Biodegradable Polymers to Biomedical and Packaging Applications

Biodegradable Polymers to Biomedical and Packaging Applications

Beata Kaczmarek-Szczepańska
Marcin Wekwejt

Basel • Beijing • Wuhan • Barcelona • Belgrade • Novi Sad • Cluj • Manchester

Editors
Beata Kaczmarek-Szczepańska
Department of Biomaterials
and Cosmetics Chemistry
Nicolaus Copernicus
University in Toruń
Toruń
Poland

Marcin Wekwejt
Department of Biomaterials
Technology
Gdańsk University
of Technology
Gdańsk
Poland

Editorial Office
MDPI AG
Grosspeteranlage 5
4052 Basel, Switzerland

This is a reprint of articles from the Special Issue published online in the open access journal *Polymers* (ISSN 2073-4360) (available at: www.mdpi.com/journal/polymers/special_issues/bio_poly_biomedical_packaging).

For citation purposes, cite each article independently as indicated on the article page online and as indicated below:

Lastname, A.A.; Lastname, B.B. Article Title. *Journal Name* **Year**, *Volume Number*, Page Range.

ISBN 978-3-7258-2464-9 (Hbk)
ISBN 978-3-7258-2463-2 (PDF)
doi.org/10.3390/books978-3-7258-2463-2

© 2024 by the authors. Articles in this book are Open Access and distributed under the Creative Commons Attribution (CC BY) license. The book as a whole is distributed by MDPI under the terms and conditions of the Creative Commons Attribution-NonCommercial-NoDerivs (CC BY-NC-ND) license.

Contents

About the Editors vii

Preface ix

Chengyu Wang, Long Mao, Bowen Zheng, Yujie Liu, Jin Yao and Heping Zhu
Development of Sustainable and Active Food Packaging Materials Composed by Chitosan, Polyvinyl Alcohol and Quercetin Functionalized Layered Clay
Reprinted from: *Polymers* 2024, *16*, 727, doi:10.3390/polym16060727 1

Sylwia Grabska-Zielińska, Judith M. Pin, Beata Kaczmarek-Szczepańska, Ewa Olewnik-Kruszkowska, Alina Sionkowska and Fernando J. Monteiro et al.
Scaffolds Loaded with Dialdehyde Chitosan and Collagen—Their Physico-Chemical Properties and Biological Assessment
Reprinted from: *Polymers* 2022, *14*, 1818, doi:10.3390/polym14091818 16

Mariia Leonovich, Viktor Korzhikov-Vlakh, Antonina Lavrentieva, Iliyana Pepelanova, Evgenia Korzhikova-Vlakh and Tatiana Tennikova
Poly(lactic acid) and Nanocrystalline Cellulose Methacrylated Particles for Preparation of Cryogelated and 3D-Printed Scaffolds for Tissue Engineering
Reprinted from: *Polymers* 2023, *15*, 651, doi:10.3390/polym15030651 28

Sara Pérez-Davila, Laura González-Rodríguez, Raquel Lama, Miriam López-Álvarez, Ana Leite Oliveira and Julia Serra et al.
3D-Printed PLA Medical Devices: Physicochemical Changes and Biological Response after Sterilisation Treatments
Reprinted from: *Polymers* 2022, *14*, 4117, doi:10.3390/polym14194117 49

Valentia Rossely Santoso, Rianita Pramitasari and Daru Seto Bagus Anugrah
Development of Indicator Film Based on Cassava Starch–Chitosan Incorporated with Red Dragon Fruit Peel Anthocyanins–Gambier Catechins to Detect Banana Ripeness
Reprinted from: *Polymers* 2023, *15*, 3609, doi:10.3390/polym15173609 69

Samantha Islam and Jonathan M. Cullen
Criteria for Assessing Sustainability of Lignocellulosic Wastes: Applied to the Cellulose Nanofibril Packaging Production in the UK
Reprinted from: *Polymers* 2023, *15*, 1336, doi:10.3390/polym15061336 87

Uruchaya Sonchaeng, Phanwipa Wongphan, Wanida Pan-utai, Yupadee Paopun, Wiratchanee Kansandee and Prajongwate Satmalee et al.
Preparation and Characterization of Novel Green Seaweed Films from *Ulva rigida*
Reprinted from: *Polymers* 2023, *15*, 3342, doi:10.3390/polym15163342 110

Nisma Agha, Arshad Hussain, Agha Shah Ali and Yanjun Qiu
Performance Evaluation of Hot Mix Asphalt (HMA) Containing Polyethylene Terephthalate (PET) Using Wet and Dry Mixing Techniques
Reprinted from: *Polymers* 2023, *15*, 1211, doi:10.3390/polym15051211 127

Ganna Kovtun, David Casas and Teresa Cuberes
Influence of Glycerol on the Surface Morphology and Crystallinity of Polyvinyl Alcohol Films
Reprinted from: *Polymers* 2024, *16*, 2421, doi:10.3390/polym16172421 148

Dengbang Jiang, Junchao Chen, Minna Ma, Xiushuang Song, Huaying A and Jingmei Lu et al.
Poly(1,3-Propylene Glycol Citrate) as a Plasticizer for Toughness Enhancement of Poly-L-Lactic Acid
Reprinted from: *Polymers* **2023**, *15*, 2334, doi:10.3390/polym15102334 . . . . **170**

About the Editors

Beata Kaczmarek-Szczepańska

Dr. Beata Kaczmarek-Szczepańska is a respected researcher and lecturer at Nicolaus Copernicus University in Toruń, affiliated with the Department of Biomaterials and Cosmetic Chemistry. Since earning her Ph.D. in chemical sciences in 2019, she has made substantial contributions to her field, particularly in cosmetics.

Her expertise lies in biomaterials, focusing on developing and modifying natural polymers such as chitosan, collagen, and hyaluronic acid for medical use. Her research involves synthesizing and characterizing bioactive materials, often incorporating tannic acid and other phenolic compounds to enhance biodegradability, mechanical strength, and biocompatibility. Her work is published in leading journals and covers areas like wound healing, tissue engineering, and antimicrobial properties of biomaterials.

Dr. Kaczmarek-Szczepańska has led projects funded by the National Science Center and the National Centre for Research and Development, respectively. She has completed internships at CURAM in Ireland and VIT in India over the past five years. Her team has been recognized by the local government authorities and the "Welconomy in Toruń 2021" forum. In 2022, she was honored by the Polish Minister of Science and Higher Education, and in 2023, she received the "Scientist of the Future" award at the Smart Development Forum. She collaborates with companies in the cosmetic, polymeric, and medical industries.

Her national and international collaborations contribute to interdisciplinary biomaterials science projects. Her work on biopolymer-based materials, integrating tannic acid and other bioactive compounds, facilitates significant research initiatives. These partnerships enhance the impact and reach of her research, advancing tissue engineering, wound healing, and sustainable material development.

Marcin Wekwejt

Dr. Marcin Wekwejt is a prominent researcher and lecturer at the Department of Biomaterials Technology, Gdańsk University of Technology, where he has been contributing since 2018. He earned his Ph.D. in materials engineering with a dissertation on modifying bone cement for long-term antibacterial protection, a work that earned him high honors and established his career in biomedical research.

Dr. Wekwejt's research focuses on developing biofunctional composite materials with antibacterial and anti-osteoporotic properties, particularly dual-setting bone cements. He has led and contributed to groundbreaking projects, including the creation of injectable ceramic–polymer bone cements. These innovations address critical challenges in biomedical engineering, such as enhancing performance, mechanical strength, and biological compatibility, which are crucial for successful medical applications.

He has completed several international research internships, including a postdoctoral fellowship at CHU de Québec–Université Laval. This experience has refined his expertise and fostered collaborations with global leaders in biomedical engineering. His partnerships with leading institutions worldwide amplify the impact of his work and ensure that his research remains at the forefront of the field.

Dr. Wekwejt's contributions have been recognized with several prestigious awards, underscoring his commitment to advancing science and translating research into clinical solutions. His work has not only advanced scientific understanding but also led to medical materials that improve patient outcomes, particularly in bone repair and regeneration. Beyond the lab, Dr. Wekwejt is dedicated to interdisciplinary collaboration, bridging the gap between research and real-world applications.

Preface

This reprint presents a curated collection of research papers on the innovative use of biodegradable polymers in biomedical and packaging applications. With the growing need for sustainable and eco-friendly materials, this Special Issue highlights the critical role of biodegradable polymers in areas such as healthcare and packaging industries. Their ability to degrade naturally and reduce environmental impact makes them a vital focus of scientific research and industrial development.

The motivation behind this publication stems from the increasing demand for materials that meet the rigorous standards of modern industries while minimizing environmental harm. Biodegradable polymers have emerged as key players in this space, offering solutions that align with the global push toward more sustainable practices. The scope of this reprint covers a wide range of topics, from polymer synthesis and physicochemical characterization to their application in two- and three-dimensional forms. The research included demonstrates the latest advancements in enhancing polymer properties to make them more effective in practical applications. These studies reflect a multidisciplinary approach, combining chemistry, materials science, biotechnology, and environmental science to address pressing challenges.

Our aim was to gather cutting-edge research that offers new insights into the development of biodegradable materials for real-world applications. This reprint is intended for a broad audience, including researchers, industry professionals, and students in the fields of biomaterials, polymers, and environmental science. It serves as a valuable resource for those looking to expand their knowledge of biodegradable polymers and their uses in key sectors such as healthcare and packaging.

The contributing authors are leading experts in their respective fields, bringing a wealth of knowledge and experience to this volume. Their innovative research has been instrumental in advancing the science and technology of biodegradable materials, and we are grateful for their contributions. We would also like to extend our deepest grattitude to the reviewers for their insightful comments, which have helped refine the research presented in this collection. Special recognition goes to the editorial team at MDPI for their unwavering support and to the institutions and colleagues who provided the necessary resources and encouragement throughout this project.

We hope that this reprint will inspire further research and innovation in the field of biodegradable polymers and contribute to the ongoing quest for sustainable materials that benefit both society and the environment.

Beata Kaczmarek-Szczepańska and Marcin Wekwejt
Editors

Article

Development of Sustainable and Active Food Packaging Materials Composed by Chitosan, Polyvinyl Alcohol and Quercetin Functionalized Layered Clay

Chengyu Wang [1], Long Mao [1,2,*], Bowen Zheng [2], Yujie Liu [2], Jin Yao [1] and Heping Zhu [1,*]

[1] Key Laboratory of Advanced Packaging Materials and Technology of Hunan Province, Hunan University of Technology, Zhuzhou 412007, China; d21080500002@hut.edu.cn (C.W.); yaojin@hut.edu.cn (J.Y.)
[2] Fujian Provincial Key Laboratory of Functional Materials and Applications, Xiamen University of Technology, Xiamen 361024, China; 13959280885@163.com (B.Z.); lyj1055348378@163.com (Y.L.)
* Correspondence: maolong@xmut.edu.cn (L.M.); zhuheping@hut.edu.cn (H.Z.)

Abstract: In order to solve the problems of insufficient active functions (antibacterial and antioxidant activities) and the poor degradability of traditional plastic packaging materials, biodegradable chitosan (CS)/polyvinyl alcohol (PVA) nanocomposite active films reinforced with natural plant polyphenol-quercetin functionalized layered clay nanosheets (QUE-LDHs) were prepared by a solution casting method. In this study, QUE-LDHs realizes a combination of the active functions of QUE and the enhancement effect of LDHs nanosheets through the deposition and complexation of QUE and copper ions on the LDHs. Infrared and thermal analysis results revealed that there was a strong interface interaction between QUE-LDHs and CS/PVA matrix, resulting in the limited movement of PVA molecules and the increase in glass transition temperature and melting temperature. With the addition of QUE-LDHs, the active films showed excellent UV barrier, antibacterial, antioxidant properties and tensile strength, and still had certain transparency in the range of visible light. As QUE-LDHs content was 3 wt%, the active films exhibited a maximum tensile strength of 58.9 MPa, representing a significant increase of 40.9% compared with CS/PVA matrix. Notably, the UV barrier (280 nm), antibacterial (*E. coli*) and antioxidant activities (DPPH method) of the active films achieved 100.0%, 95.5% and 58.9%, respectively. Therefore, CS/PVA matrix reinforced with QUE-LDHs has good potential to act as an environmentally and friendly active packaging film or coating.

Keywords: active packaging; chitosan; quercetin; layered clay; polyvinyl alcohol

1. Introduction

In order to maintain food quality and protect food from extraneous contamination, food packaging plays a crucial role, as it ensures the food industry provides safe edible products [1]. Packaging provides more than just basic physical protection and barrier properties [2,3]; however, the packaging materials commonly applied in the food industry are made from non-degradable polymers that lack active functions (such as antibacterial and antioxidant activities), leading to environmental pollution or even food spoilage [4]. With the increasingly serious environmental problems, researchers have begun to pay attention to biodegradable polymers that can extend the shelf life of food. Therefore, various environmentally friendly active packaging materials have emerged and attracted widespread attention [3].

As the only cationic polysaccharide in biomass resources, chitosan (CS) ranks as the second-largest natural polymer in nature after cellulose [5]. CS exhibits excellent film-forming abilities, broad-spectrum antibacterial activity, and easy availability [6]. However, CS also demonstrates a moisture absorption ability and brittleness [7]. Blending CS with other synthetic polymer materials can effectively improve its shortcomings [8–10]. Polyvinyl alcohol (PVA) has found widespread use in packaging, biomedicine, and other fields due to its good film-forming properties, water solubility, mechanical strength, transparency, and degradability [7,9]. The development of biodegradable active food packaging materials with excellent performance through the blending of PVA and CS has been extensively studied [11–13], and this approach takes advantage of the broad spectrum of antibacterial activity and availability of CS, as well as the excellent film forming properties and stability of PVA [7,14].

Combined with literature reports [9,13], to further enhance the comprehensive performance of CS/PVA matrix, it is generally necessary to modify them by introducing a third-phase with active functions. Natural plant polyphenols, such as tannins, anthocyanin, quercetin, tea polyphenols, etc., exhibit excellent biological activity as phenolic secondary metabolites in plants [15,16]. These plant polyphenols demonstrate active functions, including antibacterial, antioxidant, and UV barrier properties, and have been applied to the modification of CS/PVA matrix [7,9]. Koosha et al. [9] studied CS/PVA composite active films doped with black carrot anthocyanin and layered clay (bentonite). They explored the application potential of anthocyanin as a natural pH indicator and bentonite as a nanofiller in the field of active packaging. The results showed that the addition of bentonite alone significantly reduces the mechanical properties of CS/PVA composite films. Although the mechanical and antibacterial properties of CS/PVA composite films were improved by the further addition of anthocyanins, the gas barrier properties of CS/PVA composite films were significantly affected. Haghighi et al. [13] prepared CS/PVA composite films enriched with ethyl lauroyl arginate (LAE) for food packaging applications. Although the addition of LAE improved the antibacterial activity and UV barrier properties of CS/PVA composite films, the mechanical and water vapor barrier properties were deteriorated to a certain extent. Therefore, to improve the comprehensive performance of the CS/PVA matrix, it is important to realize the multi-functionality of the third phase.

Inspired by the super-adhesion and versatility of marine adhesion proteins, the functionalization of material surfaces (films, nanoparticles, etc.) can be achieved by using catechol compounds similar in structure to adhesion proteins to form functional coatings (such as polydopamine (PDA) and tannin-metalion (TA-metal) coordination compounds) [17,18]. In our previous work [19,20], a simple and green method has been reported to prepare natural plant polyphenols containing active catechol functionalized layered clay nanosheets (Layered double hydroxides, LDHs), such as LDHs@TA-Fe^{3+} and LDHs@anthocyanin-Cu^{2+}. This biomimetic modification method, based on mussels, can not only increase the interfacial compatibility between the LDHs and polymer matrix, thereby better utilizing the natural barrier properties and enhancing effects of LDHs nanosheets, but also achieve effective loading of natural active substances, endowing LDHs with antibacterial and antioxidant activities [20]. Quercetin (QUE) in natural plant polyphenols contains rich catechol groups, and it is currently yields the highest content of dietary flavonoids and polyphenols in fruits and vegetables, with good biological activity and medicinal value [21]. Therefore, the surface functionalization of LDHs using QUE can effectively combine the functions of LDHs and QUE, improving the versatility of the third phase (antibacterial, antioxidant activity, enhancement, etc.).

Based on the above analysis, this study attempted to use QUE to functionalize the LDHs to realize a combination of the active functions of QUE and the enhancement effect of LDHs nanosheets. QUE functionalized LDHs (QUE-LDHs) were synthesized by adsorption and complexation of QUE and Cu^{2+} on the surface of LDHs. Further, CS/PVA matriices reinforced with QUE-LDHs were prepared for the first time by the solution casting method. The influence of QUE-LDHs on the chemical structure, thermal, crystallization, mechanical, optical, antibacterial and antioxidant properties of CS/PVA matriices was evaluated to determine their feasibility for food active packaging.

2. Materials and Methods

2.1. Materials and Chemicals

QUE (purity ≥ 97%), copper chloride dihydrate ($CuCl_2·2H_2O$, purity ≥ 99.99%), CS (M_W = 30,000 Da, 85% deacetylation), 2,2-diphenyl-1-picrylhydrazyl (DPPH, purity ≥ 96%), ethanol (AR), methanol (AR) were provided by Shanghai Makclin Biochemical Technology Co., Ltd. (Shanghai, China). Ethylene-modified PVA (EXCEVAL™ AQ-4104, alcoholization degree 98–99 mol%) was purchased by Kuraray Co., Ltd. (Osaka, Japan). 2,2'-azinobis(3-ethylbenzothiazoline-6-sulfonic acid ammonium salt) (ABTS, purity = 98%), potassium persulfate (purity = 99.99%) were provided by Shanghai Aladdin Biochemical Technology Co., Ltd. (Shanghai, China). Phosphate buffer solution (PBS, 0.1 M, pH = 7.4) was purchased from Phygene Biotechnology Co., Ltd. (Fuzhou, China). LDHs was prepared by the hydrothermal reaction of metal salts with urea (particle size: ~1500 nm, thickness: ~40 nm) [20]. All chemical reagents are used directly without undergoing secondary treatment or purification.

2.2. Synthesis of QUE-Functionalized LDHs (QUE-LDHs)

LDHs (50 mg) was evenly dispersed in ethanol (100 mL) by ultrasonic treatment. Subsequently, QUE (50 mg) was added to the LDHs dispersion at room temperature and the dissolution of QUE was accelerated by mechanical stirring. Next, $CuCl_2·2H_2O$ (28 mg) was dissolved in deionized water (50 mL). The prepared $CuCl_2$ aqueous solution was then slowly poured into the LDHs dispersion, and the reaction was carried out at room temperature for 4 h. After the reaction, the reaction liquid was centrifuged with ethanol once and with deionized water twice. Finally, dark brown QUE-LDHs was obtained by a freeze-drying method. The characterization methods of QUE-LDHs have been given in the Supporting Information.

2.3. Preparation of QUE-LDHs/CS/PVA Nanocomposite Active Films

To enhance the water solubility and interfacial binding of CS, this study employed active catechol groups to further functionalize CS by carbodiimide coupling [22]. The composition and abbreviation of QUE-LDHs/CS/PVA nanocomposite active films are presented in Table 1. A certain amount of QUE-LDHs was dispersed in the deionized water (14 mL) and sonicated to disperse evenly. Then, CS (70 mg) was dissolved in the QUE-LDHs dispersion and the dissolution of CS was accelerated by mechanical stirring. Subsequently, PVA powder (630 mg) was added to the QUE-LDHs dispersion and heated to 95 °C under magnetic stirring until PVA powder was completely dissolved to obtain the active film-forming solution. The active film-forming solution was then sonicated to remove air bubbles. Finally, the above active film-forming solution was poured into a horizontally placed PTFE mold and transferred to an oven at 45 °C for drying for 24 h to obtain the homogeneous QUE-LDHs/CS/PVA nanocomposite active films. Before the subsequent tests, all the films were stored at 24 °C and 50% humidity for 48 h. For further information regarding the characterization of QUE-LDH/CS/PVA nanocomposite active films, please refer to the Supporting Information.

Table 1. The composition and abbreviation of QUE-LDHs/CS/PVA nanocomposite active films.

Samples	PVA/g	CS/g	QUE-LDHs/g	QUE-LDHs/wt%
CS/PVA	0.63	0.07	0	0
LQCP-0.5%	0.63	0.07	0.0035	0.5%
LQCP-1%	0.63	0.07	0.0071	1%
LQCP-3%	0.63	0.07	0.0216	3%
LQCP-5%	0.63	0.07	0.0368	5%
LQCP-7%	0.63	0.07	0.0526	7%

3. Results and Discussion

3.1. Chemical Structure of QUE-LDHs

Figure 1 shows the FT-IR and UV-Vis spectra of QUE-LDHs, QUE and LDHs. In Figure 1a, the absorption peaks at 3384, 1578, and 1352 cm^{-1} in LDHs are due to the stretching and bending vibration of O-H bonds and the stretching vibration of interlayer carbonate ions, respectively [20]. Compared with LDHs, new absorption peaks of C=C groups (marked in red) and C-O groups (marked in yellow) in the QUE-LDHs are attributed to the characteristic absorption of QUE [21,23], suggesting the successful loading of QUE on the LDHs. In Figure 1b, LDHs show no characteristic absorption peaks in the UV-Vis region, whereas QUE-LDHs show obvious characteristic absorption peaks at 290 and 434 nm. Compared with QUE, the UV-Vis characteristic absorption (at 290 and 434 nm) of QUE-LDHs shows an obvious redshift, which is due to the formation of QUE-Cu coordination compounds [24,25]. Moreover, Cu^{2+} is further confirmed to be involved in the formation of coordination compounds using XPS, and the relevant analysis is supplemented in the Supporting Information.

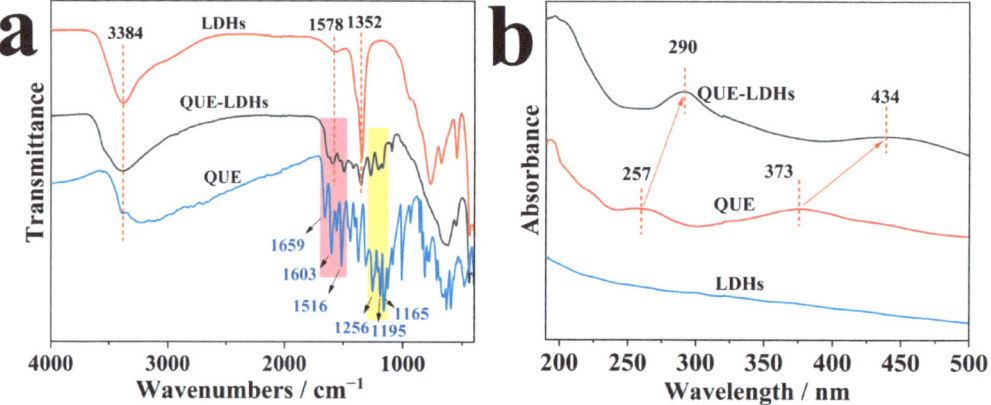

Figure 1. (a) FT-IR spectra and (b) UV-Vis spectra of QUE-LDHs, QUE and LDHs.

3.2. Microscopic Morphology of QUE-LDHs

The microscopic morphology of QUE-LDHs was characterized using TEM and SEM, as shown in Figure 2. In Figure 2a,b, the original LDHs exhibit a typical hexagonal morphology with a smooth surface. After being functionalized with QUE-Cu coordination compounds, LDHs becomes significantly rougher and the edges and corners become more rounded. Furthermore, QUE-LDH nanosheets exhibit uniform dispersion without agglomeration. In order to further confirm the existence of the functional layer, TEM was applied to study the microscopic morphology of QUE-LDHs. In Figure 2e, QUE-LDHs still exhibit a typical hexagonal structure and rough surface. In Figure 2f, a functional layer with a thickness of ~20 nm is formed on the surface of LDHs through adsorption and complexation to form QUE-Cu^{2+} coordination compounds.

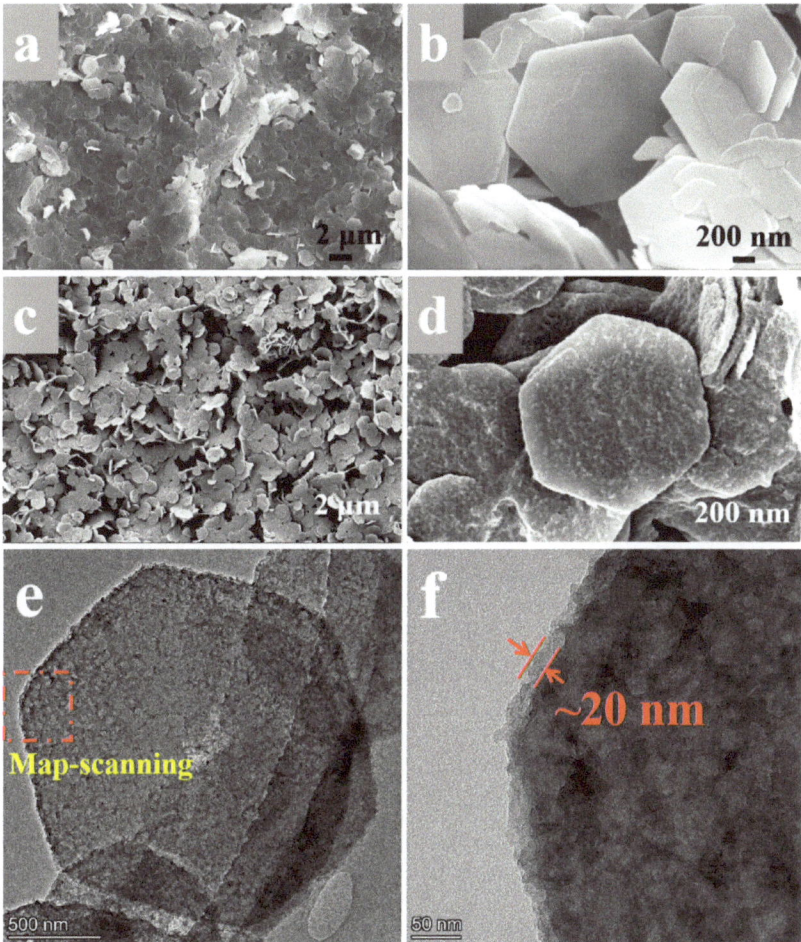

Figure 2. SEM images of (**a**,**b**) LDHs and (**c**,**d**) QUE-LDHs. (**e**,**f**) TEM images of QUE-LDHs.

3.3. Antibacterial Activity of QUE-LDHs

QUE, as a kind of plant secondary metabolite, affects the formation of microbial biofilm via its polyphenol structure, hindering the synthesis of enzymes related to metabolism, thereby increasing the permeability of biofilm, which ultimately leads to cell rupturing [21,23]. Figure 3 shows the antibacterial activity against *Escherichia coli* (*E. coli*) of QUE-LDHs, QUE and LDHs. As shown in Figure 3, the antibacterial activity of LDHs is only 24.2%, and there is no obvious antibacterial activity. After the functionalization of QUE-Cu coordination compounds, the antibacterial activity of QUE-LDHs reaches 99.4%. QUE-LDHs exhibit excellent antibacterial activity due to the natural broad-spectrum antibacterial activity of QUE and copper ions [26].

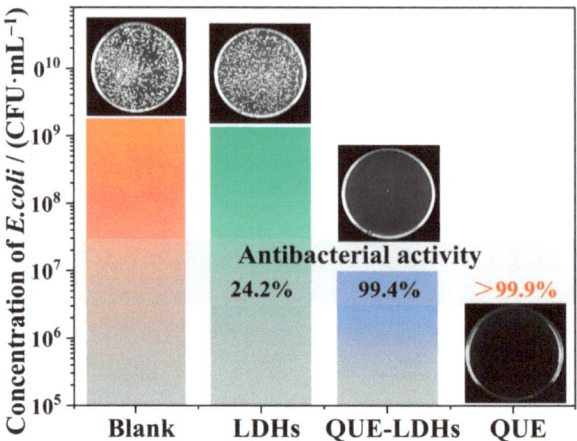

Figure 3. Antibacterial activity against *E. coli* of QUE-LDHs, QUE and LDHs.

3.4. Chemical Structure of QUE-LDHs/CS/PVA Nanocomposite Active Films

The FT-IR spectra of QUE-LDHs/CS/PVA nanocomposite active films are illustrated in Figure 4. The absorption peak at 3273 cm^{-1} in the spectrum of the CS/PVA matrix is attributed to the stretching vibration of -OH and -NH groups involved in hydrogen bonds in the CS/PVA matrix [9]. The peaks at 2929, 2852, and 1415 cm^{-1} are due to the stretching and deformation vibrations of C-H bonds [7]. Additionally, the absorption peaks at 1087 and 1019 cm^{-1} correspond to the stretching vibration of C-O bonds [5,13]. After the addition of QUE-LDHs, compared with CS/PVA matrix, there is no significant change in the position and intensity of the characteristic absorption peak for QUE-LDHs/CS/PVA nanocomposite active films. However, subtle changes are observed in the infrared spectral fingerprint range of 1200–950 cm^{-1} (marked in green). In particular, the relative intensity of the absorption peaks at 1087 and 1019 cm^{-1} changes, which reveals that the formation of the hydrogen bond between C-O groups (in the CS/PVA matrix) and phenol hydroxyl groups (in the QUE) causes the change of the infrared peak for C-O groups [27,28]. These changes suggest a strong interaction between QUE-LDHs and CS/PVA matrix, leading to the formation of new hydrogen bonds in the QUE-LDHs/CS/PVA nanocomposite active films.

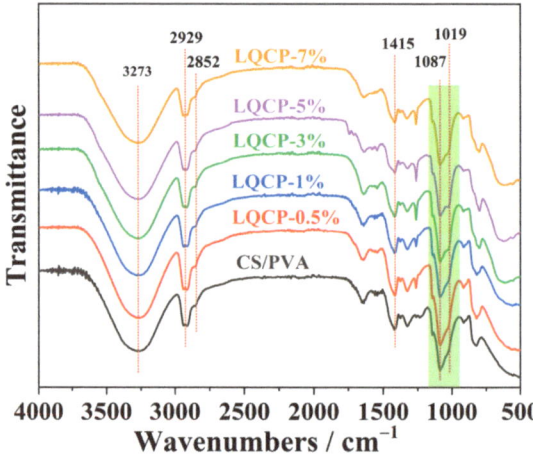

Figure 4. FT-IR spectra of QUE-LDHs/CS/PVA nanocomposite active films.

3.5. Thermal Stability of QUE-LDHs/CS/PVA Nanocomposite Active Films

The thermogravimetric curves and thermal stability parameters of the QUE-LDHs/CS/PVA nanocomposite active films are illustrated in Table 2 and Figure 5, respectively. It is observed from Table 2 that, with the addition of QUE-LDHs, the initial decomposition temperature ($T_{-5\%}$) of QUE-LDHs/CS/PVA nanocomposite active films initially increases and then decreases. This phenomenon is attributed to the enhanced insulation provided by the layered structure of QUE-LDHs [29]. When the addition of QUE-LDHs reaches 5 wt%, $T_{-5\%}$ reaches a maximum of 101.6 °C. Although $T_{-5\%}$ starts to decrease with the further addition of QUE-LDHs, it remains higher than that of CS/PVA matrix. In contrast to $T_{-5\%}$, the temperature at 50% weight loss ($T_{-50\%}$) consistently decreases with the increase in QUE-LDHs. When the addition of QUE-LDHs is only 0.5 wt%, $T_{-50\%}$ decreases from 377.3 °C (CS/PVA matrix) to 356.5 °C, showing a decrease of 20.8 °C. As the addition of QUE-LDHs increases from 0.5 wt% to 7 wt%, the decrease in $T_{-50\%}$ is only 10.8 °C.

Table 2. Thermal analysis results of QUE-LDHs/CS/PVA nanocomposite active films.

Sample	$T_{-5\%}$/°C	$T_{-50\%}$/°C	T_g/°C	T_m/°C	T_c/°C	ΔH_m/(J/g)	χ/%
CS/PVA	86.6	377.3	78.1	186.7	156.6	23.56	16.1%
LQCP-0.5%	87.3	356.5	81.5	189.5	154.5	29.40	20.1%
LQCP-1%	92.9	355.2	81.8	191.4	157.6	29.69	20.4%
LQCP-3%	95.8	355.6	81.3	190.4	156.6	28.21	19.8%
LQCP-5%	101.6	351.4	79.7	199.1	166.9	30.11	21.6%
LQCP-7%	90.6	345.7	79.8	204.5	173.5	27.50	20.2%

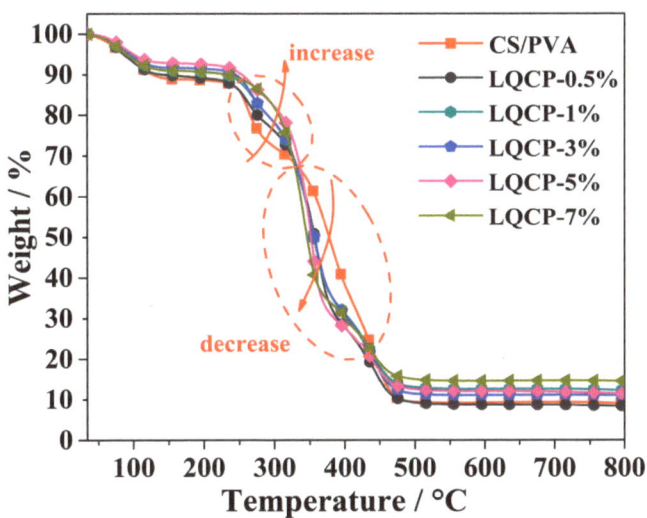

Figure 5. TG curves of QUE-LDHs/CS/PVA nanocomposite active films.

Figure 5 indicates that all the films exhibit similar thermal degradation behaviors with two obvious weight loss steps. From room temperature to ~150 °C, QUE-LDHs/CS/PVA nanocomposite active films lose free water and adsorbed water. When the temperature is between ~150 °C and ~230 °C, the weight of QUE-LDHs/CS/PVA nanocomposite active films remains constant. When the temperature is between ~230 °C and 500 °C, the weight loss of QUE-LDHs/CS/PVA nanocomposite active films increases rapidly, indicating the main thermal degradation stage. It is worth noting that the weight loss of QUE-LDHs/CS/PVA nanocomposite active films is always lower than that of CS/PVA matrix before being heated to ~325 °C, which is related to the barrier effect of QUE-LDHs [30]. However, between ~325 °C and ~435 °C, the weight loss of QUE-LDHs/CS/PVA nanocomposite active films surpasses that of CS/PVA matrix, indicating a faster thermal degradation rate, which is due to the co-thermal degradation of QUE and LDHs [21,31]. When the temperature exceeds 435 °C, the weight loss of QUE-LDHs/CS/PVA nanocomposite active films begins to be lower than that of CS/PVA matrix, resulting in a final residual (at 800 °C) being higher than that of CS/PVA matrix. This is because most of the thermal degradation of QUE-LDHs has been completed [32]. Therefore, an appropriate amount (1–5 wt%) of QUE-LDHs can enhance the initial thermal stability of CS/PVA matrix.

3.6. Thermal and Crystalline Properties of QUE-LDHs/CS/PVA Nanocomposite Active Films

The thermal and crystallization properties of QUE-LDHs/CS/PVA nanocomposite active films are characterized by DSC and XRD, as illustrated in Figure 6 and Table 2. In Figure 6a, the glass transition temperature (T_g) and melting temperature (T_m) of QUE-LDHs/CS/PVA nanocomposite active films gradually increase with the addition of QUE-LDHs. This indicates a strong interfacial interaction between QUE-LDHs and CS/PVA matrix, which weakens the mobility of PVA molecules [33,34]. Additionally, combined with a thermal stability analysis of the TG test, the barrier effect of QUE-LDHs prolongs the melting time of CS/PVA matrix. Figure 6b and Table 2 illustrate that the crystallization temperature (T_c) and melting enthalpy (ΔH_m) of all the films show a gradual increase. The addition of QUE-LDHs promotes the rearrangement and stacking of PVA molecules, playing a certain degree of heterogeneous nucleation. However, with the increase in QUE-LDHs, the crystallinity (χ) varies from 19.8% to 21.6%. This indicates that the overall crystallinity remains relatively stable, which may be due to the comprehensive effect of heterogeneous nucleation and interfacial interaction [35,36]. Combined with Figure 6a,b, the advance of crystallization indicates that PVA molecules can form a large number of crystallization nucleation sites on the surface of QUE-LDHs, resulting in an increase in T_c. However, the strong interaction between QUE-LDHs and CS/PVA matrix limits the growth of crystal nuclei, reducing heterogeneous nucleation efficiency [29]. In Figure 6c, characteristic diffraction peaks of (003) and (006) for QUE-LDHs appear in QUE-LDHs/CS/PVA nanocomposite active films around 2θ = 11.7° and 23.5° [37]. The intensity of these peaks increases with the addition of QUE-LDHs, which indicates that the mass fraction of QUE-LDHs increases. Additionally, all the films exhibit a characteristic diffraction peak of (101) at θ = 19.6°, which is attributed to X-ray diffraction of PVA [38]. However, the (101) crystal plane of PVA does not change significantly with the addition of QUE-LDHs, which is similar to the crystallinity change in the DCS analysis.

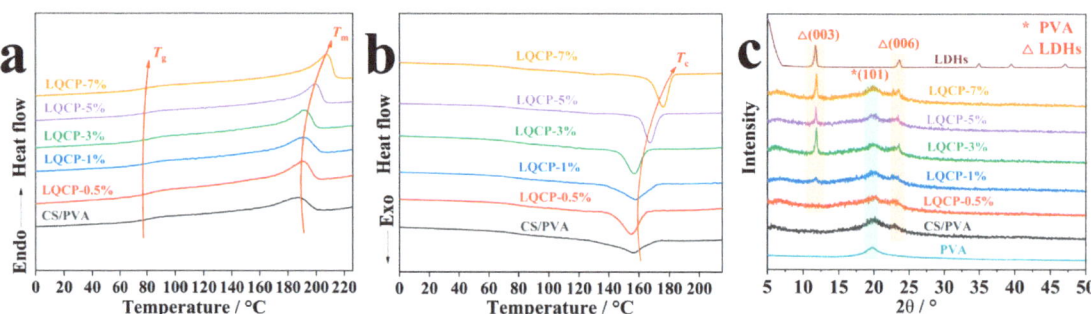

Figure 6. (a) DSC second heating curves, (b) DSC cooling curves, (c) XRD patterns of QUE-LDHs/CS/PVA nanocomposite active films.

3.7. Mechanical Properties of QUE-LDHs/CS/PVA Nanocomposite Active Films

Figure 7 illustrates the mechanical properties and stress–strain curves of QUE-LDHs/CS/PVA nanocomposite active films. In Figure 7a, the tensile strength and elongation at break of CS/PVA matrix are 41.8 MPa and 126.2%, respectively. When the addition of QUE-LDHs reaches 3 wt%, the tensile strength achieves the maximum value of 58.9 MPa, representing a significant 40.9% increase compared with CS/PVA matrix (41.8 MPa). Although the tensile strength begins to decrease continuously with the addition of QUE-LDHs, the tensile strength of LQCP-7% (42.1 MPa) does not change significantly compared with CS/PVA matrix. In contrast to the changes in the tensile strength, the elongation at break gradually decreases with the increase in QUE-LDHs. When the molecular chain slides under stress, the chain mobility of CS/PVA matrix is decreased due to the limitation of strong interface interactions caused by hydrogen bondings, ultimately leading to premature fracture of the film [5,39]. Meanwhile, according to DSC test results, the existence of QUE-LDHs increases χ of CS/PVA matrix, further decreases the mobility of molecular chains.

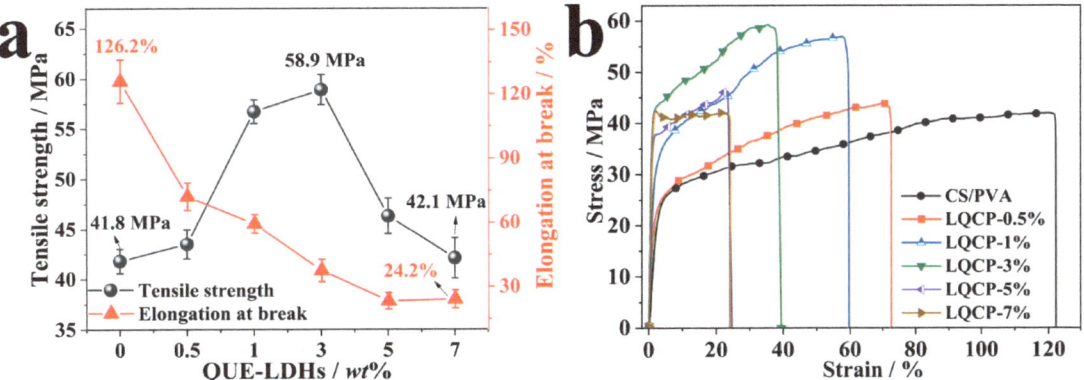

Figure 7. (**a**) Mechanical properties and (**b**) stress–strain curves of QUE-LDHs/CS/PVA nanocomposite active films.

As shown in Figure 7b, the stress–strain curve of CS/PVA matrix are similar to those of pure PVA, with no obvious yield point and strain softening stage [40]. With the addition of QUE-LDHs (no more than 3 wt%), the yield strength and breaking strength of QUE-LDHs/CS/PVA nanocomposite active films show an increasing trend, which indicates that QUE-LDHs play a significant enhancement effect. When the addition of QUE-LDHs reaches 5 wt%, the strength of QUE-LDHs/CS/PVA nanocomposite active films begins to decrease significantly, which is related to the decrease in enhancement efficiency and agglomeration of QUE-LDHs. The strain decreases with the increase in QUE-LDHs, which reveals that the existence of QUE-LDHs weakens the interactions between CS and PVA.

In order to further analyze the reasons for the changes of the mechanical properties, the microscopic morphology of the fracture surface was investigated, as shown in Figure 8. As shown in Figure 8a,b, QUE-LDHs is uniformly dispersed at nanometer size in CS/PVA matrix without obvious agglomeration, and there are no visible defects or gaps between QUE-LDHs and CS/PVA matrix. This reveals that QUE-LDHs and CS/PVA matrix have good interface compatibility, which helps to enhance the strength of CS/PVA matrix. The EDS element analysis of the marked region in Figure 8b, as presented in Figure 8c,d, further confirms the presence of QUE-LDHs.

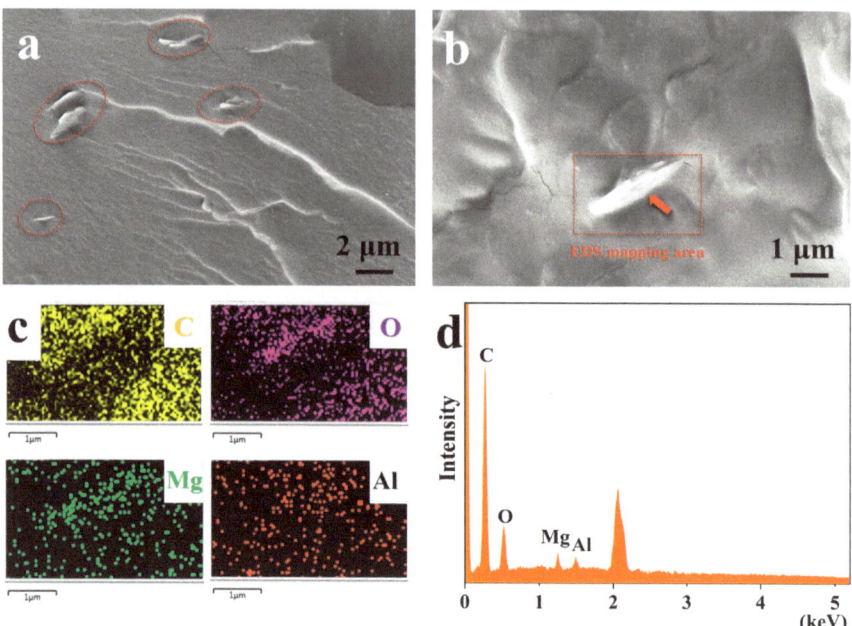

Figure 8. (**a**,**b**) SEM images for fracture surface, (**c**) EDS map-scanning, (**d**) EDS spectrum of LQCP-3%.

3.8. Optical Properties of QUE-LDHs/CS/PVA Nanocomposite Active Films

Figure 9 displays the UV-Vis light transmittance and digital photographs of QUE-LDHs/CS/PVA nanocomposite active films. Good UV barrier properties can effectively delay the deterioration of food under UV radiation, ensuring the quality of food during storage and transportation, and prolonging the shelf life of food [41,42]. As shown in Figure 9a, the UV transmittance of CS/PVA matrix at 280 nm and 400 nm is only 25.3% and 53.2%, demonstrating certain UV barrier properties. It indicates that CS/PVA matrix already exhibits significant absorption in the UV light region (190–400 nm), which is caused by the catechol structure and unsaturated bonds in CS [22]. With the addition of QUE-LDHs, the UV absorption of QUE-LDHs/CS/PVA nanocomposite active films significantly increases, which is attributed to the large amount of active catechol groups in QUE-LDHs. When the addition of QUE-LDHs is only 1 wt%, the UV transmittance of LQCP-1% at 280 nm and 400 nm reduces to 0.1% and 2.9%, which is a 99.6% and 94.5% lower than that of CS/PVA matrix, indicating excellent UV barrier properties. In our previous study [35], when the same amount (1 wt%) of curcumin-functionalized LDHs was added to CS/PVA matrix, the UV transmittance at 280 nm and 400 nm was reduced by only 70.4% and 35.3%. This indicates that QUE-LDHs contains a large number of active catechol groups, which show better UV barrier properties, while curcumin-functionalized LDHs only contain a large number of active phenolic hydroxyl groups. With further addition of QUE-LDHs (3–7 wt%), the UV barrier properties reach 100% at 280 nm. In the visible light region (400–780 nm), the light transmittance at 600 nm of CS/PVA film is 70.5%. With the addition of QUE-LDHs, the light transmittance decreases gradually. When the addition of QUE-LDHs reaches 7 wt%, the reduction in light transmittance at 600 nm reaches 60.7% compared with CS/PVA matrix. When the same amount (7 wt%) of curcumin-functionalized LDHs was added to CS/PVA matrix, the light transmittance at 600 nm is reduced by only 29.8% [35]. This suggests that QUE-LDHs have a greater effect on visible light transmittance. However, the absorption of UV light is much greater than that of visible light. Meanwhile, the addition of QUE-LDHs can significantly enhance the high-energy blue light (400–500 nm) barrier properties of CS/PVA matrix. The barrier of high-energy blue light is also crucial for food preservation, as it helps delay the photooxidation of organic compounds and the degradation of vitamins and other pigments [43].

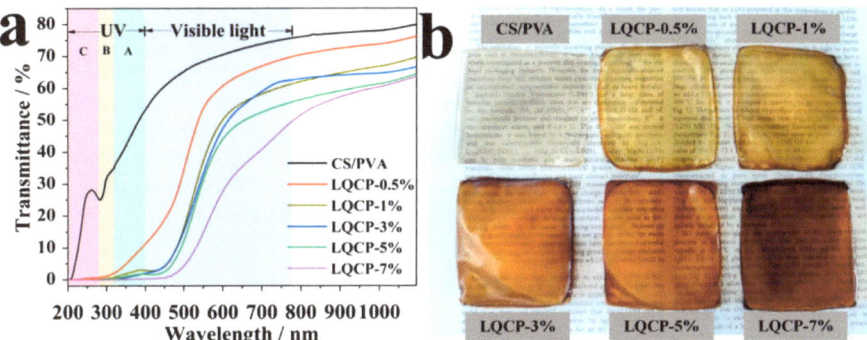

Figure 9. (**a**) UV-Vis light transmittance and (**b**) digital photographs of QUE-LDHs/CS/PVA nanocomposite active films.

Table 3 presents the optical properties of QUE-LDHs/CS/PVA nanocomposite active films. As shown in Table 3, the opacity of QUE-LDHs/CS/PVA nanocomposite active films gradually decreases with the addition of QUE-LDHs. When the addition of QUE-LDHs does not exceed 5 wt%, the opacity is less than five, indicating that QUE-LDHs/CS/PVA nanocomposite active films exhibit transparency. Providing a clear view of food and packaging conditions while providing good UV and high-energy blue light barrier properties are prerequisite for excellent food packaging film materials [13]. As shown in Figure 9b, the color of QUE-LDHs/CS/PVA nanocomposite active films shows a trend from colorless to brown. Despite the deepening color, the text behind QUE-LDHs/CS/PVA nanocomposite active films (the addition of QUE-LDHs does not exceed 5 wt%) remains clearly visible, demonstrating a certain degree of transparency, which is consistent with the opacity results. The colorimeter was used to further accurately analyze the color change of QUE-LDHs/CS/PVA nanocomposite active films, as shown in Table 3. It is evident that with the increase in QUE-LDHs, the L^* (brightness) significantly decreases, which is correlating with the decrease in visible light transmittance. Influenced by the color of QUE-LDHs, the values of a^* and b^* continue to increase, indicating that QUE-LDHs/CS/PVA nanocomposite active films exhibit a trend of reddening and yellowing. Due to the growing color difference between the films and standard whiteboard, ΔE^* increases from 21.6 to 84.8. In our previous study [35], the addition of curcumin-functionalized LDHs resulted in an increase in ΔE^* from 21.6% to 59.1%. The results indicate that the addition of QUE-LDHs has a significant effect on the color of CS/PVA matrix.

Table 3. Optical properties of QUE-LDHs/CS/PVA nanocomposite active films.

Sample	T_{280}/% [1]	T_{400}/% [1]	T_{600}/% [1]	Thickness/mm	Abs_{600} [2]	Opacity	L^*	a^*	b^*	ΔE
CS/PVA	25.3	53.2	70.5	0.091	0.152	1.670	81.9 ± 0.1	1.9 ± 0.2	15.7 ± 0.2	21.6 ± 0.4
LQCP-0.5%	0.6	11.3	60.8	0.073	0.216	2.959	76.8 ± 0.2	6.0 ± 0.1	40.4 ± 0.4	45.4 ± 0.3
LQCP-1%	0.1	2.9	50.1	0.071	0.300	4.225	73.4 ± 0.1	6.9 ± 0.3	42.2 ± 0.2	46.9 ± 0.3
LQCP-3%	0	2.0	47.3	0.075	0.325	4.333	56.5 ± 0.2	21.0 ± 0.4	48.5 ± 0.4	66.4 ± 0.5
LQCP-5%	0	1.7	44.0	0.072	0.357	4.958	55.8 ± 0.3	26.0 ± 0.5	52.2 ± 0.6	71.3 ± 0.5
LQCP-7%	0	0.20	27.7	0.075	0.558	7.440	41.0 ± 0.5	24.6 ± 0.3	59.3 ± 0.2	84.8 ± 0.4

[1] Light transmittance at 280, 400, 600 nm. [2] Absorbance at 600 nm.

3.9. Antibacterial Activity of QUE-LDHs/CS/PVA Nanocomposite Active Films

Figure 10 shows the antibacterial activity against *E. coli* of QUE-LDHs/CS/PVA nanocomposite active films. The antibacterial activity of CS/PVA matrix reaches 91.1%, showing good antibacterial ability, which is due to the excellent broad-spectrum antibacterial activity of CS [6]. According to literature reports, the antibacterial mechanism for CS is due to the strong interaction between positively charged amino groups of CS and negatively charged membranes of bacterial, resulting in increased membrane permeability and the leakage of intracellular components, and ultimately leading to bacterial death [6,44]. The antibacterial activity of QUE-LDHs/CS/PVA nanocomposite active films exhibits a gradual increase trend upon the increase in QUE-LDHs. When the addition of QUE-LDHs reaches 7 wt%, the antibacterial activity of LQCP-7% reaches the maximum of 97.3%, showing excellent antibacterial ability. These results indicate that the presence of active QUE-LDHs can enhance the antibacterial ability of CS/PVA matrix.

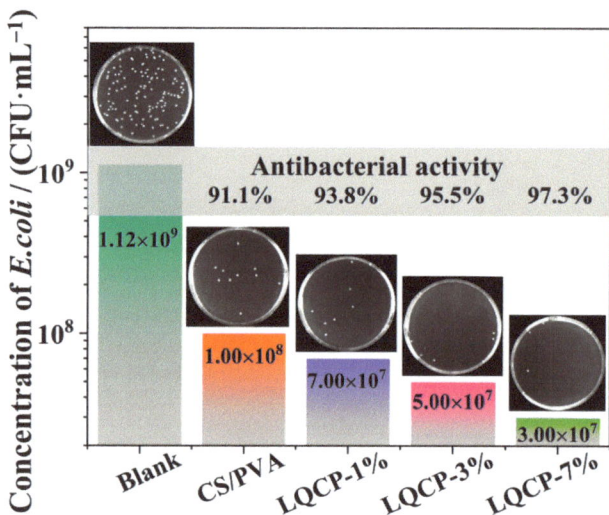

Figure 10. Antibacterial activity against *E. coli* of QUE-LDHs/CS/PVA nanocomposite active films.

3.10. Antioxidant Activity of QUE-LDHs/CS/PVA Nanocomposite Active Films

The antioxidant ability of food packaging materials can effectively prevent the oxidation of nutrients in food and reduce the loss of food quality [45]. DPPH and ABTS radical scavenging activities were applied to evaluate the antioxidant ability of QUE-LDHs/CS/PVA nanocomposite active films, as shown in Figure 11. In Figure 11a, DPPH radical scavenging activity of QUE reaches 96.6%, showing excellent antioxidant activity, which is due to the large number of active groups in QUE effectively trapping radicals and preventing the oxidation reaction of nutrients [21,23]. CS/PVA matrix does not have sufficient active groups, therefore it cannot effectively scavenge DPPH radicals, and its antioxidant activity is only 14.2%. With the addition of QUE-LDHs, its antioxidant activity gradually increases. When the amount of QUE-LDHs reaches 7 wt%, the antioxidant activity reaches the maximum value of 78.3%. ABTS radical scavenging activity experiment also shows the same trend. However, it can be seen from radical scavenging experiments that the scavenging activity of DPPH radical is significantly higher than that of ABST radical under the same amount of QUE-LDHs. This is because the solvent used in DPPH radical scavenging activity experiment is alcohol solution, while ABTS radical scavenging activity experiment is aqueous solution. Additionally, QUE is almost insoluble in water, but soluble in ethanol.

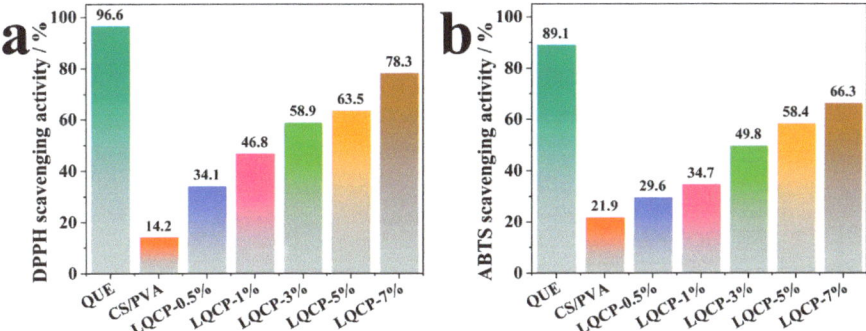

Figure 11. Antioxidant activity of QUE-LDHs/CS/PVA nanocomposite active films ((**a**), DPPH mehod. (**b**), ABTS method).

4. Conclusions

In this study, QUE-LDHs realizes a combination of active functions and enhancement effect through the deposition and complexation of QUE and copper ions on the LDHs' surface. CS/PVA matrix, reinforced with active QUE-LDHs, was prepared by a solution casting method. The results show that QUE-Cu^{2+} coordination compounds successfully to functionalize the LDHs nanosheets with a functional layer thickness of ~20 nm. Infrared and thermal analysis results revealed that there was a strong interface interaction between QUE-LDHs and CS/PVA matrix, resulting in the limited movement of PVA molecular chains and the increase in glass transition temperature and melting temperature. The thermal stability analysis revealed that the addition of QUE-LDHs can increase the initial decomposition temperature to a certain extent, but it still increased the thermal degradation rate of CS/PVA matrix in the later stage. After the addition of QUE-LDHs, QUE-LDHs/CS/PVA nanocomposite active films showed an excellent UV barrier, antibacterial, antioxidant properties and tensile strength, and still had certain transparency in the range of visible light. As the QUE-LDH content was 3 wt%, the active films exhibited the maximum tensile strength of 58.9 MPa, representing a significant increase in 40.9% compared with CS/PVA matrix. Notably, the UV barrier (280 nm) and antibacterial (*E. coli*) and antioxidant activities (DPPH method) of the active films achieved 100.0%, 95.5% and 58.9%, respectively. Therefore, LDHs@QUE-Cu/CS/PVA nanocomposite active films show excellent active packaging functions and have good potential in extending the shelf life of food.

Supplementary Materials: The following supporting information can be downloaded at: https://www.mdpi.com/article/10.3390/polym16060727/s1. Figure S1: XPS spectra of LDHs and QUE-LDHs. References [46–48] are cited in the supplementary materials.

Author Contributions: Conceptualization, C.W. and L.M.; methodology, L.M.; software, C.W. and L.M.; validation, B.Z.; formal analysis, L.M. and Y.L.; investigation, B.Z. and Y.L.; resources, J.Y.; data curation, C.W.; writing—original draft, C.W. and L.M.; writing—review and editing, L.M.; visualization, L.M.; supervision, L.M. and H.Z.; project administration, L.M. and H.Z.; funding acquisition, L.M. All authors have read and agreed to the published version of the manuscript.

Funding: This study was supported by National Natural Science Foundation of China (No. 52103098), Natural Science Foundation of Fujian province, China (No. 2021J05265).

Institutional Review Board Statement: Not applicable.

Data Availability Statement: Data are contained within the article and Supplementary Materials.

Conflicts of Interest: All authors declare no conflicts of interest.

References

1. Mohamad, N.; Mazlan, M.M.; Tawakkal, I.S.M.A.; Talib, R.A.; Kian, L.K.; Jawaid, M. Characterization of Active Polybutylene Succinate Films Filled Essential Oils for Food Packaging Application. *J. Polym. Environ.* **2022**, *30*, 585–596. [CrossRef]
2. Singh, A.K.; Kim, J.Y.; Lee, Y.S. Phenolic Compounds in Active Packaging and Edible Films/Coatings: Natural Bioactive Molecules and Novel Packaging Ingredients. *Molecules* **2022**, *27*, 7513. [CrossRef]
3. Lai, W.; Wong, W. Design and Practical Considerations for Active Polymeric Films in Food Packaging. *Int. J. Mol. Sci.* **2022**, *23*, 6295. [CrossRef]
4. Jayakumar, A.; Radoor, S.; Kim, J.T.; Rhim, J.W.; Nandi, D.; Parameswaranpillai, J.; Siengchin, S. Recent innovations in bionanocomposites-based food packaging films—A comprehensive review. *Food Packag. Shelf.* **2022**, *33*, 100877. [CrossRef]
5. Panda, P.K.; Dash, P.; Yang, J.; Chang, Y. Development of chitosan, graphene oxide, and cerium oxide composite blended films: Structural, physical, and functional properties. *Cellulose* **2022**, *29*, 2399–2411. [CrossRef]
6. Flórez, M.; Guerra-Rodríguez, E.; Cazón, P.; Vázquez, M. Chitosan for food packaging: Recent advances in active and intelligent films. *Food Hydrocolloid.* **2022**, *124*, 107328. [CrossRef]
7. Abdelghany, A.M.; Menazea, A.A.; Ismail, A.M. Synthesis, characterization and antimicrobial activity of Chitosan/Polyvinyl Alcohol blend doped with *Hibiscus sabdariffa* L. extract. *J. Mol. Struct.* **2019**, *1197*, 603–609. [CrossRef]
8. Chenwei, C.; Zhipeng, T.; Yarui, M.; Weiqiang, Q.; Fuxin, Y.; Jun, M.; Jing, X. Physicochemical, microstructural, antioxidant and antimicrobial properties of active packaging films based on poly(vinyl alcohol)/clay nanocomposite incorporated with tea polyphenols. *Prog. Org. Coat.* **2018**, *123*, 176–184. [CrossRef]
9. Koosha, M.; Hamedi, S. Intelligent Chitosan/PVA nanocomposite films containing black carrot anthocyanin and bentonite nanoclays with improved mechanical, thermal and antibacterial properties. *Prog. Org. Coat.* **2019**, *127*, 338–347. [CrossRef]
10. Yang, W.; Fortunati, E.; Bertoglio, F.; Owczarek, J.S.; Bruni, G.; Kozanecki, M.; Kenny, J.M.; Torre, L.; Visai, L.; Puglia, D. Polyvinyl alcohol/chitosan hydrogels with enhanced antioxidant and antibacterial properties induced by lignin nanoparticles. *Carbohyd. Polym.* **2018**, *181*, 275–284. [CrossRef] [PubMed]
11. Wu, J.; Ooi, C.W.; Song, C.P.; Wang, C.; Liu, B.; Lin, G.; Chiu, C.; Chang, Y. Antibacterial efficacy of quaternized chitosan/poly (vinyl alcohol) nanofiber membrane crosslinked with blocked diisocyanate. *Carbohyd. Polym.* **2021**, *262*, 117910. [CrossRef]
12. Ali, A.; Ahmed, S. Eco-friendly natural extract loaded antioxidative chitosan/polyvinyl alcohol based active films for food packaging. *Heliyon* **2021**, *7*, e6550.
13. Haghighi, H.; Leugoue, S.K.; Pfeifer, F.; Siesler, H.W.; Licciardello, F.; Fava, P.; Pulvirenti, A. Development of antimicrobial films based on chitosan-polyvinyl alcohol blend enriched with ethyl lauroyl arginate (LAE) for food packaging applications. *Food Hydrocolloid.* **2020**, *100*, 105419. [CrossRef]
14. Amalraj, A.; Haponiuk, J.T.; Thomas, S.; Gopi, S. Preparation, characterization and antimicrobial activity of polyvinyl alcohol/gum arabic/chitosan composite films incorporated with black pepper essential oil and ginger essential oil. *Int. J. Biol. Macromol.* **2020**, *151*, 366–375. [CrossRef]
15. Yong, H.; Liu, J. Active packaging films and edible coatings based on polyphenol-rich propolis extract: A review. *Compr. Rev. Food Sci. F* **2021**, *20*, 2106–2145. [CrossRef]
16. Soltani Firouz, M.; Mohi-Alden, K.; Omid, M. A critical review on intelligent and active packaging in the food industry: Research and development. *Food Res. Int.* **2021**, *141*, 110113. [CrossRef] [PubMed]
17. Lee, H.; Dellatore, S.M.; Miller, W.M.; Messersmith, P.B. Mussel-inspired surface chemistry for multifunctional coatings. *Science* **2007**, *318*, 426–430. [CrossRef] [PubMed]
18. Filippidi, E.; Cristiani, T.R.; Eisenbach, C.D.; Waite, J.H.; Israelachvili, J.N.; Ahn, B.K.; Valentine, M.T. Toughening elastomers using mussel-inspired iron-catechol complexes. *Science* **2017**, *358*, 502–505. [CrossRef] [PubMed]
19. Mao, L.; Wang, C.; Yao, J.; Lin, Y.; Liao, X.; Lu, J. Design and fabrication of anthocyanin functionalized layered clay/poly(vinyl alcohol) coatings on poly(lactic acid) film for active food packaging. *Food Packag. Shelf.* **2023**, *35*, 101007. [CrossRef]
20. Mao, L.; Wu, H.; Liu, Y.; Yao, J.; Bai, Y. Enhanced mechanical and gas barrier properties of poly(ε-caprolactone) nanocomposites filled with tannic acid-Fe(III) functionalized high aspect ratio layered double hydroxides. *Mater. Chem. Phys.* **2018**, *211*, 501–509. [CrossRef]
21. Luzi, F.; Pannucci, E.; Santi, L.; Kenny, J.M.; Torre, L.; Bernini, R.; Puglia, D. Gallic Acid and Quercetin as Intelligent and Active Ingredients in Poly(vinyl alcohol) Films for Food Packaging. *Polymers* **2019**, *11*, 1999. [CrossRef]
22. Lei, Y.; Mao, L.; Yao, J.; Zhu, H. Improved mechanical, antibacterial and UV barrier properties of catechol-functionalized chitosan/polyvinyl alcohol biodegradable composites for active food packaging. *Carbohyd. Polym.* **2021**, *264*, 117997. [CrossRef]
23. de Barros Vinhal, G.L.R.R.; Silva-Pereira, M.C.; Teixeira, J.A.; Barcia, M.T.; Pertuzatti, P.B.; Stefani, R. Gelatine/PVA copolymer film incorporated with quercetin as a prototype to active antioxidant packaging. *J. Food Sci. Technol.* **2021**, *58*, 3924–3932. [CrossRef] [PubMed]
24. Wang, Q.; Li, P.; Xue, J.; Zhao, W.; Wu, D. Complexing Reactions of Quercetin with Cu(II) and Al(III) Studied by UV-vis Absorption Spectroscopy. *J. Light Scatt.* **2009**, *21*, 174–177.
25. Ji, J. XPS Study on Cu^{2+}-Chitosan Chelate and Adsorption Mechanism of Chitosan for Cu^{2+}. *Chin. J. Appl. Chem.* **2000**, *17*, 115–116.
26. Aytac, Z.; Ipek, S.; Durgun, E.; Uyar, T. Antioxidant electrospun zein nanofibrous web encapsulating quercetin/cyclodextrin inclusion complex. *J. Mater. Sci.* **2018**, *53*, 1527–1539. [CrossRef]
27. He, Y.; Zhu, B.; Inoue, Y. Hydrogen bonds in polymer blends. *Prog. Polym. Sci.* **2004**, *29*, 1021–1051. [CrossRef]

28. Lan, W.; Zhang, R.; Ahmed, S.; Qin, W.; Liu, Y. Effects of various antimicrobial polyvinyl alcohol/tea polyphenol composite films on the shelf life of packaged strawberries. *LWT-Food Sci. Technol.* **2019**, *113*, 108297. [CrossRef]
29. Mao, L.; Liu, Y.; Wu, H.; Chen, J.; Yao, J. Poly(ε-caprolactone) filled with polydopamine-coated high aspect ratio layered double hydroxide: Simultaneous enhancement of mechanical and barrier properties. *Appl. Clay Sci.* **2017**, *150*, 202–209. [CrossRef]
30. Kiliaris, P.; Papaspyrides, C.D. Polymer/layered silicate (clay) nanocomposites: An overview of flame retardancy. *Prog. Polym. Sci.* **2010**, *35*, 902–958. [CrossRef]
31. Nagendra, B.; Das, A.; Leuteritz, A.; Gowd, E.B. Structure and crystallization behaviour of syndiotactic polystyrene/layered double hydroxide nanocomposites. *Polym. Int.* **2016**, *65*, 299–307. [CrossRef]
32. Qiu, L.; Gao, Y.; Lu, P.; O'Hare, D.; Wang, Q. Synthesis and properties of polypropylene/layered double hydroxide nanocomposites with different LDHs particle sizes. *J. Appl. Polym. Sci.* **2018**, *135*, 46204. [CrossRef]
33. Ramaraj, B.; Nayak, S.K.; Yoon, K.R. Poly(vinyl alcohol) and layered double hydroxide composites: Thermal and mechanical properties. *J. Appl. Polym. Sci.* **2010**, *116*, 1671–1677. [CrossRef]
34. Ghaderi, J.; Hosseini, S.F.; Keyvani, N.; Gómez-Guillén, M.C. Polymer blending effects on the physicochemical and structural features of the chitosan/poly(vinyl alcohol)/fish gelatin ternary biodegradable films. *Food Hydrocolloid.* **2019**, *95*, 122–132. [CrossRef]
35. Yao, J.; Mao, L.; Wang, C.; Liu, X.; Liu, Y. Development of chitosan/poly (vinyl alcohol) active films reinforced with curcumin functionalized layered clay towards food packaging. *Prog. Org. Coat.* **2023**, *182*, 107674. [CrossRef]
36. Bian, J.; Han, L.; Wang, X.; Wen, X.; Han, C.; Wang, S.; Dong, L. Nonisothermal crystallization behavior and mechanical properties of poly(butylene succinate)/silica nanocomposites. *J. Appl. Polym. Sci.* **2010**, *116*, 902–912. [CrossRef]
37. Valente, J.S.; Sánchez-Cantú, M.; Lima, E.; Figueras, F. Method for Large-Scale Production of Multimetallic Layered Double Hydroxides: Formation Mechanism Discernment. *Chem. Mater.* **2009**, *21*, 5809–5818. [CrossRef]
38. Balavairavan, B.; Saravanakumar, S.S. Characterization of Ecofriendly Poly(vinyl alcohol) and Green Banana Peel Filler (GBPF) Reinforced Bio-Films. *J. Polym. Environ.* **2021**, *29*, 2756–2771. [CrossRef]
39. Shokuhi Rad, A.; Ebrahimi, D. Improving the Mechanical Performance and Thermal Stability of a PVA-Clay Nanocomposite by Electron Beam Irradiation. *Mech. Compos. Mater.* **2017**, *53*, 373–380. [CrossRef]
40. Marangoni, R.; Gardolinski, J.E.F.D.; Mikowski, A.; Wypych, F. PVA nanocomposites reinforced with Zn2Al LDHs, intercalated with orange dyes. *J. Solid State Electr.* **2011**, *15*, 303–311. [CrossRef]
41. Hajji, S.; Chaker, A.; Jridi, M.; Maalej, H.; Jellouli, K.; Boufi, S.; Nasri, M. Structural analysis, and antioxidant and antibacterial properties of chitosan-poly(vinyl alcohol) biodegradable films. *Environ. Sci. Pollut. Res.* **2016**, *23*, 15310–15320. [CrossRef] [PubMed]
42. Bhowmik, S.; Agyei, D.; Ali, A. Bioactive chitosan and essential oils in sustainable active food packaging: Recent trends, mechanisms, and applications. *Food Packag. Shelf.* **2022**, *34*, 100962. [CrossRef]
43. Yu, J.; Wei, D.; Li, S.; Tang, Q.; Li, H.; Zhang, Z.; Hu, W.; Zou, Z. High-performance multifunctional polyvinyl alcohol/starch based active packaging films compatibilized with bioinspired polydopamine nanoparticles. *Int. J. Biol. Macromol.* **2022**, *210*, 654–662. [CrossRef] [PubMed]
44. Babaei-Ghazvini, A.; Acharya, B.; Korber, D.R. Antimicrobial Biodegradable Food Packaging Based on Chitosan and Metal/Metal-Oxide Bio-Nanocomposites: A Review. *Polymers* **2021**, *13*, 2790. [CrossRef] [PubMed]
45. Vidal, O.L.; Barros Santos, M.C.; Batista, A.P.; Andrigo, F.F.; Baréa, B.; Lecomte, J.; Figueroa-Espinoza, M.C.; Gontard, N.; Villeneuve, P.; Guillard, V.; et al. Active packaging films containing antioxidant extracts from green coffee oil by-products to prevent lipid oxidation. *J. Food Eng.* **2022**, *312*, 110744. [CrossRef]
46. Fortunati, E.; Luzi, F.; Dugo, L.; Fanali, C.; Tripodo, G.; Santi, L.; Kenny, J.M.; Torre, L.; Bernini, R. Effect of hydroxytyrosol methyl carbonate on the thermal, migration and antioxidant properties of PVA-based films for active food packaging. *Polym. Int.* **2016**, *65*, 872–882. [CrossRef]
47. Dai, W.; Sun, Q.; Deng, J.; Wu, D.; Sun, Y. XPS studies of $Cu/ZnO/Al_2O_3$ ultra-fine catalysts derived by a novel gel oxalate co-precipitation for methanol synthesis by CO_2+H_2. *Appl. Surf. Sci.* **2001**, *177*, 172–179. [CrossRef]
48. Bai, Y.; Li, Y.; Wang, E.; Wang, X.; Lu, Y.; Xu, L. A novel reduced α-Keggin type polyoxometalate coordinated to two and a half copper complex moieties: $[Cu(2,2'-bipy)2][PMoVI8MoV4O40\{Cu(2,2'-bipy)\}2.5]\cdot H_2O$. *J. Mol. Struct.* **2005**, *752*, 54–59. [CrossRef]

Disclaimer/Publisher's Note: The statements, opinions and data contained in all publications are solely those of the individual author(s) and contributor(s) and not of MDPI and/or the editor(s). MDPI and/or the editor(s) disclaim responsibility for any injury to people or property resulting from any ideas, methods, instructions or products referred to in the content.

Article

Scaffolds Loaded with Dialdehyde Chitosan and Collagen—Their Physico-Chemical Properties and Biological Assessment

Sylwia Grabska-Zielińska [1,*,†], Judith M. Pin [2,†], Beata Kaczmarek-Szczepańska [3], Ewa Olewnik-Kruszkowska [1], Alina Sionkowska [3], Fernando J. Monteiro [4,5,6], Kerstin Steinbrink [2] and Konrad Kleszczyński [2]

1. Department of Physical Chemistry and Physicochemistry of Polymers, Faculty of Chemistry, Nicolaus Copernicus University, Gagarin 7, 87-100 Toruń, Poland; olewnik@umk.pl
2. Department of Dermatology, University of Münster, Von-Esmarch-Str. 58, 48149 Münster, Germany; j_pin001@uni-muenster.de (J.M.P.); kerstin.steinbrink@ukmuenster.de (K.S.); konrad.kleszczynski@ukmuenster.de (K.K.)
3. Department of Biomaterials and Cosmetics Chemistry, Faculty of Chemistry, Nicolaus Copernicus University, Gagarin 7, 87-100 Toruń, Poland; beata.kaczmarek@umk.pl (B.K.-S.); alinas@umk.pl (A.S.)
4. i3S—Instituto de Investigação e Inovação em Saúde, Universidade do Porto, 4200-180 Porto, Portugal; fjmont@i3s.up.pt
5. INEB—Instituto de Engenharia Biomédica, Universidade do Porto, 4200-180 Porto, Portugal
6. FEUP—Faculdade de Engenharia, Universidade do Porto, 4200-465 Porto, Portugal
* Correspondence: sylwiagrabska91@gmail.com
† These authors contributed equally to this work.

Citation: Grabska-Zielińska, S.; Pin, J.M.; Kaczmarek-Szczepańska, B.; Olewnik-Kruszkowska, E.; Sionkowska, A.; Monteiro, F.J.; Steinbrink, K.; Kleszczyński, K. Scaffolds Loaded with Dialdehyde Chitosan and Collagen—Their Physico-Chemical Properties and Biological Assessment. *Polymers* 2022, 14, 1818. https://doi.org/10.3390/polym14091818

Academic Editor: Brian G. Amsden

Received: 30 March 2022
Accepted: 27 April 2022
Published: 29 April 2022

Publisher's Note: MDPI stays neutral with regard to jurisdictional claims in published maps and institutional affiliations.

Copyright: © 2022 by the authors. Licensee MDPI, Basel, Switzerland. This article is an open access article distributed under the terms and conditions of the Creative Commons Attribution (CC BY) license (https:// creativecommons.org/licenses/by/ 4.0/).

Abstract: In this work, dialdehyde chitosan (DAC) and collagen (Coll) scaffolds have been prepared and their physico-chemical properties have been evaluated. Their structural properties were studied by Fourier Transform Infrared Spectroscopy with Attenuated Internal Reflection (FTIR–ATR) accompanied by evaluation of thermal stability, porosity, density, moisture content and microstructure by Scanning Electron Microscopy—SEM. Additionally, cutaneous assessment using human epidermal keratinocytes (NHEK), dermal fibroblasts (NHDF) and melanoma cells (A375 and G-361) was performed. Based on thermal studies, two regions in DTG curves could be distinguished in each type of scaffold, what can be assigned to the elimination of water and the polymeric structure degradation of the materials components. The type of scaffold had no major effect on the porosity of the materials, but the water content of the materials decreased with increasing dialdehyde chitosan content in subjected matrices. Briefly, a drop in proliferation was noticed for scaffolds containing 20DAC/80Coll compared to matrices with collagen alone. Furthermore, increased content of DAC (50DAC/50Coll) either significantly induced the proliferation rate or maintains its ratio compared to the control matrix. This delivery is a promising technique for additional explorations targeting therapies in regenerative dermatology. The using of dialdehyde chitosan as one of the main scaffolds components is the novelty in terms of bioengineering.

Keywords: chitosan dialdehyde; collagen; bioengineering; cutaneous cells; scaffolds

1. Introduction

Collagen (Coll) is one of the most important biopolymers, belonging to the group of proteins [1,2]. It is found in the skin, bone and cartilage tissue, tendons, endothelial vessels, and in the extracellular matrix (ECM) [3–5]. The collagen family consists of 29 distinct collagen types. They are divided into four classes based on existence of various α chains, isoforms of particles, supermolecular structures of each collagen type, differences in the expressions of genes involved in protein biosynthesis and post-translational modifications of collagens [2,3]. Thus, the class of collagen depends on its structural and composition properties [2].

Collagen is widely used in many areas including biomaterials, tissue engineering, drug delivery systems, cosmetology, pharmacy, and the food industry [4,6–8]. As a biomaterial, collagen exerts numerous advantages such as biodegradability and bioresorbability, non-toxicity and biocompatibility, non-antigenicity, as well as synergy with bioactive components. Moreover, the possibility of its formulation in a number of different forms, it is easily modifiable to produce materials as desired by utilizing its functional groups, compatibility with synthetic polymers [7]. Additionally, it also has some disadvantages, i.e., a high cost of pure type I collagen, variability of isolated collagen (e.g., crosslink density, fiber size, trace impurities, etc.), hydrophilicity which leads to swelling and a more rapid release of substances incorporated to material, and variability in enzymatic degradation rate as compared with hydrolytic degradation. Furthermore, complex handling properties, side effects such as bovine spongeform encephalopathy (BSF), mineralization, low stability under high temperature or presence of enzymes [4,7]; therefore, the structure of pure collagen requires stabilization and modification [4,7].

The most commonly used methods to modify collagen materials are mixing with natural (chitosan [9], hyaluronic acid [10], silk fibroin [11], elastin [12], keratin [13]) or synthetic polymers (poly(vinyl pyrrolidone) [14], poly(vinyl alcohol) [15], poly(ethylene glycol) [16], poly(ethylene oxide) [16]), and cross-linking with chemical (EDC/NHS [17], dialdehyde starch [18], glutaraldehyde [19], genipin [20]), physical (temperature [21], UV light [22]) or enzymatic (microbial transglutaminase [4,23]) factors.

Dialdehyde chitosan is the compound obtained from chitosan by oxidation with sodium or potassium periodate. The process of periodate oxidation endows chitosan with multiple functional aldehyde groups. The dialdehydes, including dialdehyde chitosan, are considered as a safe additives and green cross-linking agents for various biomaterials and nanomaterials. They also can be used as a biological tissue fixation and tanning agents [24–27]. The use of dialdehyde compounds for materials' modification is a good route since dialdehydes react readily with functional groups from other polymers. Dialdehyde starch [18,28,29], dialdehyde carboxymethyl cellulose [30], dialdehyde nanocellulose [31], dialdehyde cellulose [32], dialdehyde alginate [33,34] and dialdehyde chitosan [24,27] have been used to cross-link collagen materials. Thus, Figure 1 shows the mechanism of the reaction between collagen and a dialdehyde compound.

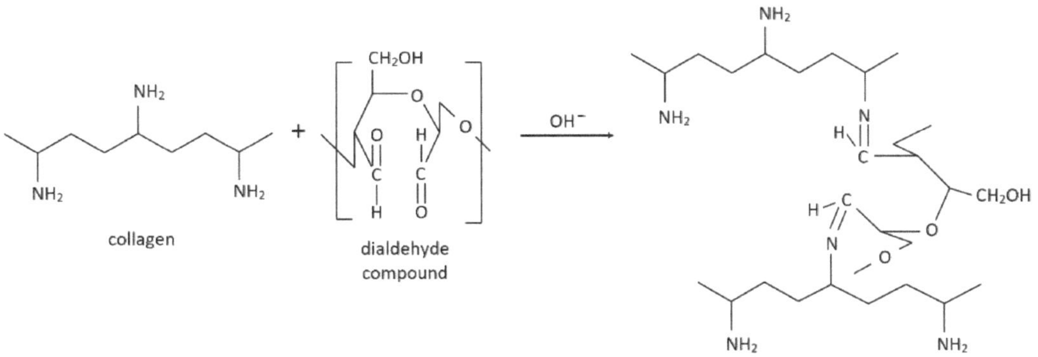

Figure 1. The mechanism of collagen and dialdehyde compound reaction (based on collagen and dialdehyde cellulose example from Pietrucha and Safandowska's work [32]).

Pietrucha and Safandowska [32] have studied the physicochemical properties of silver carp collagen modified by dialdehyde cellulose, and they reported a marked increase in thermostability of collagen structure and the improvement of the mechanical strength of the modified materials. Hu et al. [33] described the interaction between collagen and alginate dialdehyde (ADA) used as naturally derived cross-linker, and they indicated that the ADA addition could improve the properties of collagen-based materials such as thermal stability and hydrophilicity. Additionally, the dialdehyde compound resulted in the aggregation of collagen molecules, not destroying the triple helix conformation of collagen and it has a positive effect on cells proliferation at a certain content of ADA [33]. Yu et al. [34] also used dialdehyde alginate to modify collagen and proved that ADA significantly improving swelling, rheological behaviors and capability to resist against type I collagenase. Concerning dialdehyde chitosan as a cross-linking agent, Wanli et al. [35] used it for collagen fibers cross-linking. They reported that the thermal denaturation temperature of collagen fiber rose with increasing oxidation degree of chitosan dialdehyde, and the porosity of collagen fiber was reduced accordingly [35]. Liu et al. [27] also worked with dialdehyde chitosan as cross-linking agent for collagen materials and concluded that introducing DAC into collagen may be favorable for cellular growth, adhesion and proliferation. According to their report, chitosan dialdehyde might be an ideal cross-linking agent for the chemical fixation of collagen [27]. On the other hand, Bam et al. [24] designed biostable scaffolds based on collagen cross-linked with dialdehyde chitosan in the presence of gallic acid (GA), and observed that the formed stable Schiff's base between collagen and DAC with GA had significant effects in improving microstructural integrity. Additionally, the texture, thermal and structural properties, biostability, swelling and water uptake have been improved after the introduction of DAC in the scaffolds [24].

In this work, we decided to use dialdehyde chitosan as one of the scaffolds components. Usually, as we considered above, dialdehydes generally were used as additives to mixtures of biopolymers or to pure biopolymers, that is as cross-linking agents [24–27], not as one of the main components of scaffolds. The aim of this work was to obtain and characterize materials based on collagen with chitosan dialdehyde (DAC) in various compositions (DAC/Coll: 80/20, 50/50 and 20/80). The use of dialdehyde chitosan as a component of collagen materials, not a cross-linking agent is a novelty. There is only one report, where dialdehyde chitosan is mixed as component of scaffold. Mixtures with hyaluronic acid and the results of their physico-chemical characterization were described in our previous reports [36,37].

2. Materials and Methods

2.1. Materials

Collagen (Coll) and dialdehyde chitosan (DAC) were obtained in-house. Collagen was prepared from tail tendons of young rats following our previously reported method [38]. Dialdehyde chitosan was obtained by one-step synthesis following the method described by Bam et al. [24] with slight modifications. The synthesis of dialdehyde chitosan was previously described by this research group [36]. Reagents purchased form Sigma-Aldrich (St. Louis, MO, USA): chitosan (DD = 78%), acetic acid, HCl, acetone, isopropanol, sodium periodate, Minimum Essential Medium Eagle (MEM) (1000 mg/L), 1% penicillin-streptomycin solution, 3-(4,5-dimethylthiazol-2-yl)-2,5-diphenyltetrazolium bromide (MTT), L-glutamine (200 mM), and 0.05% trypsin/0.53 mM EDTA solution. Fetal bovine serum was purchased from Thermo Fisher Scientific (Waltham, MA, USA). Human epidermal keratinocytes (NHEKs) and human dermal fibroblasts (NHDFs) were supplied by PromoCell (Heidelberg, Germany) and American Type Culture Collection (ATCC) (Manassas, VA, USA), respectively. Human melanoma cell i.e., amelanotic A375 and G-361 cell lines supplied by ATCC (Manassas, VA, USA).

2.2. Obtaining the Scaffolds

Chitosan dialdehyde was dissolved in water and collagen was dissolved in 0.1 M acetic acid at 1% concentration separately. They were mixed in different weight ratios: 80/20, 50/50 and 20/80. The scaffolds based on pure collagen were treated as control samples. Solutions were mixed with a magnetic stirrer for 1 h and the obtained mixtures were poured into 24-well polystyrene culture plates, frozen, and lyophilized (−20 °C, 100 Pa, 48 h, ALPHA 1–2 LDplus, CHRIST, Ostreode am Harz, Germany).

2.3. Structural Studies—Attenuated Total Reflectance–Fourier Transform Infrared Spectroscopy (FTIR-ATR)

Nicolet iS10 spectrometer equipped with an attenuated total reflectance (FTIR–ATR) device with a germanium crystal (Nicolet iS10, Thermo Fisher Scientific, Waltham, MA, USA) was used to analyze the chemical structure of the obtained scaffolds. The spectra were evaluated in the range of 600–4000 cm^{-1}. All spectra were recorded with the resolution of 4 cm^{-1} with 64 scans.

2.4. Thermal Stability

Thermogravimetric analyses were performed at a heating rate of 10 °C/min (20–600 °C) under nitrogen atmosphere, by using TA Instruments SDT 2960 Simultaneous TG-DTG (TA Instruments Manufacturer, Eschborn, Germany). From thermogravimetric curves, the characteristic temperature at a maximum decomposition rate of the investigated composites was determined.

2.5. Determination of Density, Porosity and Water Content

The liquid displacement method with isopropanol was used to measure density and porosity of the scaffolds. A fragment of the sample with a known weight was immersed in a cylinder with a known volume of isopropanol for 3 min. The density was calculated using the following Equation (1):

$$d\left[\frac{mg}{cm^3}\right] = \frac{W}{V_2 - V_3} \cdot 100\% \qquad (1)$$

where W—weight of sample (mg), V_2—total volume of isopropanol with the isopropanol impregnated sample (cm^3), and V_3—volume of isopropanol after scaffold removal (cm^3). The porosity was calculated using following Equation (2):

$$\varepsilon\,[\%] = \frac{V_1 - V_3}{V_2 - V_3} \times 100\% \qquad (2)$$

where V_1—initial volume of isopropanol (cm^3), and V_2, V_3—as above.

Gravimetric analysis was used to determine the water content of the samples. The water content of scaffolds was measured by drying samples at 105 °C until they reached a constant weight. The results were expressed as grams of water per 100 g of dry sample.

2.6. Scanning Electron Microscopy Imaging

The morphology of the samples was studied using a Scanning Electron Microscope (LEO Electron Microscopy Ltd., Cambridge, UK). Scaffolds were frozen in liquid nitrogen for a few minutes, cut with a razor blade and gold coated prior to the observation.

2.7. Cell Culture and Proliferation Ratio Assessment

Human epidermal keratinocytes (NHEKs) and human dermal fibroblasts (NHDFs) were supplied by PromoCell (Heidelberg, Germany); amelanotic A375 and G-361 cell lines were supplied by the American Type Culture Collection (ATCC) (Manassas, VA, USA), respectively. NHEKs were grown in Keratinocyte Growth Medium 2 supplemented with 1% penicillin-streptomycin solution while NHDFs were maintained in MEM medium supplemented with 10% (v/v) heat-inactivated fetal bovine serum, 2 mM of L-glutamine, and 1% (v/v) streptomycin-penicillin solution. Melanoma cells were maintained in MEM medium supplemented with 10% (v/v) heat-inactivated fetal bovine serum, 2 mM of L-glutamine, and 1% (v/v) streptomycin-penicillin solution. Cells were seeded in 24-wells plates at the density of 0.5×10^5 cells/well and were allowed to attach to the surface of the scaffolds for 24 h. Afterwards, cells were cultured in supplemented culture medium in a humidified 5% CO_2 atmosphere at 37 °C for 96 h while the culture medium was exchanged every 48 h. Differences in cell viability were assessed using the MTT assay. MTT (5 mg/mL in 1 × PBS) was prepared in the respective culture medium (the final dilution, 1:10), 100 µL of assay reagent was added to each well, and cells were subsequently incubated for 3 h in a humidified atmosphere of 5% CO_2 at 37 °C. The resultant formazan crystals were dissolved using 100 µL isopropanol/0.04 N HCl, absorbance was measured at λ = 595 nm using the BioTek ELx808™ microplate reader (BioTek Instruments, Inc., Winooski, VT, USA), and the results were normalized to the control cells.

2.8. Statistics

Statistical analysis of the data was completed using commercial software (GraphPad Prism 8.0.1.244, GraphPad Software, San Diego, CA, USA). The results were presented as a mean ± standard deviation (S.D.) and were statistically analyzed using one-way analysis of variance (one-way ANOVA). Multiple comparisons between the means were performed with the statistical significance set at $p \leq 0.05$. Results from mechanical tests, density, porosity and water content measurements were subjected to statistical analysis.

For cell culture and proliferation ratio assessment, data were expressed as pooled means + S.D. of six independent experiments ($n = 6$). Statistically significant differences between results were determined by the univariate analysis of variance (ANOVA) or the Student's t-test and appropriate post hoc analysis (Tukey or Dunnett tests, accordingly). All the analysis are presented as percentage of the control sample and a $p < 0.05$ was considered as statistically significant.

3. Results and Discussion

3.1. Structural Studies—FTIR-ATR

To evaluate the molecular structure of DAC/Coll scaffolds and confirm possible interactions between dialdehyde chitosan and collagen, FTIR-ATR analysis was performed. Additionally, FTIR analysis allows to observe the presence of functional groups, that may be used to identify compounds and interactions between material components. The spectra of the obtained materials are shown in Figure 2. The structure of collagen [37–41] and dialdehyde chitosan [25–27,42,43] have often been analyzed and described in the literature (Figure S1 in the Supplementary Materials). Herein, it was decided to record only the spectra of mixtures based on dialdehyde chitosan and collagen, as presented via the positions of their characteristic bands (Table 1). It may be observed that the position of Amide II and Amide III does not depend on the DAC/Coll weight ratio. However, the Amide A and Amide B band positions are shifted. This suggests that they participate in the formation of hydrogen bonds between these materials. Additionally, the C–OH peak was not observed in the spectra with the lowest DAC content. The –C=N peak was noticed only for the material composition with the highest DAC content. In the other two materials (20DAC/80Coll and 50DAC/50Coll), this peak of stretching vibration (around 1630–1640 cm^{-1}) formed by Schiff base reaction overlapped with the C=O stretching in Amide I [44]. Based on these observations, our results indicated that dialdehyde chitosan was successfully introduced into the collagen matrix.

Table 1. The positions of characteristic bands of DAC/Coll based scaffolds.

Specimen	Amide A	Amide B	CH_3	C–OH	Amide I	C=N	Amide II	Amide III
20DAC/80Coll	3313	3077	2932	—	1656	—	1556	1241
50DAC/50Coll	3306	3078	2934	1730	1656	—	1556	1241
80DAC/20Coll	3322	3086	2938	1731	1658	1632	1556	1240

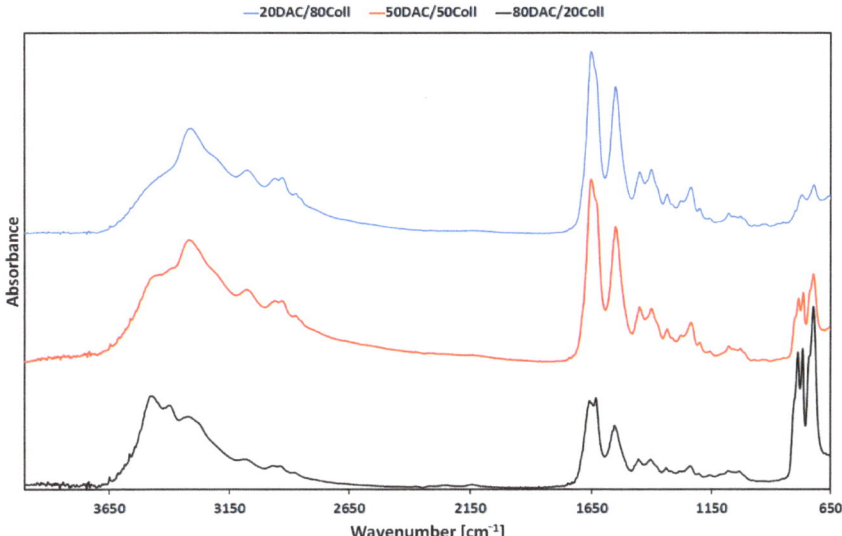

Figure 2. The FTIR-ATR spectra of DAC/Coll based materials.

3.2. Thermal Stability

The biopolymers are characterized by low denaturation temperature, and their thermal properties should be considered (Table 2). TG-DTG curves of collagen, 20DAC/80Coll and 80DAC/20Coll scaffolds are shown in Figure S2 and S3 in the Supplementary Materials.

Table 2. Results of DTG analysis with temperatures of maximum peaks.

Specimen	T_{max} (1) (°C)	T_{max} (2) (°C)
100Coll	47.5	324.3
20DAC/80Coll	147.0	326.7
50DAC/50Coll	147.1	314.3
80DAC/20Coll	147.6	307.3

Two regions in DTG curves could be distinguished in each type of scaffolds. The first one may be correlated with the elimination of water molecules present in the scaffolds [36]. For scaffolds made of pristine collagen, it was 47.5 °C. For materials based on dialdehyde chitosan and collagen mixtures, the region responsible for water elimination, were observed at much higher temperature, namely 147.0 °C for 20DAC/80Coll scaffolds and approximately 147 °C for 50/50 and 80/20 DAC/Coll materials. Thus, no significant differences between scaffolds with different DAC/Coll compositions were observed, but a significant improvement in thermal stability was noticed for DAC/Coll based materials over pristine collagen-based matrices. The second region in DTG curves may be assigned to the degradation of the polymeric structure of the materials components [36] and the fast volatilization of the polymer segment due to the thermal scission of the polymer backbone [44]. T_{max} (2) for collagen materials was comparable with T_{max} (2) for DAC/Coll samples, and it was within the limits of 307.3 to 326.7 °C. Similar results were obtained by Liu et al. [27] who showed that the addition of oxidized chitosan resulted in the increase in maximum temperature of the first and second stages. Additionally, the addition of chitosan dialdehyde to cellulose increased the thermal stability of the obtained films [44]. In summary, for materials made of dialdehyde chitosan and collagen-based mixtures, the temperature below 146 °C is safe and does not cause material degradation. The Schiff base reactions (what was reported in Sections 1 and 3.1. Structural studies—FTIR–ATR) that occurred between dialdehyde chitosan and collagen significantly improved the thermal stability of the scaffolds (Figure 1) [44].

3.3. Density, Porosity and Water Content

Parameters such as density, porosity or moisture content are important from the point of view of using the material in tissue engineering. They were determined and the results are shown in Figure 3.

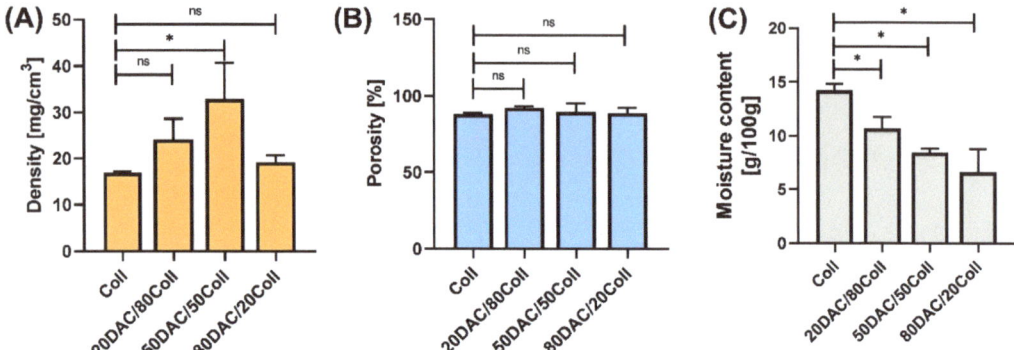

Figure 3. Results for (**A**) density, (**B**) porosity and (**C**) moisture content of DAC/Coll scaffolds. Statistically significant differences versus pristine collagen scaffolds were indicated as follows: * $p < 0.05$; ns—not significant.

Namely, the porosity (Figure 3B) of investigated materials ranged from 88.08 ± 0.64% (Coll) to 92.27 ± 0.99% (20DAC/80Coll). No statistically significant differences between control (Coll) and other materials were observed. Potential material targeting tissue engineering should be characterized by porosity of about 90% [45–47]. Here, porosity of our assessed matrices was higher or almost 90% and this means that they met the above requirement. The content of dialdehyde chitosan in the scaffolds had no major influence on the porosity of the materials but the addition of dialdehyde chitosan to collagen may affect the density of the material, which is in agreement with Pietrucha and Safandowska [32].

Furthermore, the highest density was observed for 50DAC/50Coll scaffolds. No statistically significant differences between pristine collagen scaffolds and 20DAC/80Coll and 80DAC/20Coll have been noticed (Figure 3A). With increasing dialdehyde chitosan to collagen, the material density increased until at the weight ratio 50DAC/50Coll.

Concerning moisture content (Figure 3C), it may be seen that it decreased with increasing dialdehyde chitosan to collagen ratio, and it reached 14.22 ± 0.34 g/100 g for collagen and 6.62 ± 1.27 g/100 g for 80DAC/20Coll scaffold. The moisture content of pure collagen materials was the same as in our previous study, where collagen, collagen and silk fibroin as well as collagen, silk fibroin and chitosan scaffolds were cross-linked with dialdehyde starch [47]. In the above-mentioned study, it was concluded that in the case of materials modified with a dialdehyde compound, the moisture content decreased in comparison with the moisture content found in the unmodified material [47]. It suggests that collagen has more polar character than chitosan as it has more hydrophilic groups for a given chain length. Only with pristine collagen the moisture content for the modified and unmodified material was 14.17 ± 1.36 g/100 g and 13.10 ± 0.79 g/100 g, respectively. Nevertheless, such a small difference could be due to the fact that in that study only a 10% addition of dialdehyde compound to the polymer was used [47].

3.4. Morphological Studies—SEM

SEM imaging was used to assess the morphological structure of scaffolds. The SEM images show the inner region of the scaffold after cutting it with a scalpel, as shown in

Figure 4. Namely, the heteroporous structure of collagen and dialdehyde chitosan/collagen materials were well-ordered. After freeze-drying process where solvents (water and 0.1 M acetic acid) act as pore-forming agents (porogens), the materials were characterized by pores with variable size and geometry. The scaffold morphology was related to the feed ratio of dialdehyde chitosan to collagen. As for 80DAC/20Coll and 20DAC/80Coll materials, the sponges had structures very similar to that of native collagen. 80DAC/20Coll materials presented the best microstructure. Pores were most regular and most similar to each other in shape and size that diameter was approximately 140 μm for collagen, 133 μm for 20DAC/80Coll, 90 μm for 50DAC/50Coll, and 75 μm for 80DAC/20Coll.

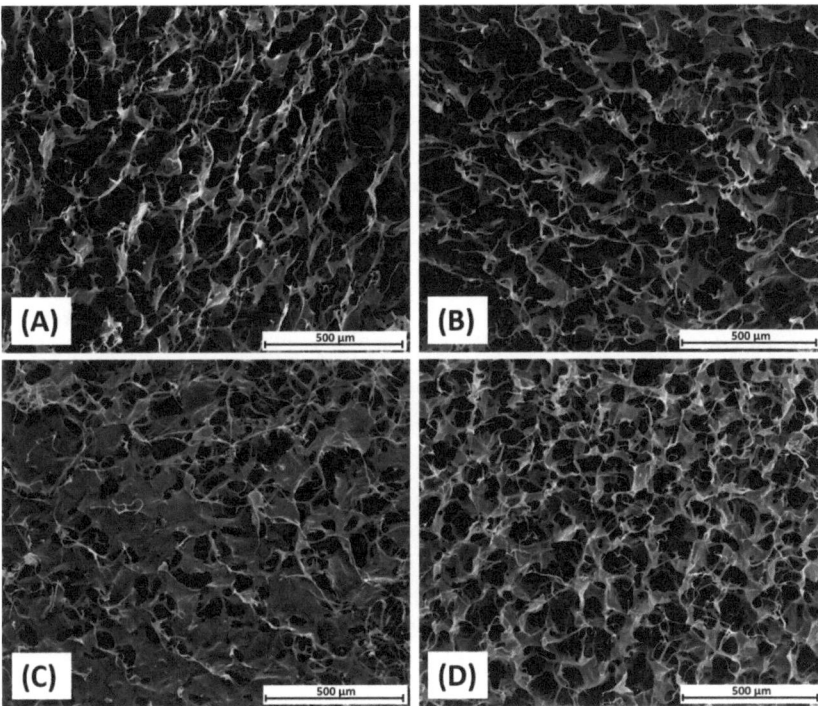

Figure 4. The SEM pictures of collagen (**A**); 20DAC/80Coll (**B**); 50DAC/50Coll (**C**); 80DAC/20Coll (**D**) scaffolds. Magnification 200×.

About 50DAC/50Coll, the material was characterized by a heterogenous structure, and it could be a result of the excessively fast cross-linking between collagen and dialdehyde chitosan. In this case, the pores were irregular, more closed and unequal. A similar observation was reported by Ding et al. [48] where collagen materials were cross-linked by dialdehyde cellulose. This is consistent with the results of density (Figure 3), e.g., the highest density was visible, and it was 32.89 ± 4.50 mg/cm^3.

Bam et al. [24] reported a little bit more flattened structure of DAC modified collagen than pure collagen scaffold. Mu et al. [18] studied several physico-chemical properties of collagen materials modified with dialdehyde starch (DAS) with weight ratios of DAS/Coll: 1:100, 1:70 and 1:10, and they observed that fibrous surface structures were more pronounced at low DAS content, while integrally lamellar aggregates appeared as DAS increased. One may say that this coincides with our results if we look at the materials 50DAC/50Coll (C) and 80DAC/20Coll (D). They also reported [18] that the thickness of the pore wall increased with the content of DAS, and their cryogels became more solid with higher DAS content.

3.5. Cellular Assessments Using Cutaneous Models

We performed the proliferation rate assessment (Figure 5) using selected cutaneous cell models, i.e., human epidermal keratinocytes and dermal fibroblasts providing the new insight into a considerable improvement of wound dressing, re-epithelization or therapeutic approaches for skin. Comparatively, we also tested melanoma cell lines used as reference cellular models to confirm the responses of primary cell lines as it was presented in our latest studies [36,49,50]. Within all investigated cell lines, we observed similar pattern of regulation where 20DAC/80Coll scaffolds caused prominent drop in cell proliferation, ranging from 29% to 46% versus scaffolds containing only collagen. Furthermore, increased content of DAC (50DAC/50Coll) either significantly induced the proliferation rate or maintained its ratio compared to the control matrix. A similar response was noticed when comparing scaffolds loaded only with DAC and those with respective addition of collagen. Namely, 50DAC/50Coll scaffolds enhanced proliferation rate versus DAC alone by 40% (NHEKs, Figure 5A), 33% (NHDFs, Figure 5B), 30% (A375, Figure 5C), 37% (G361, Figure 5D) while cells cultured on 80DAC/20Coll matrices revealed slight but not statistically significant increase versus DAC. The composition of 50DAC/50Coll provides the most suitable environment for cells as for them the cell viability is the highest for each type of cell line. It may be assumed that both collagen and dialdehyde chitosan are valuable as matrices.

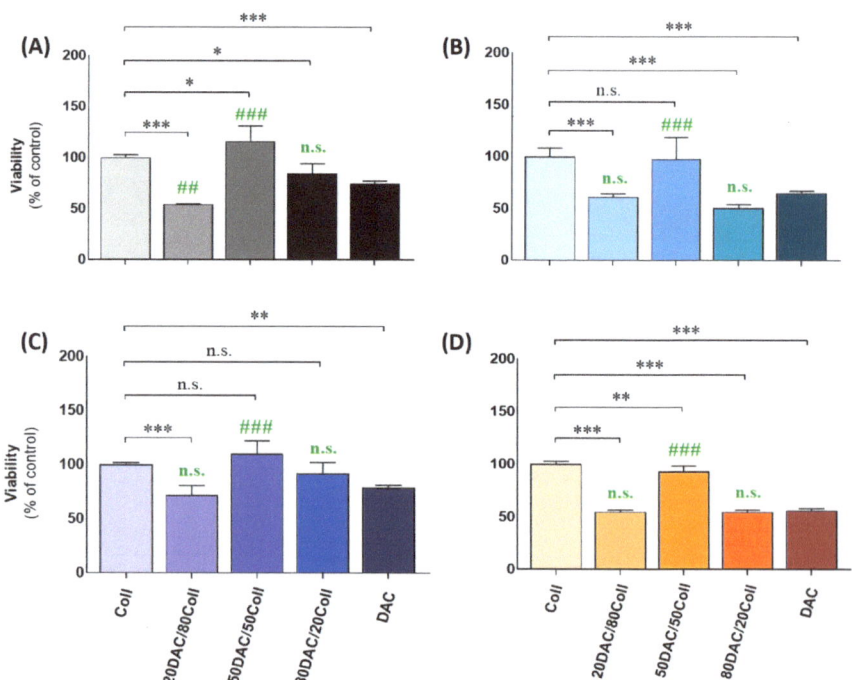

Figure 5. Assessment of proliferation rate in human cutaneous cell lines. Human cells from epidermal keratinocytes (NHEKs, **A**), dermal fibroblasts (NHDFs, **B**) as well as human amelanotic melanoma models, i.e., A375 (**C**) and G-361 (**D**) were seeded on subjected scaffolds loaded with respective components, cultured for 96 h, and viability was assessed using the MTT viability assay as described in Section 2. Data are presented as mean S.E.M. (n = 6), expressed as a percentage of the control cells cultured on matrices with collagen. Statistically significant differences versus collagen-contained scaffolds were indicated as follows: * $p < 0.05$, ** $p < 0.01$, *** $p < 0.001$ while changes versus scaffolds with DAC alone were indicated as ## $p < 0.01$ and ### $p < 0.01$ and selected with green color; n.s.—not significant.

Obtained results are in line with other physico-chemical properties enclosed in this study. Nevertheless, further assessments are utmost needed to understand the correlation between cell proliferation and the composition of scaffolds targeted in wound healing.

4. Conclusions

In conclusion, we have successfully prepared dialdehyde chitosan/collagen scaffolds. The dialdehyde chitosan have been prepared by a fast, not complicated, and cheap one step process. The materials have been obtained by a freeze-drying method.

The obtained scaffolds have been characterized by two regions in DTG curves, which were responsible for water molecules elimination and polymeric structure degradation in polymeric materials. The scaffold had a porous structure with a porosity around 90%, which is adequate for bioengineering applications. The highest material density was observed for 50DAC/50Coll scaffold, which is in agreement with SEM micrographs, where irregular and more closed structure of scaffold have been shown. Human epidermal keratinocytes (NHEK), dermal fibroblasts (NHDF) and reference melanoma cells (A375 and G-361) have been used for biological assessment of the obtained materials and that test led to the conclusion that increased content of dialdehyde chitosan either maintained or significantly increased the proliferation rates when compared to collagen scaffold. Thereby, it may be assumed that the most suitable material is in the composition 50DAC/50Coll.

Further tests to characterize dialdehyde compounds/collagen materials could be performed as a next step of this area of research. However, it is believed that the addition of chitosan dialdehyde to collagen may find potential use in the preparation of biopolymeric scaffolds for biological applications.

Supplementary Materials: The following supporting information can be downloaded at: https://www.mdpi.com/article/10.3390/polym14091818/s1, Figure S1. The FTIR spectra of (A) collagen and (B) dialdehyde chitosan, Figure S2. The TGA-DTA curves of collagen, Figure S3. The STG curves of 20DAC/80Coll and 80DAC/20Coll scaffolds.

Author Contributions: S.G.-Z. and B.K.-S. designed and prepared collagen/chitosan dialdehyde materials, designed and performed the experiments, analyzed obtained data, carried out statistical assessment, and together with K.K. and J.M.P. wrote the first draft of the manuscript. J.M.P. and K.K. performed experiments in terms of cutaneous cell lines, analyzed the results, and carried out statistical assessment. S.G.-Z., B.K.-S., K.S. and K.K. obtained respective funding. Finally, S.G.-Z., B.K.-S., J.M.P. and K.K. evaluated all the results, made their interpretation and together with E.O.-K., A.S., F.J.M. and K.S. drafted and approved the final version of the manuscript. All authors have read and agreed to the published version of the manuscript.

Funding: This research was funded by the Polish National Agency for Academic Exchange, Iwanowska project PPN/IWA/2018/1/00064/U/00001 (S.G.-Z.), and some part by the National Science Centre, PRELUDIUM grant number 2018/31/N/ST8/01391 (S.G.-Z.), Nicolaus Copernicus University, Toruń, grant number: 282/2021 IDUB SD (B.K.-S.), and German Research Foundation (Deutsche Forschungsgemeinschaft [DFG]), grant numbers: KL2900/2-1 (K.K.), TR156/C05-246807620 (K.S.), SFB1009/B11-194468054 (K.S.), SFB1066/B06-213555243 (K.S.), SFB1450/C06-431460824 (K.S.).

Institutional Review Board Statement: The study was conducted according to the guidelines of the Declaration of Helsinki, and approved by the Local Ethics Committee for Animal Experiments at the University of Life Sciences in Lublin, Poland (62/2016, 24 October 2016).

Informed Consent Statement: Not applicable.

Data Availability Statement: The data presented in this study are available on request from the corresponding author. The data are not publicly available due to projects realization.

Acknowledgments: The authors would like to acknowledge Medical University of Lublin, Poland for the materials to obtain collagen; and members of the Biocomposites Group (INEB, i3S, Porto, Portugal) for their support and help during correspondence author's stay in i3S Institute.

Conflicts of Interest: The authors declare no conflict of interest.

References

1. Ricard-Blum, S. The collagen family. *Cold Spring Harb. Perspect. Biol.* **2011**, *3*, a004978. [CrossRef] [PubMed]
2. Troy, E.; Tilbury, M.A.; Power, A.M.; Wall, J.G. Nature-based biomaterials and their application in biomedicine. *Polymers* **2021**, *13*, 3321. [CrossRef] [PubMed]
3. Sionkowska, A.; Adamiak, K.; Musial, K.; Gadomska, M. Collagen based materials in cosmetic applications: A review. *Materials* **2020**, *13*, 4217. [CrossRef] [PubMed]
4. Adamiak, K.; Sionkowska, A. Current methods of collagen cross-linking: Review. *Int. J. Biol. Macromol.* **2020**, *161*, 550–560. [CrossRef] [PubMed]
5. Shoulders, M.D.; Raines, R.T. Collagen structure and stability. *Annu. Rev. Biochem.* **2009**, *78*, 929–958. [CrossRef] [PubMed]
6. Sionkowska, A. Current research on the blends of natural and synthetic polymers as new biomaterials: Review. *Prog. Polym. Sci.* **2011**, *36*, 1254–1276. [CrossRef]
7. Lee, C.H.; Singla, A.; Lee, Y. Biomedical applications of collagen. *Int. J. Pharm.* **2001**, *221*, 1–22. [CrossRef]
8. Seal, B.L.; Otero, T.C.; Panitch, A. Polymeric biomaterials for tissue and organ regeneration. *Mater. Sci. Eng. Rep.* **2001**, *34*, 147–230. [CrossRef]
9. Kaczmarek, B.; Sionkowska, A. Chitosan/collagen blends with inorganic and organic additive-a review. *Adv. Polym. Technol.* **2018**, *37*, 2367–2376. [CrossRef]
10. Sionkowska, A.; Gadomska, M.; Musiał, K.; Piatek, J. Hyaluronic acid as a component of natural polymer blends for biomedical applications: A review. *Molecules* **2020**, *25*, 4035. [CrossRef]
11. Sionkowska, A.; Grabska, S.; Lewandowska, K.; Andrzejczyk, A. Polymer films based on silk fibroin and collagen-the physicochemical properties. *Mol. Cryst. Liq. Cryst.* **2016**, *640*, 13–20. [CrossRef]
12. Skopinska-Wisniewska, J.; Sionkowska, A.; Kaminska, A.; Kaznica, A.; Jachimiak, R.; Drewa, T. Surface characterization of collagen/elastin based biomaterials for tissue regeneration. *Appl. Surf. Sci.* **2009**, *255*, 8286–8292. [CrossRef]
13. Sionkowska, A. Collagen blended with natural polymers: Recent advances and trends. *Prog. Polym. Sci.* **2021**, *122*, 101452. [CrossRef]
14. Sionkowska, A. Interaction of collagen and poly(vinyl pyrrolidone) in blends. *Eur. Polym. J.* **2003**, *39*, 2135–2140. [CrossRef]
15. Sionkowska, A.; Skopińska, J.; Wisniewski, M. Photochemical stability of collagen/poly(vinyl alcohol) blends. *Polym. Degrad. Stab.* **2004**, *83*, 117–125. [CrossRef]
16. Sionkowska, A.; Skopinska-Wisniewska, J.; Wisniewski, M. Collagen–synthetic polymer interactions in solution and in thin films. *J. Mol. Liq.* **2009**, *145*, 135–138. [CrossRef]
17. Yang, C. Enhanced physicochemical properties of collagen by using EDC/NHS-crosslinking. *Bull. Mater. Sci.* **2012**, *35*, 913–918. [CrossRef]
18. Mu, C.; Liu, F.; Cheng, Q.; Li, H.; Wu, B.; Zhang, G.; Lin, W. Collagen cryogel cross-linked by dialdehyde starch. *Macromol. Mater. Eng.* **2010**, *295*, 100–107. [CrossRef]
19. Nimni, M.E.; Cheung, D.; Strates, B.; Kodama, M.; Sheikh, K. Chemically modified collagen: A natural biomaterial for tissue replacement. *J. Biomed. Mater. Res.* **1987**, *21*, 741–771. [CrossRef]
20. Sundararaghavan, H.G.; Monteiro, G.A.; Lapin, N.A.; Chabal, Y.J.; Miksan, J.R.; Shreiber, D.I. Genipin-induced changes in collagen gels: Correlation of mechanical properties to fluorescence. *J. Biomed. Mater. Res. A* **2008**, *87*, 308–320. [CrossRef]
21. Sionkowska, A.; Kozłowska, J. Properties and modification of porous 3-D collagen/hydroxyapatite composites. *Int. J. Biol. Macromol.* **2013**, *52*, 250–259. [CrossRef] [PubMed]
22. Weadock, K.S.; Miller, E.J.; Bellincampi, L.D.; Zawadsky, J.P.; Dunn, M.G. Physical crosslinking of collagen fibers: Comparison of ultraviolet irradiation and dehydrothermal treatment. *J. Biomed. Mater. Res.* **1995**, *29*, 1373–1379. [CrossRef] [PubMed]
23. Chen, R.N.; Ho, H.O.; Sheu, M.T. Characterization of collagen matrices crosslinked using microbial transglutaminase. *Biomaterials* **2005**, *26*, 4229–4235. [CrossRef] [PubMed]
24. Bam, P.; Bhatta, A.; Krishnamoorthy, G. Design of biostable scaffold based on collagen crosslinked by dialdehyde chitosan with presence of gallic acid. *Int. J. Biol. Macromol.* **2019**, *130*, 836–844. [CrossRef] [PubMed]
25. Wegrzynowska-Drzymalska, K.; Grebicka, P.; Mlynarczyk, D.T.; Chelminiak-Dudkiewicz, D.; Kaczmarek, H.; Goslinski, T.; Ziegler-Borowska, M. Crosslinking of chitosan with dialdehyde chitosan as a new approach for biomedical applications. *Materials* **2020**, *13*, 3413. [CrossRef]
26. He, X.; Tao, R.; Zhou, T.; Wang, C.; Xie, K. Structure and properties of cotton fabrics treated with functionalized dialdehyde chitosan. *Carbohydr. Polym.* **2014**, *103*, 558–565. [CrossRef]
27. Liu, X.; Dan, N.; Dan, W.; Gong, J. Feasibility study of the natural derived chitosan dialdehyde for chemical modification of collagen. *Int. J. Biol. Macromol.* **2016**, *82*, 989–997. [CrossRef]
28. Langmaier, F.; Mládek, M.; Mokrejš, P. Hydrogels of collagen hydrolysate cross-linked with dialdehyde starch. *J. Therm. Anal. Calor.* **2009**, *98*, 807–812. [CrossRef]
29. Jayakumar, G.; Kanth, S.; Rao, J.R.; Nair, B.U. A molecular level investigation of dialdehyde starch interaction with collagen for eco-friendly stabilization. *J. Am. Leath. Chem. Assoc.* **2015**, *110*, 145–151.
30. Tan, H.; Wu, B.; Li, C.; Mu, C.; Li, H.; Lin, W. Collagen cryogel cross-linked by naturally derived dialdehyde carboxymethyl cellulose. *Carbohydr. Polym.* **2015**, *129*, 17–24. [CrossRef]

31. Lu, T.; Li, Q.; Chen, W.; Yu, H. Composite aerogels based on dialdehyde nanocellulose and collagen for potential applications as wound dressing and tissue engineering scaffold. *Comp. Sci. Technol.* **2014**, *94*, 132–138. [CrossRef]
32. Pietrucha, K.; Safandowska, M. Dialdehyde cellulose-crosslinked collagen and its physicochemical properties. *Proc. Biochem.* **2015**, *50*, 2105–2111. [CrossRef]
33. Hu, Y.; Liu, L.; Gu, Z.; Dan, W.; Dan, N.; Yu, X. Modification of collagen with a natural derived cross-linker, alginate dialdehyde. *Carbohydr. Polym.* **2014**, *102*, 324–332. [CrossRef]
34. Yu, X.; Yuan, Q.; Yang, M.; Liu, R.; Zhu, S.; Li, J.; Zhang, W.; You, J.; Xiong, S.; Hu, Y. Development of biocompatible and antibacterial collagen hydrogels via dialdehyde polysaccharide modification and tetracycline hydrochloride loading. *Macromol. Mater. Eng.* **2019**, *304*, 1800755. [CrossRef]
35. Wanli, H.; Zihan, Y.; Yanan, W.; Bi, S. Preparation of dialdehyde chitosan and its application for crosslinking collagen fiber. *Leath. Sci. Eng.* **2021**, *31*, 1–5.
36. Grabska-Zielińska, S.; Sosik, A.; Małkowska, A.; Olewnik-Kruszkowska, E.; Steinbrink, K.; Kleszczyński, K.; Kaczmarek-Szczepańska, B. The characterization of scaffolds based on dialdehyde chitosan/hyaluronic acid. *Materials* **2021**, *14*, 4993. [CrossRef] [PubMed]
37. Lewandowska, K.; Sionkowska, A.; Grabska, S.; Kaczmarek, B.; Michalska, M. The miscibility of collagen/hyaluronic acid/chitosan blends investigated in dilute solutions and solids. *J. Mol. Liq.* **2016**, *220*, 726–730. [CrossRef]
38. Sionkowska, A.; Grabska, S. Preparation and characterization of 3D collagen materials with magnetic properties. *Polym. Test.* **2017**, *62*, 382–391. [CrossRef]
39. Sionkowska, A.; Kaczmarek, B.; Lewandowska, K. Modification of collagen and chitosan mixtures by the addition of tannic acid. *J. Mol. Liq.* **2014**, *199*, 318–323. [CrossRef]
40. Sionkowska, A.; Wiśniewski, M.; Skopińska, J.; Mantovani, D. Effects of solar radiation on collagen-based biomaterials. *Int. J. Phot.* **2006**, *2006*, 1–6. [CrossRef]
41. Sionkowska, A.; Kozłowska, J.; Skorupska, M.; Michalska, M. Isolation and characterization of collagen from the skin of brama australis. *Int. J. Biol. Macromol.* **2015**, *80*, 605–609. [CrossRef] [PubMed]
42. Keshk, S.M.A.S.; Ramadan, A.M.; Al-Sehemi, A.G.; Irfan, A.; Bondock, S. An unexpected reactivity during periodate oxidation of chitosan and the affinity of its 2,3-di-aldehyde toward sulfa drugs. *Carbohydr. Polym.* **2017**, *175*, 565–574. [CrossRef] [PubMed]
43. Lv, Y.; Long, Z.; Song, C.; Dai, L.; He, H.; Wang, P. Preparation of dialdehyde chitosan and its application in green synthesis of silver nanoparticles. *BioResources* **2013**, *8*, 6161–6172. [CrossRef]
44. Gao, C.; Wang, S.; Liu, B.; Yao, S.; Dai, Y.; Zhou, L.; Qin, C.; Fatehi, P. Sustainable chitosan-dialdehyde cellulose nanocrystal film. *Materials* **2021**, *14*, 5851. [CrossRef] [PubMed]
45. Kaczmarek, B.; Nadolna, K.; Owczarek, A. The physical and chemical properties of hydrogels based on natural polymers. In *Hydrogels Based on Natural Polymers*; Chen, Y., Ed.; Elsevier: Cambridge, MA, USA, 2020; pp. 151–172.
46. Carey, S.P.; Kraning-Rush, C.M.; Williams, R.M.; Reinhart-King, C.A. Biophysical control of invasive tumor cell behavior by extracellular matrix microarchitecture. *Biomaterials* **2012**, *33*, 4157–4165. [CrossRef]
47. Grabska-Zielińska, S.; Sionkowska, A.; Reczyńska, K.; Pamuła, E. Physico-chemical characterization and biological tests of collagen/silk fibroin/chitosan scaffolds cross-linked by dialdehyde starch. *Polymers* **2020**, *12*, 372. [CrossRef]
48. Ding, C.; Zhang, Y.; Yuan, B.; Yang, X.; Shi, R.; Zhang, M. The preparation of nano-SiO_2/dialdehyde cellulose hybrid materials as a novel cross-linking agent for collagen solutions. *Polymers* **2018**, *10*, 550. [CrossRef]
49. Kaczmarek, B.; Miłek, O.; Michalska-Sionkowska, M.; Zasada, L.; Twardowska, M.; Warżyńska, O.; Kleszczyński, K.; Osyczka, A.M. Novel eco-friendly tannic acid-enriched hydrogels-preparation and characterization for biomedical application. *Materials* **2020**, *13*, 4572. [CrossRef]
50. Kaczmarek-Szczepańska, B.; Ostrowska, J.; Kozłowska, J.; Szota, Z.; Brożyna, A.A.; Dreier, R.; Reiter, R.J.; Slominski, A.T.; Steinbrink, K.; Kleszczyński, K. Evaluation of polymeric matrix loaded with melatonin for wound dressing. *Int. J. Mol. Sci.* **2021**, *22*, 5658. [CrossRef]

Article

Poly(lactic acid) and Nanocrystalline Cellulose Methacrylated Particles for Preparation of Cryogelated and 3D-Printed Scaffolds for Tissue Engineering

Mariia Leonovich [1], Viktor Korzhikov-Vlakh [1,*], Antonina Lavrentieva [2], Iliyana Pepelanova [2], Evgenia Korzhikova-Vlakh [1,3] and Tatiana Tennikova [1,*]

[1] Institute of Chemistry, Saint Petersburg State University, Peterhoff, Universitetskii pr. 26, 198504 Saint Petersburg, Russia
[2] Institute of Technical Chemistry, Gottfried-Wilhelm-Leibniz University of Hannover, 30167 Hannover, Germany
[3] Institute of Macromolecular Compounds, Russian Academy of Sciences, 199004 Saint Petersburg, Russia
* Correspondence: v.korzhikov-vlakh@spbu.ru (V.K.-V.); tennikova@mail.ru (T.T.)

Abstract: Different parts of bones possess different properties, such as the capacity for remodeling cell content, porosity, and protein composition. For various traumatic or surgical tissue defects, the application of tissue-engineered constructs seems to be a promising strategy. Despite significant research efforts, such constructs are still rarely available in the clinic. One of the reasons is the lack of resorbable materials, whose properties can be adjusted according to the intended tissue or tissue contacts. Here, we present our first results on the development of a toolbox, by which the scaffolds with easily tunable mechanical and biological properties could be prepared. Biodegradable poly(lactic acid) and nanocrystalline cellulose methacrylated particles were obtained, characterized, and used for preparation of three-dimensional scaffolds via cryogelation and 3D printing approaches. The composition of particles-based ink for 3D printing was optimized in order to allow formation of stable materials. Both the modified-particle cytotoxicity and the matrix-supported cell adhesion were evaluated and visualized in order to confirm the perspectives of materials application.

Keywords: poly(lactic acid); nanocrystalline cellulose; methacrylation; particles; 3D printing; scaffolds; tissue engineering

1. Introduction

Bone loss and fracture caused by disease or injury re considered major health problems [1]. Today, the most common method to address bone defects is bone grafting, but unfortunately, it is still problematic for clinicians to choose between autografts, allografts or engineered tissues [2]. Recently, bone tissue engineering (BTE) has gained more attention as a potential treatment for bone defects. The strategy for bone tissue engineering involves the following steps: (1) identification of a suitable cell source, isolation of cells and expansion of cells to sufficient amounts; (2) obtaining of a biocompatible material that can be used as a cell substrate and processed into the required shape (scaffold); (3) seeding of the scaffold with cells, which can then be cultivated in bioreactors; (4) placement of the material-cell construct into the target site in vivo [2]. In this regard, scaffolds are the key component of BTE, as they regulate bone healing and mimic the extracellular matrix (ECM) function in the bone tissue [2,3]. An ideal scaffold for bone tissue repair should have a three-dimensional porous structure with a highly interconnected pore network, being biocompatible and controllably bioresorbable [3,4]. The ideal strategy for regenerative bone therapies is to use a tunable scaffolding material that can modulate the process of healing while providing mechanical support [2,5].

Tuning of scaffolding material properties requires a certain level of versatility, which could be provided by polymers—the intelligent choice of macromolecules—and their combinations allow the formation of materials with a wide variety of both mechanical and biological properties [6]. The development of composites allows further upgrades the level of the materials' versatility [7]. However, despite the great progress in the preparation of various materials for scaffold formation, there are still unsolved problems, such as control over spatial and temporal distribution of bioactive molecules, as well as the formation of gradients [8–10]. It is now obvious that successful scaffold application requires control over its micro/nanotopology as well as providing the scaffold with the ability to promote and guide cell-induced tissue regeneration through the regulation of local microenvironment by exposing the appropriate signals at the desired site for the required time frame [11]. The development of such complex scaffolds is essential, because the combination/distribution of various materials could resemble the complexity of natural bone tissue. These issues require higher levels of material versatility, which could be provided by the application of combinations of polymeric particles of different origins with different properties as "ink" for the 3D printing of scaffolds. Particles possessing different chemical natures and various densities and surface properties can be aggregated into supermacroporous matrices using different combinations of such particles, which should allow the tuning of the mechanical properties and surface topology of the pores [12,13]. It could be supposed, for example, that increasing the number of hard particles and the density of cross-linking should result in greater mechanical strength, while the addition of soft particles should favor the formation of gel-like materials. Particles with different properties could be combined into one material to mimic the tissue contacts; for example, osteochondral scaffolds [14]. Moreover, the useful peculiarities of the particles are their potential for surface modification and their ability to entrap different substances into their inner volume, which allow the spatial distribution of osteogenic or angiogenic factors and their temporally controlled release [15]. The release of various drugs could also be very useful in the cases when bone regeneration is associated with surgical interventions due to various diseases [16].

There are many methods that were developed in order to obtain porous materials applicable for supporting 3D cell growth: melt-molding, fiber bonding, porogen leaching, gas foaming and phase separation [17]. Pore diameter greater than 50 μm, pore uniformity and interconnectivity are the main goals for these methods. However, not all these approaches are suitable for preparation of particles-based scaffolds. One of the interesting techniques that is useful for the preparation of supermacroporous matrices is cryogelation. This method is based on solid–liquid phase separation, and includes freezing of the solution or preformed gel, cross-linking interactions between macromolecules or particles and removal of the solvent [18]. As the solution is frozen, the solvent crystals grow until they come into contact with other crystals. Thus, the solvent crystals form a common extended structure. The phase containing the polymer or particles is concentrated around the crystals and is called the unfrozen liquid microphase (ULMP). In fact, crystals of water push the polymer molecules or particles together. This process is called "cryo-concentration" [18]. In such concentrated conditions, different chemical cross-linking reactions, such as cryo-polymerization, could run very efficiently, even at temperatures below 0 °C [19]. The last stage of cryogel formation is the removal of the solvent crystals via freeze-drying, leaving the supermacroporous structure, which is useful for cell seeding and proliferation (see the scheme and images in [19]).

Previously developed 3D printing techniques are also very good candidates for the preparation of scaffolds by particle controllable aggregation. Modern 3D printers allow precise control over the material composition as well as simultaneous combination of different printing techniques: [20–23]. However, 3D organization of particles into scaffolds via direct ink writing (DIW) and stereolithography (SLA) is still an unexplored area [24]. Wide application of such 3D printing techniques in clinics with utilization of particles as versatile material for scaffold formation could help to solve the problems of auto- and allograft applications, which comprise a shortage of bone, donor site morbidity and additional operations on the patients [2].

Poly(lactic acid) (PLA) is one of the most widely used polymers for biomedical applications [25]. This polyester is often a good selection in the case of scaffolds for BTE [23,26]. PLA is biocompatible and its application in medicine is approved by the Food and Drug Administration (FDA) agency [25,27]. One of the most valuable features of PLA is its controllable biodegradability [27]. PLA is a thermoplastic polymer that is non-soluble in water and relatively hydrophobic [25]. On one hand, this allows the formation of stable biomaterials with different geometries, but on the other hand, this could cause unpredictable in vivo biointeractions between PLA-based materials and the surrounding tissues [28]. In order to improve the mechanical properties and hydrophilicity of PLA-based scaffolds, the PLA is commonly combined with various particles, such as hydroxyapatite [29], chitosan [30,31], graphene oxide [32] etc., in order to form composite scaffolds. Recently it was recognized that among the prospective additives for PLA are micro- or nanocrystalline cellulose (MCC/NCC) particles, which are very biocompatible and can significantly increase scaffold mechanical performance and hydrophilicity [33]. NCC particle surfaces could be also chemically modified with special molecules in order to affect the biological properties of the polyester-based composite scaffold; for example, inducing calcification and bone tissue formation around the implanted material [34,35]. Such composite scaffolds could be considered as "particle-incorporating scaffolds", in which particles are dispersed into a PLA-based continuous phase. This type of composite scaffold is often prepared via 3D printing, which uses fused deposition modelling (FDM) technology [24,34]. Despite significant research in this area, the approach presents challenges with regard to control over biomolecule delivery, cell infiltration and viability within the scaffold matrix and clinical handling [36]. Moreover, it is difficult to use FDM technology only for spatial control of component distribution.

Another possible type of particle organization in materials for tissue engineering is the "particle-based scaffold". In such a structure, PLA microspheres themselves or their combinations with other particles represent the building blocks for scaffold framework formation [37,38]. It is well-known that crude PLA can be easily converted into microspheres via emulsion protocols [39]. In previous studies, microspheres based on PLA or other polyesters were used as blocks for preparation of scaffolds via interfusion of particles, which was provoked by thermal sintering [14,40] or solvent-induced sintering [41] protocol. Additionally, the selective laser sintering (SLS) technique could be successfully applied to produce nanocomposite PLA-hydroxyapatite scaffolds [42]. The mentioned techniques of particle-based scaffolds preparation are very similar to the procedures of powder-based technologies. In fact, none of these techniques allow reliable preservation of encapsulated biomolecule stability, due to the application of elevated temperatures, laser, or organic solvents. In this regard, the simultaneous application of soft DIW and SLA 3D printing techniques is of interest because it allows the formation of a scaffold with spatial distribution of different particle types within the scaffold (Figure 1). These techniques also allow the formation of so-called gradient scaffolds.

To the best of our knowledge, no studies on PLA and NCC particle controllable organization into scaffolds via covalent cross-linking has been previously published. Thus, the purpose of this study was investigation of the unexplored possibility of PLA and NCC nanoparticle (NPs) covalent interaction to allow the production of 3D matrices, which could be used as scaffolds for tissue engineering. To provide the possibility of their chemical cross-linking, both PLA and NCC NPs were methacrylated (PLA-methacrylate—PLA-MA and NCC-methacrylate—NCC-MA). The effect of particle modification on the viability of cells was tested. Then, we aimed to investigate the possibility of chemical interparticle cross-linking without external cross-linkers, as well as in the presence of bifunctional (poly(ethylene glycol) dimethacrylate) and multifunctional (gelatin methacrylate) macromolecular cross-linkers. The ability of methacrylated particles to form 3D structures was firstly analyzed via rheological measurements, and then optimized mixtures were used for porous matrix formation. Cryogelation and 3D-printing approaches were applied for this purpose, allowing comparison of materials obtained by these methods. The effect of composition used for matrix preparation on cell attachment was finally tested.

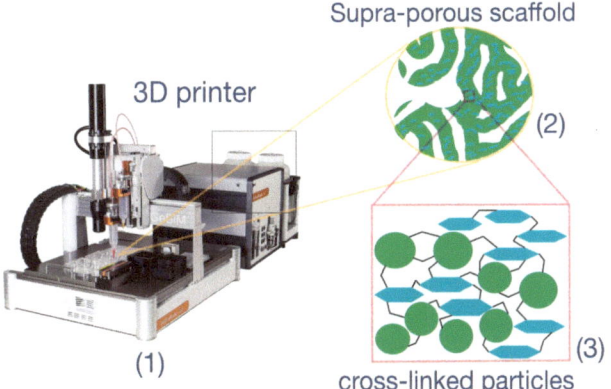

Figure 1. The idea of particle-based scaffold formation with application of DIW + SLA 3D-printing technologies. (**1**) 3D-printer supplied with extruder-type printing head and UV-lamp; (**2**) super-macroporous matrix with interconnected pores, which could serve as scaffold; (**3**) covalently cross-linked particles. The green circles represent PLA particles, and the turquoise hexagons indicate NCC particles.

2. Materials and Methods

2.1. Materials

All monomers, initiators, and agents for modification, as well as nonionic poly(oxyethylene-*b*-oxypropylene) (Pluronic F-68, M_W = 138.16), sodium dodecyl sulfate (SDS), poly(vinyl alcohol) (PVA, M_W = 70,000), 2-hydroxy-4′-(2-hydroxyethoxy)-2-methylpropiophenone (Irgacure 2959) and poly(ethylene glycol) dimethacrylate (PEGDA, M_W = 700) were purchased from Sigma-Aldrich (Munich, Germany). NCC was a product of Blue Goose Biorefineries Inc. (Saskatoon, SK, Canada). Organic solvents (toluene, chloroform, methanol, etc.) used for polymer synthesis and modification were from Vecton (St. Petersburg, Russia). All organic solvents were distilled before application. Poly(L-lactic acid) (PLA, M_W = 18,500; Đ = 1.23) was synthesized by ring-opening polymerization of L-lactide (Sigma-Aldrich, Munich, Germany) as previously described [43]. Methacrylated gelatin (GelMA) was synthesized as described in [44].

Dulbecco's Modified Eagle Medium (DMEM-F12) (Biolot, Saint Petersburg, Russia)/10% fetal bovine serum (FBS) (Biowest, Riverside, MO, USA)/50 IU/mL penicillin/50 μg/mL streptomycin (Biolot, Saint Petersburg, Russia) was used for cell culture experiments.

2.2. Cells

MSCs (bone marrow mesenchymal stem cells) and NIH 3T3 (mouse fibroblast cells) cell lines were obtained from the German Collection of Microorganisms and Cell Culture (DSMZ, Braunschweig, Germany).

2.3. Methods

2.3.1. PLA-Based Particle Formation

PLA-based particles were obtained by single emulsion method as previously described [39,45]. Shortly, 2 mL solution of PLA in dichloromethane (50 mg/mL) was dispersed into 20 mL of ice-cold water phase, in which 0.2 g of SDS (1 wt%) and 0.6 g of Pluronic F-68 (3 wt%) were preliminarily dissolved. The emulsion was formed via simultaneous application of ultrasound homogenizer (Sonopuls HD2070, Bandelin,

Berlin, Germany) and magnetic stirrer (MR Hei-Mix S, Heidolph, Schwabach, Germany) at 700 rpm during 10 min. The resulting emulsion was diluted in 80 mL of ice-cold 1 wt% PVA aqueous solution. Dichloromethane was removed by evaporation using rotary evaporator (Hei-VAP Precision ML/G3B, Heidolph, Schwabach, Germany) at 100 mbar for 2–3 h. The obtained particles were isolated by centrifugation (Sigma 2–16 KL, Sigma, Darmstadt, Germany) at 10,000 g and washed three times with distilled water.

2.3.2. PLA Particle Modification with 2-Aminoethyl Methacrylate

First, the available reagent 2-aminoethyl methacrylate in hydrochloride form (2-AEMA•HCl) was converted to a free base by treatment with diisopropylethylamine. For that 0.3 g of 2-AEMA•HCl was placed in a 25 mL vial, 15 mL of diethyl ether was added and 450 µL of diisopropylethylamine was added at stirring. The reaction proceeded for 2 h under vigorous stirring at room temperature. Then the precipitate was separated by centrifugation, and the supernatant containing 2-AEMA was concentrated on a rotary evaporator. As a result, the free base 2-AEMA as an oil was isolated from the crystalline hydrochloride form of 2-AEMA•HCl.

The PLA particle modification scheme is shown in Figure 2A. A quantity of 500 mg of PLA particles were treated with 0.1 M NaOH solution (50 mL) to generate carboxyl groups (Figure 2(A1)) on the particle surface [46]. The hydrolysis was conducted for 30 min at room temperature, under stirring (350 rpm). Then, the particles were purified 5 times by Vivaspin-column dialysis (filter MWCO 10,000; 25 mL; Merck, Darmstadt, Germany) against 20 mL of ice-cold grade water (MQ water) and 3 times against 0.05 M buffer solution of 2-(N-morpholino)ethanesulfonic acid, pH 5.6 (MES buffer).

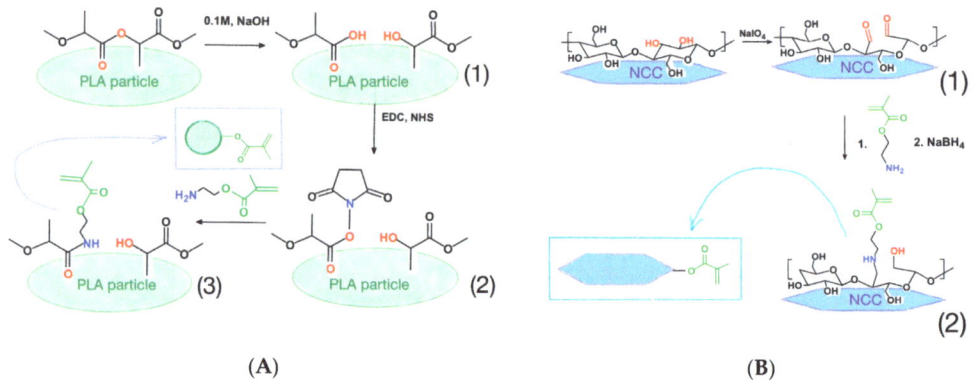

Figure 2. The schemes of methacrylated particles formation via reaction with 2-aminoethylmethacrylate: (**A**)—modification of PLA particles; (**B**)—modification of NCC particles.

The suspension of particles in 0.05 M MES with a pH of 5.6 was precooled down to 5 °C, then the solution of 30 mg N-hydroxysuccinimide (NHS) and 20 mg of (N-3-dimethylaminopropyl)-N-ethylcarbodiimide hydrochloride (EDC) in the same buffer were added consecutively to activate the carboxyl groups (Figure 2(A2)) on the particle surface. After that, the particles were purified 5 times by Vivaspin-column dialysis (MWCO 10,000) against MQ water and 3 times against 0.1 M borate buffer solution (BBS, pH 8.4).

Next, 200 mg of AEMA in a 0.1 M borate buffer solution (pH 8.4) was added to the particle suspension. The reaction was performed for 3 h at room temperature while stirring at 500 rpm to obtain the final product—methacrylated PLA particles (PLA-MA, Figure 2(A3)). Then, the particles were purified by Vivaspin-column dialysis (MWCO 1000) against ultrapure water.

2.3.3. Nanocrystalline Cellulose Modification with AEMA

The PLA particle modification scheme is shown on Figure 2B. A quantity of 1 g of 8% stock nanocrystalline cellulose (NCC) suspension was diluted 10 times with MQ water and cooled to 5 °C while stirring for 30 min. Then, 500 mg of sodium metaperiodate in 5 mL of water was added into the reaction mixture. The periodate-containing mixture was wrapped in aluminum foil to avoid light exposure. The reaction mixture was gently stirred at 5 °C in the dark for 24 h. Thereafter, the obtained 2,3-dialdehyde NCC (DA-NCC, Figure 2(B1)) was purified by Vivaspin-column dialysis (MWCO 1000) against MQ water.

Then, 300 mg of AEMA in 6 mL of MQ water was added to 100 mL of DA-NCC (1.0 g) suspension in 0.01 M BBS (pH 8.4). The reaction was carried out for 24 h at 20 °C under vigorous stirring at 500 rpm. After that, 300 mg of sodium borohydride was added to the reaction mixture to reduce the remaining aldehyde groups and imine bonds. The reaction was carried out for 24 h at 20 °C under stirring at 500 rpm. The product (Figure 2(B2)) was purified by Vivaspin-column dialysis (MWCO 1000) against MQ water.

2.3.4. Particle Characterization

The introduction of methacrylate (MA) groups onto the particle surfaces was proved by: (1) ^{1}H NMR spectra, which were recorded with equipment of Magnetic Resonance Research Centre of St. Petersburg State University (Bruker Avance spectrometer, 400.13 MHz) in $CDCl_3$; (2) FTIR spectra, which were recorded from 4000–400 cm^{-1} in KBr tablets with application of equipment of Chemical Analysis and Materials Research Centre of St. Petersburg State University (IRAffinity-1, Shimadzu Corporation, Kyoto, Japan).

Particle hydrodynamic diameter and ζ-potential were measured with Zetasizer Nano ZS (Malvern, Enigma Business Park, UK) equipped with a He–Ne laser beam at λ = 633 nm. Nanoparticle tracking analysis (NTA) was performed with a NanoSight NS300 particle analyzer (Malvern, Enigma Business Park, UK).

The morphology of particles was detected by TEM with equipment from the Centre for Molecular and Cell Technologies of St. Petersburg State University—Jeol JEM-1400 STEM (Tokyo, Japan). SEM was used for the investigation of particles and matrix morphology and this equipment was provided by Interdisciplinary Resource Centre for Nanotechnology of St. Petersburg State University—Zeiss Supra 40VP (Carl Zeiss MicroImaging GmbH, Oberkochen, Germany).

2.3.5. Rheological Measurements of Particle Cross-Linking

The UV photo-curing of methacrylated particle suspensions was investigated at 25 °C using a MCR 302 Modular Rheometer (Anton Paar, Graz, Austria) equipped with a plate-plate geometry (20 mm diameter, 0.3 mm gap size) and by performing in situ UV cross-linking by irradiation with a UV lamp (365 nm, 75 W/cm^2, Delolux 80, Delo, Windach, Germany) from below. The crosslinking kinetics was recorded with a time sweep oscillatory test under constant strain amplitude of 1% and at a constant frequency of 1 Hz, which is within the linear viscoelastic (LVE) region.

A series of samples of nanoparticle suspensions of different compositions (PLA-MA/NCC-MA: 80:20, 60:40, 40:60, 20:80 wt/wt; PLA-MA/NCC-MA 80:20 wt/wt + 20 wt% PEGDA or GelMA) were prepared. A quantity of 1 mg of photoinitiator (0.1 wt/vol %, Irgacure 2959) was added to the 1 mL of prepared mixture. The concentration of particles in the suspension was varied (10, 50 and 100 mg/mL). A 400 µL sample was placed between the two rheometer plates using an automatic pipette. The sample was irradiated with UV radiation in situ; the irradiation time was 10 min. Then, the rheological measurements were continued for a further 10 min without irradiation.

2.3.6. Cryogelation

Matrices were formed in a 1 mL syringe. A quantity of 0.8 mL of particle aqueous suspension containing 64 mg of PLA-MA and 16 mg of NCC-MA was prepared, and

0.15 mL of solution containing 216 mg of cross-linking agent (20 wt% from the total mass of reaction mixture, PEGDA or GelMA) was added to the particle suspension and homogenized with a thermoshaker (TS-100, Biosan, Riga, Latvia) at 500 rpm during 5 min. Then, solutions of 1.3 mg of initiator—ammonium peroxodisulfate (APS)—and 1.5 mg of reaction activator N,N,N′,N′-tetramethylethylenediamine (TMEDA) in 250 µL of ultrapure water were prepared. First APS and then TMEDA solutions were added under vigorous stirring to the precooled to 5 °C suspensions of particles with cross-linker. After that, the reaction mixture was immediately transferred into the syringe and placed in the freezer at −13 °C for 24 h. After this period, the samples were allowed to thaw, and the matrices were washed by passing a 100-fold of the volume of MQ water through them. Then, syringes were gently cut open to release the resulting matrix, which was further washed with MQ water over 48 h at 150 rpm stirring with thermoshaker.

2.3.7. 3D Printing

3D CAD models of the desired object for printing were designed and the necessary settings were lined up using Autodesk Inventor Professional 2015 software (Autodesk Inc., San Rafael, CA, USA). 3D printer Allevi 1 (Allevi Inc., Philadelphia, PA, USA) and GeSim Bioscaffolder 3D printer (Radeberg, Germany) were used for 3D printing. These printers were equipped with extrusion-type print head and LED 365 nm (1.5 W/cm^2) UV lamp for photo-cross-linking of the material.

Matrices were printed in the wells of 24-well plates using a mixture of particle suspension with added cross-linking in a composition similar to that described above (see Section 2.3.6). Printing parameters were optimized: printing needle diameter, pressure, and printing speed. In order to avoid gel formation, the optimal printing suspension temperature of 37 °C was also selected when working with the sample containing GelMA as an additional cross-linking agent. Uniform extrusion was observed at the printing needle diameter 0.254 and length 6.35 mm, pressure (0.5 PSI or 3.45 kPa). The optimum printing speed was 5 mm/s.

Before printing 10 µL (11 µg/µL) of photoinitiator (Irgacure 2959) solution in ethanol was added to the mixture. The suspension was loaded into an extruder cartridge with pre-screwed print heads of the required diameter. The cross-linking of the sample was performed in layers: after the printing of the first level of the matrix, the UV lamp was turned on, the sample was irradiated for 10 min, and then the printing process was resumed. Finally, printed samples (length/width/height = 1/1/0.4 cm) were left for an hour in order to allow the radical cross-linking reaction to proceed, as well as for drying of the samples.

2.3.8. Study of Mechanical Properties

The samples of cryogels and 3D printed materials were prepared in the form of unified discs 5 mm in diameter and 5 mm thick. The mechanical analysis of the samples was performed by uniaxial compression tests on a Shimadzu EZTest EZ-L (Shimadzu Corporation, Kyoto, Japan) tensile testing machine. To ensure tight surface contact with the sample, the initial force was 0.1 N; the rate of motion was 2 mm/min. The series of parallel measurements (n = 3) were performed for all cryogel specimens.

2.3.9. Cell Culture Experiments

Cytotoxicity of particles. The cytotoxicity of the particles was studied using the CellTiter-Blue test (CTB), which is based on the ability of living cells to reduce the blue nonfluorescent reagent CTB-resazurin (7-hydroxy-3H-phenoxazine-3-one-10-oxide) to the pink fluorescent resorufin.

A total of 8×10^3 cells (MSCs) in 100 µL of culture medium were seeded in each well of a 96-well plate and cultured for 24 h (37 °C, 5% CO_2, 80% humidity). Then, the culture medium was removed using a sterile glass Pasteur pipette and 200 µL of culture medium containing nanoparticles in the concentration range of 1–10 mg/mL was added.

The cells were incubated in a CO_2 incubator for 24/48/72 h at 37 °C, then the medium was removed and 100 μL of CTB solution in basal medium (1:10 (vol/vol)) was added to each well. The cells were incubated for another 2 h at 37 °C. The formation of a fluorescent product proportional to the number of viable cells was monitored fluorometrically (λ_{ex} = 544 nm and λ_{emm} = 590 nm). The data were normalized as a percentage of the control (wells containing cells incubated without test substances). The analysis was repeated five times for each concentration.

Cells adhesion. Before the cell culture experiments, the samples obtained by 3D-printing and cryogelation were washed with ultrapure water while stirring to remove unreacted components. Then they were sterilized with UV light (wavelength range: 200–295 nm) for 30 min on each side.

MSCs were used in the experiments. A total of 4×10^4 cells were seeded on the surface of matrices in 500 μL of nutrient medium in each well. The plate was then placed in a CO_2 incubator (37 °C, 5% CO_2, 80% humidity) for a predetermined period of time. Then cells attached to the cryogel surface were visualized. Staining was performed using DAPI fluorescent dye. For this purpose, the matrices were washed twice with warm 0.01 M phosphate-salt buffer solution (pH 7.4) for 30 min. Then, 200 μL of cold 4% paraformaldehyde in phosphate-salt buffer solution was added to the sample to fix the cells, and was left for 20 min at room temperature. After fixation, the samples were washed again with 0.01 M phosphate-salt buffer solution (pH 7.4), and the cell nuclei were stained for 15 min with DAPI working solution (1:1000 (vol/vol) in phosphate-salt buffer solution) at room temperature (22 °C) in the dark. The samples were analyzed using a fluorescence microscope (Olympus, BX-51 Olympus Corporation, Tokyo, Japan). The data from these experiments are presented as average ± SD (n = 5).

3. Results and Discussion

3.1. Preparation of Particles Capable of Participation in Free-Radical Polymerization-Mediated Cross-Linking

The idea of this study was to use PLA and NCC particles for the production of particle-based 3D scaffolds. PLA NPs were obtained via single emulsion protocol, while NCC particles were used as purchased. In order to provide the particles with the ability to participate in the free-radical cross-linking process, the methacrylate groups were introduced onto particle surfaces to obtain PLA-methacrylate (PLA-MA) and NCC-methacrylate (NCC-MA).

In the case of PLA, we applied the previously developed protocol of PLA particle surface partial hydrolysis [46], which results in the formation of carboxylic groups (Figure 2A). These groups were then transformed into the activated esters and put into reaction with 2-aminoethylmethacrylate (AEMA) to obtain PLA-MA particles. In order to introduce methacrylate groups on the surfaces of the NCC particles, they were partially oxidized with sodium periodate. The formed aldehyde groups easily reacted with amino groups of AEMA to give NCC-MA (Figure 2B).

The successful introduction of methacrylate moieties was proved by ^{1}H NMR and FTIR spectra for PLA-MA and NCC-MA, respectively (Figure 3). In the ^{1}H NMR spectrum of PLA-MA particles dissolved in $CDCl_3$, the signals corresponding to the presence of diastereotopic protons of methacrylate double bond (Figure 3, signals *f* and *f'*), protons of methylene groups between ester and amino groups (Figure 3, signals *d* and *c*) and protons of methacrylate methyl group (Figure 3, signal *b*) were found.

Because of the non-solubility of NCC, the presence of methacrylate groups in modified NCC was analyzed by FTIR. In the spectrum of raw NCC, the absorption bands at 3357 cm^{-1} and around 2908 cm^{-1} were attributed to the O-H and C-H stretching vibrations, respectively. The absorption band at 1642 cm^{-1} could be attributed to the O-H vibration of absorbed water [47–49]. The bands corresponding to the C-H and C-O vibrations, contained in the polysaccharide rings of cellulose, were around 1374 cm^{-1}. The vibration of C-O-C in pyranose ring was indicated by the absorption band at 1060 cm^{-1}. In the

NCC-MA spectrum, it was possible to observe the appearance of new bands. The band at 1738 cm^{-1} corresponded to the C=O stretching vibrations. It was also possible to observe an enhancement of intensity and sharpness of the band at 1617 cm^{-1}, which could be due to the C=CH$_2$ out-of-plane bending vibrations.

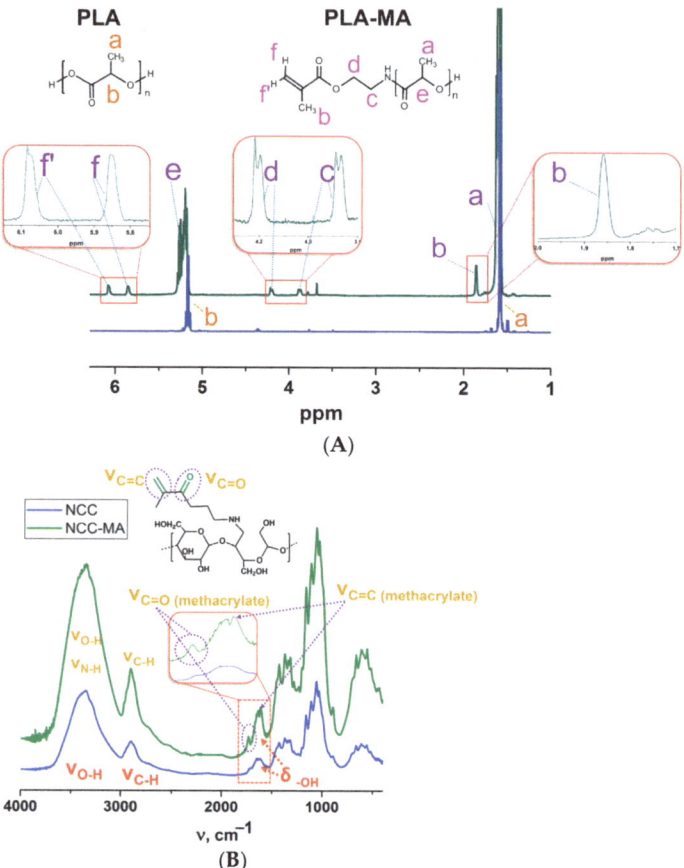

Figure 3. The evidence for successful introduction of methacrylate groups into particles under study: (**A**) ^{1}H NMR spectra of PLA and PLA-MA particles dissolved in CDCl$_3$; (**B**) FTIR spectra of NCC and NCC-MA.

The Raman spectra of PLA and PLA-MA were recorded in order to check the presence of methacrylate bonds in modified PLA particles (see Supplementary Materials, Figure S1). In the presented spectrum of the modified particles, there were valent vibrations of the C=C bond at 1650 and 1626 cm^{-1}, while in the spectrum of non-modified PLA, no signals were observed in this region.

Initial and modified particles were characterized by dynamic and electrophoretic light scattering (DLS and ELS), nanoparticle tracking analysis (NTA), and electron microscopy (TEM/SEM) (Table 1). The obtained data are presented in Table 1, and the data obtained by different methods are in good agreement with each other. One can observe that the initial PLA particles possessed greater diameter than that of raw NCC. Both PLA and NCC had negative ζ-potential. The value of this parameter was large enough to stabilize particle suspensions. Modification of particles by MA groups led to some growth in particle size and a slight decrease in surface charge. However, these changes were not drastic. It is obvious that growth of particle size could not be caused by particle enlargement due to the attachment of MA groups, but most probably resulted from particle aggregation due to the introduction of hydrophobic MA moieties. The increased width of particle size distribution after modification favors this statement (see Supplementary Materials, Figure S2). It should be noted that D_H of NCC particles is averaged as if they were spherical particles, although they were not.

Table 1. Effect of particle modification on particle diameter and surface charge.

Sample #	D_H (DLS), nm	ζ-Potential, mV	D (TEM), nm	D (NTA), nm
PLA MPs	258 ± 74	−45 ± 7	210 ± 70	263 ± 85
PLA-MA MPs	347 ± 129	−33 ± 8	290 ± 120	443 ± 181
NCC	95 ± 18	−39 ± 11	not determined	121 ± 30
NCC-MA	110 ± 24	−36 ± 5	not determined	182 ± 51

The morphology of the modified particles was analyzed by TEM and SEM. One can observe that both PLA and PLA-MA (Figure 4) possessed round-shaped particles. Modification of PLA particles with MA groups did not seem to change their morphological features. However, these particles seemed to be quite prone to aggregation, which is also supported by the data obtained by nanoparticle tracking analysis (Figure S2, Supplementary Materials).

Figure 4. Morphology of PLA particles: (**A**) TEM microphotographs of initial PLA NPs; (**B**) TEM microphotographs of PLA NPs modified with AEMA; (**C**,**D**) SEM microphotographs of PLA NPs modified with AEMA.

NCC-MA particles appear on the TEM images in the form of needles (Figure 5). The obtained sample of NCC-MA was observed by SEM, as presented in Figure 5. One can see the anisotropic character of the sample, which seems to reflect the organization of the NCC-MA particles into the fibrillary structures. The organization of NCC into such structures is well-known as also being the case for non-modified NCC [50].

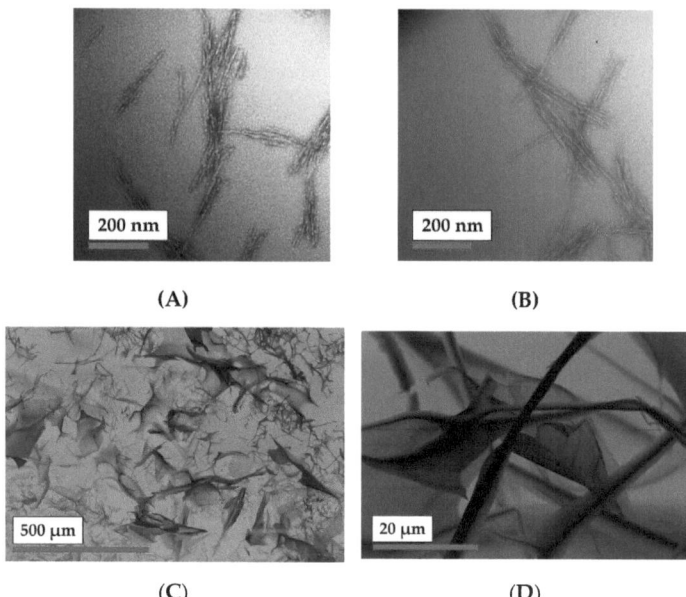

Figure 5. Morphology of NCC particles: (**A**) TEM images of non-modified NCC particles; (**B**) TEM images of NCC-MA particles; (**C**) SEM microphotographs NCC modified with AEMA; (**D**) SEM microphotographs NCC modified with AEMA.

The study of nanoparticle morphology can be summarized by the observation that the modification of nanoparticles does not affect their morphological features, but possibly could lead to their more intensive aggregation. Additionally, the size of the particles grew a little after the attachment of MA groups.

3.2. Effect of Particle Modification on Cells Viability

It is well known that PLA and NCC particles are quite biocompatible materials [35,46]. Here we were aimed to evaluate the effect of methacrylic groups attachment to the surface of such particles on their cytotoxicity. Two types of cells, which were used in this study, namely bone marrow mesenchymal stem cells (MSCs) and mouse embryonic fibroblasts (NIH/3T3), were widely used in tissue engineering related research. Different concentrations of particles were added to the cultured cells, and cell viability was evaluated via MTT assay after different periods of co-incubation (Figure 6).

The obtained results clearly show that both methacrylated types of particles (PLA-MA and NCC-MA) did not greatly affect cells viability at concentrations 1 and 5 mg/mL. There was a certain decrease in viability in the case of NIH/3T3 cells after 72 h of 10 mg/mL for both types of particles co-incubated with cells. However, no such effect was observed for MSCs.

In general, it might be concluded that modification of PLA and NCC NPs by methacrylate groups does not result in a drastic increase in their toxicity. The obtained particles could be considered as quite non-toxic and recommended for further studies on their possible application for scaffolds formation.

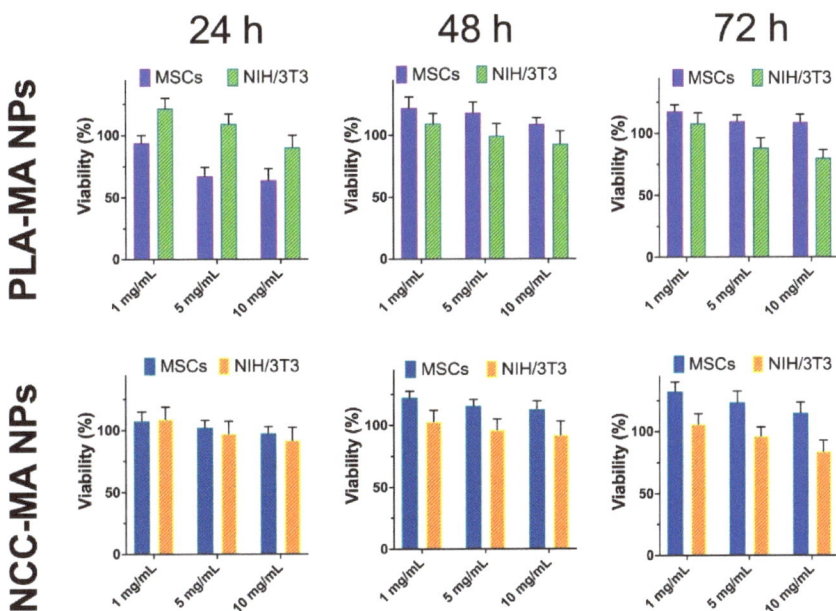

Figure 6. The effect of PLA-MA and NCC-MA particles on cell viability. MSCs and NIH/3T3 were exposed to the indicated concentrations of particles for 24, 48 and 72 h. Viability (%) is referred to the control, which was pure media without particles. Data are presented as mean ± SD ($n = 5$).

3.3. Rheological Studies of PLA and NCC Particle Cross-Linking

The study of the possibility of particle interaction under free-radical process conditions was the key stage of this research. For that, the rheology experiments were applied to detect the appropriate particle suspension concentration and PLA-MA/NCC-MA ratio at which the storage modulus (G') would increase, indicating the formation of a three-dimensional network. Particle suspensions of different compositions together with photo-initiators were placed between the two rheometer plates (Supplementary Materials, Figure S3). The rheological measurements were started simultaneously with sample UV irradiation to detect the growth of G'. The exposure to UV continued over 10 min and the measurement process after that period was conducted without irradiation. Thus, we also detected the effect of post-exposure cross-linking on the rheological properties of the system.

First, we studied the effect of particle concentrations on their ability to form cross-links between each other (Figures 7A and 8A). Figure 7A shows the possible scheme of direct cross-linking. One can observe that the concentrations of both PLA-MA and NCC-MA played important roles in the growth of storage modulus. This is logical, because the increase in concentration affects the probability of particle interaction, which could lead to cross-linking. It was shown that for the formation of cross-links between particles, a particle concentration of at least 100 mg/mL was required. It appears that at such concentrations, particles could form a three-dimensional network; however, the mechanical properties of such networks were not enough to form stable 3D matrices. Higher concentrations of particles were not applied here, because particles could aggregate, and it was highly probable 3D printer nozzle would be clogged. it is notable that investigation of the effects of the PLA-MA and NCC-MA ratio on G' growth showed that an 80:20 weight ratio of corresponding particles resulted in the best growth of storage modulus (Supplementary materials, Figure S4A). It should be also noted that significant after-exposure free-radical cross-linking was not observed.

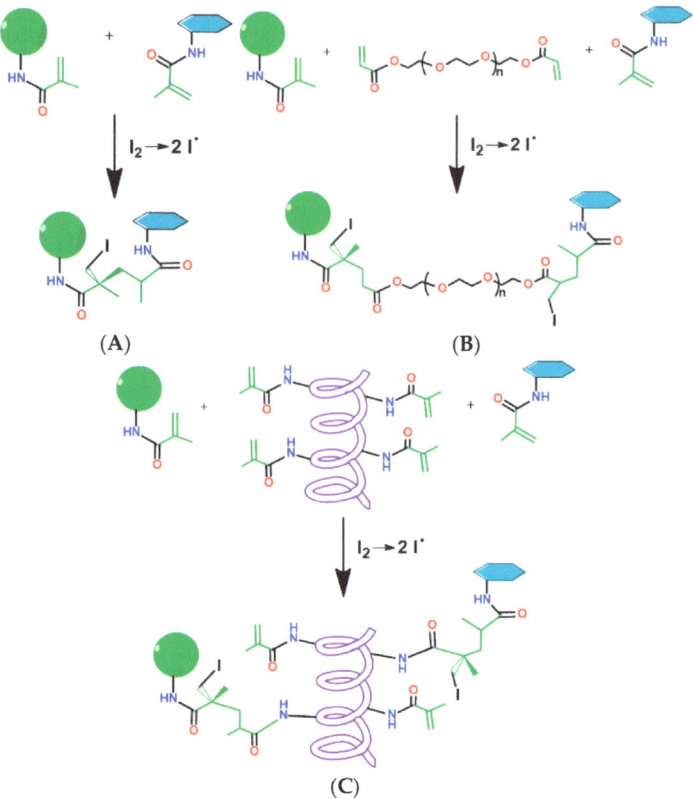

Figure 7. The possible schemes of particle cross-linking: (**A**) direct methacrylated particle cross-linking; (**B**) particle cross-linking with application of PEGDA as cross-linking agent; (**C**) particle cross-linking with application of GelMA as cross-linking agent.

Considering the abovementioned circumstances and previous experience on the application of poly(ethylene glycol) diacrylate (PEGDA) and gelatin methacrylate (GelMA) for preparation of matrices for 3D cell culturing [51,52] we made and attempted to use such macromolecules as crosslinkers for intensification of the inter-particle interaction. The schemes of cross-linking in these cases are shown on Figure 7B,C. The effect of cross-linker concentration on G' growth showed that concentration of 20 wt% GelMA was optimal, because increasing GelMA concentration did not result in better rheological properties (Supplementary Materials, Figure S4B).

One can observe an impressive difference in the growth of the storage modulus when using PEGDA and GelMA as crosslinking agents (Figure 8B). The values of the G' increased by more than two orders of magnitude after 10 min of irradiation with UV lamp when crosslinking macromolecules were added to the mixture. The materials formed under such conditions seemed to be stable 3D networks. It is notable that the application of GelMA yielded a greater increase in G' than in the case of PEGDA. This might be explained by the fact that GelMA is a multifunctional cross-linker agent, bearing several MA groups in its structure, while PEGDA is only bifunctional. Such a benefit of GelMA provides better steric accessibility of MA groups for participation in free-radical cross-linking. The greater molecular mass of GelMA as compared to that of PEGDA could be also an important factor in the observed effect.

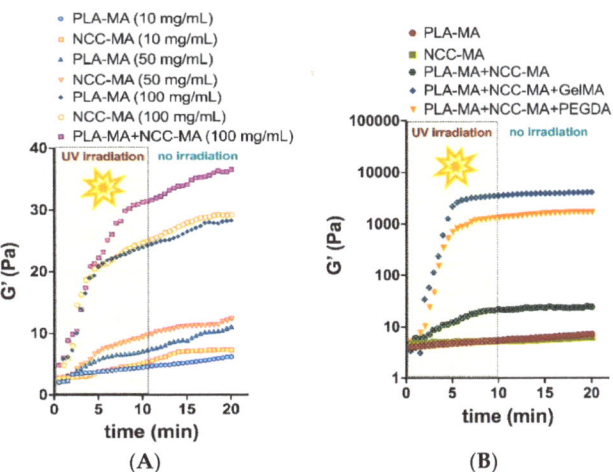

Figure 8. Results of rheological studies on crosslinking kinetics: (**A**) effect of particle concentration; (**B**) effect of cross-linkers (PEGDA and GelMA) on particle cross-linkers. Shown is G' (storage modulus) evolution with time; G" (loss modulus) is not shown for the purpose of clarity. In all samples G' > G" after exposure to UV, indicating a well-developed and solid polymer network. Samples were measured at 25 °C and were exposed to UV intensity of 1.2 J/cm^2. The curves shown are representative ones of three measurements.

The conducted rheological experiments led us to significant conclusions: the interparticle interaction provoked by free-radical cross-linking of methacrylate groups is generally possible at concentrations above 100 mg/mL, but does not lead to the formation of mechanically stable polymer networks. Thus, the formation of particle-based scaffolds requires the application of macromolecular cross-linker. Multifunctional cross-linkers with greater molecular mass result in better cross-linking than bifunctional ones.

3.4. Cryogelation and 3D-Printing: Methods for Preparation of Particle-Based Scaffolds

Rheological studies allowed us to identify the composition of the mixtures containing PLA-MA and NCC-MA particles that could be turned into reasonably stable materials. In the next stage of this study, we aimed to use such compositions for obtaining macroporous matrices via cryogelation and 3D printing.

Both methods were successfully applied for the preparation of materials (Figure 9). The composition of the particle/cross-linker mixture used for cryogelation and 3D printing that was selected showed the best performance when subjected to rheological studies, and comprised PLA-MA/NCC-MA 80:20 (wt/wt) and 20 wt% cross-linker (PEGDA or GelMA).

In the case of cryogelation, the free-radical process was initiated by ammonium persulfate (APS) and 1,2-bis(dimethylamino)ethane (TEMED) RedOx initiation system. The idea of cryogelation consists in the cryo-concentration of particles and cross-linker into unfrozen liquid microphase due to the crystallization of the solvent (water), which pushes particles and cross-linker out of the crystallizing front. The action of APS and TEMED could initiate the free-radical process below 0 °C, which leads to interparticle cross-linking with the involvement of macromolecular cross-linker (PEGDA or GelMA). The described process was performed in the syringe, leading to the formation of matrices with cylindrical geometrical forms (Figure 9A).

The CAD models of the materials were designed and the DIW 3D printing of scaffolds was performed with an extrusion-type printhead and a UV lamp for photo-cross-linking of the material. A photo-initiator (Irgacure 2959, 0.1 wt%) was added to the system to provide the possibility of free-radical cross-linking reactions. Such printing parameters as printing needle diameter, pressure, and printing speed were optimized. To avoid the formation of hard gel and printhead clogging, in the case of the sample containing thermosensitive Gel-MA as an additional crosslinking agent, the optimal printing suspension temperature was also determined to be 37 °C. Uniform extrusion was observed with a printing needle diameter of 0.254 mm and length of 6.35 mm, and with pressure of 0.5 PSI or 3.45 kPa. The optimal printing speed was 5 mm/s. Three-dimensional printing and cross-linking of the sample was performed in layers (Supplementary Materials, Figure S5): after the printing of the first level of the matrix, the UV lamp was turned on, the sample was irradiated for 10 min, and then the printing process was resumed. Fully printed samples were left for one hour for the completion of free-radical reaction and drying (Figure 9B).

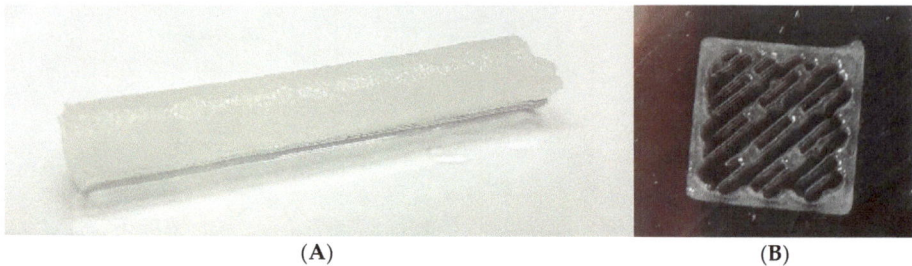

Figure 9. Materials obtained based on the PLA-MA/NCC-MA/Gel-MA mixture by: (**A**) cryogelation in 1 mL syringe; (**B**) 3D printing. The 80:20 weight ratio of PLA-MA and NCC-MA along with 20 wt% GelMA was applied for the preparation of these materials.

The morphology of samples obtained by cryogelation and 3D printing was studied by SEM (Figures 10 and 11). Particle-based cryogel samples, obtained with the application of both GelMA and PEGDA, demonstrated a macroporous nature with average pore diameter of around 200 µm (according to SEM). Additionally, matrices showed good pore interconnectivity, which is important for scaffolds. Interestingly, the examination of the pore walls under a larger magnification demonstrated the fact that the particles were incorporated into the pore walls (Figure 10B,D).

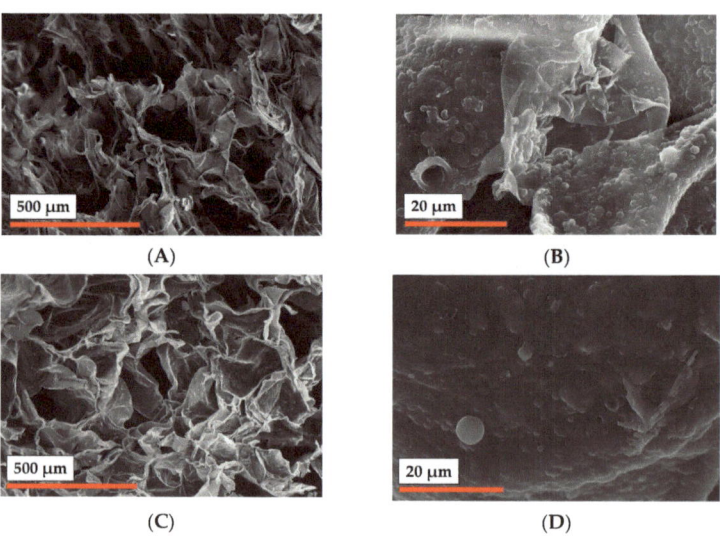

Figure 10. SEM images of matrices obtained by cryogelation of PLA-MA NPs with NCC-MA with different cross-linkers at different magnification: (**A**) Gel-MA (×100); (**B**) Gel-MA (×2000); (**C**) PEGDA (×100); (**D**) PEGDA (×2000).

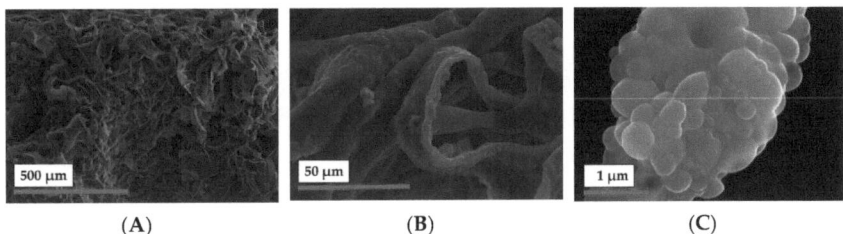

Figure 11. SEM images of PLA-MA NPs/NCC-MA 3D-printed matrix: (**A**) ×100; (**B**) ×1000; (**C**) ×30,000.

Based on the results of the study of the rheology of particle crosslinking with different cross-linkers in the presence of the photoinitiator (Figure 8B), only compositions containing GelMA were used for 3D printing. The morphology of the obtained material at different magnification was studied by SEM (Figure 11). It should be noted that in contrast to the cryogelation process, the pore structure in the 3D printing was not provided by phase separation, but was governed by the designed CAD model. However, the surface of the 3D printed hydrogel scaffold was shown to be quite rough, having pores with 10 to 100 μm in diameter (Figure 11A). As in the case of cryogelated materials, the magnification of the material surface allowed the observation of the particles incorporated into the walls of the 3D printed matrices (Figure 11B,C).

Thus, both cryogelation and 3D printing approaches could be applied to form particle-based materials with the application of PLA-MA, NCC-MA, and GelMA as cross-linkers. The application of PEGDA as a cross-linker does not result in any beneficial structure. At the same time, GelMA could be considered as a more promising material, because it bears a specific arginine-glycine-aspartic acid (RGD) cell-adhesion moiety [53], which is important for the induction of novel tissue formation [11].

3.5. Mechanical Testing

The mechanical strength of materials is an important characteristic for scaffolds, especially in the case of the tissue engineering of hard tissues [1]. Thus, we performed a comparative study of matrices obtained by cryogelation and 3D printing based on PLA-MA/NCC-MA and GelMA as cross-linkers. The strain–stress curves obtained during sample compression are presented in Figure 12. The mechanical properties of both cryogelated and 3D printed matrices were not extraordinary and more or less corresponded to those of hydrogels. At the same time, it was obvious that the 3D printed material possessed better mechanical properties than the cryogelated one. This can be explained by the thicker walls and lower overall porosity, and hence the greater density of the printed material as compared to cryogel. It also should be mentioned here that 3D printing allowed control over the thickness of formed material walls, thus the mechanical properties could be controlled to form gradients of mechanical strength. Such material formations will be presented in our future papers. Thus, based on the above-mentioned arguments, we can propose 3D printed scaffolds as better candidates for hard-tissue engineering.

3.6. Cell Adhesion on Cryogelated and 3D Printed Matrices

Aa a final step of the current study, we performed a comparative study of MSC adhesion on the surface of the cryogelated matrices and the 3D printed ones. PLA-MA and NCC-MA ratios were chosen as based on the previous results and were 80:20 (wt/wt). In the case of cryogel, PEGDA was applied as cross-linking agent, while in the case of 3D printed material it was GelMA. To test cell adhesion matrices were incubated with MSC cells on a 24-well plate for one week in MEM. Then, the cells adhered to the matrices were visualized by DAPI-staining. One can observe (Figure 13) that the number of cells on the surface of 3D printed material seem to be greater than on the surface of cryogelated matrix.

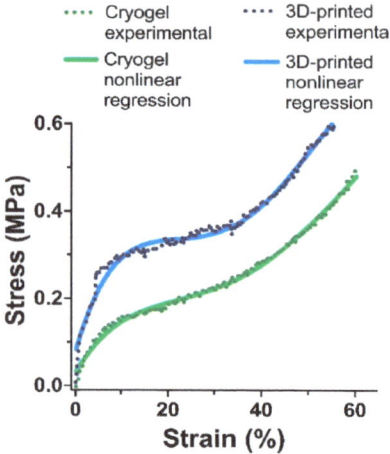

Figure 12. Strain–stress test curves for cryogelated and 3D-printed matrices. The compression stress was applied with initial force 0.1 N. The speed of compression was 2 mm/min. The data on the graph are representative of 3 measurements (n = 3).

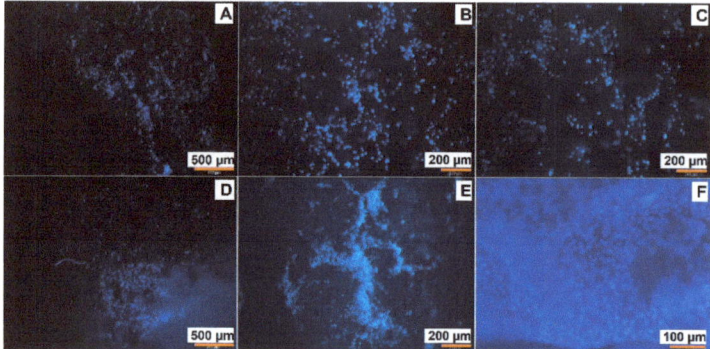

Figure 13. Fluorescent images of DAPI-stained mesenchymal stem cells (MSC) on the surface of: (**A–C**)—cryogelated matrix PLA-MA/NCC-MA + 20% PEGDA; (**D–F**) 3D-printed matrix PLA-MA/NCC-MA + 20% GelMA. Images were recorded at different magnification: A and D ×2; B and E ×4; C and F ×10.

In order to obtain quantitative results, we measured the intensity of the fluorescence on the surface of cryogelated and 3D printed materials with equal geometry (round discs). The obtained data (Figure 14) are in line with visual data on Figure 13 and clearly show that the number of cells attached to the 3D printed material is significantly greater than that in the case of cryogels. This observation is even more interesting given the apparently higher surface area of the cryogel matrices. It is most likely that the surface area factor was more than counterbalanced by the presence of GelMA in the printed matrices, which in turn contained specific cell adhesion moiety, namely RGD. Such a specific moiety initiates intensive cell adhesion due to signal transmission between cells [54].

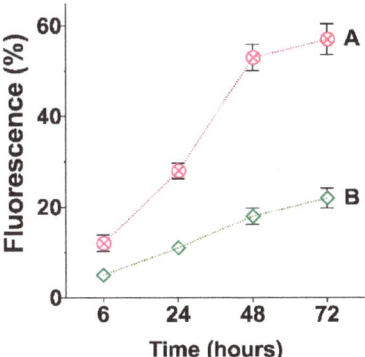

Figure 14. Results of fluorescence analysis of attached DAPI-labeled MSCs on the surface of: (**A**) 3D printed matrix PLA-MA/NCC-MA + 20% GelMA and (**B**) cryogelated matrix PLA-MA/NCC-MA + 20% PEGDA after 6, 24, 48, and 72 h after seeding.

4. Conclusions

In the presented study we were aimed to study the possibility of particle-based scaffold formation as based on PLA and NCC. In order to form covalent cross-links between these particles, we introduced methacrylate groups on their surfaces. Such modification did not greatly affect particle morphology, namely the spherical nature particles, in the case of PLA and needles in the case of NCC. Furthermore, no increasing toxicity of modified particles in relation to the MSCs and NIH/3T3 cells was observed.

The investigation of particle interaction and the ability to form three-dimensional networks under free-radical polymerization conditions showed that particles can form only very poor networks (storage modulus 30–40 Pa), when the reaction takes place only between the particles themselves. The addition of macromolecular cross-links, such as PEGDA and GelMA, in concentrations of 20 wt% significantly changed the situation, and reasonably stable networks could be formed. Such networks were characterized by a storage modulus as large as 2000–4000 Pa. The developed particles/cross-linker compositions were successfully applied for the formation of materials via cryogelation/cryopolymerization technique and DIW 3D printing with photo-curing. The study of the material morphology showed that both types of materials, namely cryogels and 3D printed matrices, contained particles in their structure. Mechanical tests showed that 3D printed materials possessed greater strength than cryogelated ones. The study of MSCs adhesion clearly showed that application of GelMA as a cross-linking agent is highly preferable if one wants to initiate specific cell adhesion. The fluorescence of DAPI-stained cells was three times higher when GelMA was used as the particle cross-linker.

The results of this study could be useful for the preparation of 3D scaffolds with the application of particles. Further attempts should be made to generate gradient scaffolds via printing with different printheads and extruders loaded with different particles. Thus, we can say that particles are another prospective component in the toolbox for creating scaffolds that will effectively support the formation of new tissue both in vitro and in vivo.

Supplementary Materials: The following supporting information can be downloaded at: https://www.mdpi.com/article/10.3390/polym15030651/s1, Figure S1. Raman spectrum of PLA (green line) and PLA-MA (blue line). Raman spectra were recorded with T64000 Horiba Scientific spectrometer (Kyoto, Japan). Figure S2. Non-methacrylated and methacrylated particles characteristics: (A)—PLA and PLA-MA; (B)—NCC and NCC-MA. Figure S3. The scheme of rheological experiment, which was applied for the testing of interparticle cross-linking. Figure S4: Results of rheological experiment: effect of PLA-MA and NCC-MA weight ratios (A) and Gel-MA concentration (B) on storage modulus increase during 10 min of exposure to UV-light and thereafter. Figure S5: The scheme of 3D printing with the application of particles as photo-curable ink.

Author Contributions: Conceptualization, V.K.-V. and T.T.; methodology, V.K.-V. and A.L.; formal analysis, V.K.-V., M.L., A.L. and E.K.-V.; investigation, M.L., I.P. and V.K.-V.; resources, A.L., E.K.-V. and T.T.; data curation, V.K.-V. and M.L.; writing—original draft preparation, V.K.-V.; writing—review and editing, E.K.-V. and T.T.; visualization, M.L. and V.K.-V.; supervision, V.K.-V. and T.T.; project administration, T.T.; funding acquisition, T.T. All authors have read and agreed to the published version of the manuscript.

Funding: This research was funded by Megagrant of the Ministry of Science and Higher Education of the Russian Federation (#075-15-2021-637).

Institutional Review Board Statement: Not applicable.

Informed Consent Statement: Not applicable.

Data Availability Statement: Data available within the article or its Supplementary Materials.

Acknowledgments: The paper is dedicated to the 300th anniversary of Saint Petersburg University. The authors are grateful to Magnetic Resonance Research Centre, Chemical Analysis and Materials Research Centre, Centre for Molecular and Cell Technologies, Interdisciplinary Resource Centre for Nanotechnology of St. Petersburg State University for NMR, FTIR, TEM and SEM measurements, correspondingly. M. Leonovich thanks the IP@Leibniz program for two-month scholarship.

Conflicts of Interest: The authors declare no conflict of interest.

References

1. Hutmacher, D.W. Scaffolds in tissue engineering bone and cartilage. *Biomaterials* **2000**, *21*, 2529–2543. [CrossRef] [PubMed]
2. Lee, S.S.; Du, X.; Kim, I.; Ferguson, S.J. Scaffolds for bone-tissue engineering. *Matter* **2022**, *5*, 2722–2759. [CrossRef]
3. Zhu, G.; Zhang, T.; Chen, M.; Yao, K.; Huang, X.; Zhang, B.; Li, Y.; Liu, J.; Wang, Y.; Zhao, Z. Bone physiological microenvironment and healing mechanism: Basis for future bone-tissue engineering scaffolds. *Bioact. Mater.* **2021**, *6*, 4110–4140. [CrossRef] [PubMed]
4. Bhat, Z.F.; Bhat, H.; Pathak, V. Prospects for In Vitro Cultured Meat – A Future Harvest. In *Principles of Tissue Engineering*, 4th ed.; Academic Press: Cambridge, MA, USA, 2014; pp. 1663–1683. [CrossRef]
5. Langer, R.; Vacanti, J. Advances in tissue engineering. *J. Pediatr. Surg.* **2016**, *51*, 8–12. [CrossRef]
6. Jafari, M.; Paknejad, Z.; Rad, M.R.; Motamedian, S.R.; Eghbal, M.J.; Nadjmi, N.; Khojasteh, A. Polymeric scaffolds in tissue engineering: A literature review. *J. Biomed. Mater. Res. Part B Appl. Biomater.* **2017**, *105*, 431–459. [CrossRef]
7. Turnbull, G.; Clarke, J.; Picard, F.; Riches, P.; Jia, L.; Han, F.; Li, B.; Shu, W. 3D bioactive composite scaffolds for bone tissue engineering. *Bioact. Mater.* **2018**, *3*, 278–314. [CrossRef]
8. Fernandez-Yague, M.A.; Abbah, S.A.; McNamara, L.; Zeugolis, D.I.; Pandit, A.; Biggs, M.J. Biomimetic approaches in bone tissue engineering: Integrating biological and physicomechanical strategies. *Adv. Drug Deliv. Rev.* **2015**, *84*, 1–29. [CrossRef]
9. Cattalini, J.P.; Roether, J.; Hoppe, A.; Pishbin, F.; Haro Durand, L.; Gorustovich, A.; Boccaccini, A.R.; Lucangioli, S.; Mouriño, V. Nanocomposite scaffolds with tunable mechanical and degradation capabilities: Co-delivery of bioactive agents for bone tissue engineering. *Biomed. Mater.* **2016**, *11*, 065003. [CrossRef] [PubMed]
10. Samorezov, J.E.; Alsberg, E. Spatial regulation of controlled bioactive factor delivery for bone tissue engineering. *Adv. Drug Deliv. Rev.* **2015**, *84*, 45–67. [CrossRef]
11. Brown, B.N.; Badylak, S.F. Extracellular matrix as an inductive scaffold for functional tissue reconstruction. *Transl. Res.* **2014**, *163*, 268–285. [CrossRef]
12. Huang, W.; Li, X.; Shi, X.; Lai, C. Microsphere based scaffolds for bone regenerative applications. *Biomater. Sci.* **2014**, *2*, 1145–1153. [CrossRef]
13. Shi, X.; Wang, Y.; Varshney, R.R.; Ren, L.; Gong, Y.; Wang, D.A. Microsphere-based drug releasing scaffolds for inducing osteogenesis of human mesenchymal stem cells in vitro. *Eur. J. Pharm. Sci.* **2010**, *39*, 59–67. [CrossRef] [PubMed]
14. Shalumon, K.; Sheu, C.; Fong, Y.; Liao, H.-T.; Chen, J.-P. Microsphere-Based Hierarchically Juxtapositioned Biphasic Scaffolds Prepared from Poly(Lactic-co-Glycolic Acid) and Nanohydroxyapatite for Osteochondral Tissue Engineering. *Polymers* **2016**, *8*, 429. [CrossRef] [PubMed]
15. Son, J.S.; Kim, S.G.; Jin, S.C.; Piao, Z.G.; Lee, S.Y.; Oh, J.S.; Kim, C.S.; Kim, B.H.; Jeong, M.A. Development and structure of a novel barrier membrane composed of drug-loaded poly(lactic-co-glycolic acid) particles for guided bone regeneration. *Biotechnol. Lett.* **2012**, *34*, 779–787. [CrossRef] [PubMed]
16. Dang, W.; Jin, Y.; Yi, K.; Ju, E.; Zhuo, C.; Wei, H.; Wen, X.; Wang, Y.; Li, M.; Tao, Y. Hemin particles-functionalized 3D printed scaffolds for combined photothermal and chemotherapy of osteosarcoma. *Chem. Eng. J.* **2021**, *422*, 129919. [CrossRef]
17. Carter, P.; Bhattarai, N. Bioscaffolds: Fabrication and Performance. In *Engineered Biomimicry*; Elsevier: Amsterdam, The Netherlands, 2013; pp. 161–188. ISBN 9780124159952.
18. Rogers, Z.J.; Bencherif, S.A. Cryogelation and cryogels. *Gels* **2019**, *5*, 46. [CrossRef]

19. Korzhikov-Vlakh, V.; Pepelanova, I. Biological, Natural, and Synthetic 3D Matrices. In *Basic Concepts on 3D Cell Culture*; Kasper, C., Egger, D., Lavrentieva, A., Eds.; Springer: Cham, Switzerland, 2021; pp. 79–104. ISBN 978-3-030-66748-1.
20. Holländer, J.; Hakala, R.; Suominen, J.; Moritz, N.; Yliruusi, J.; Sandler, N. 3D printed UV light cured polydimethylsiloxane devices for drug delivery. *Int. J. Pharm.* 2018, *544*, 433–442. [CrossRef]
21. Pahlevanzadeh, F.; Emadi, R.; Valiani, A.; Kharaziha, M.; Poursamar, S.A.; Bakhsheshi-Rad, H.R.; Ismail, A.F.; RamaKrishna, S.; Berto, F. Three-dimensional printing constructs based on the chitosan for tissue regeneration: State of the art, developing directions and prospect trends. *Materials* 2020, *13*, 2663. [CrossRef] [PubMed]
22. Nava-Medina, I.B.; Gold, K.A.; Cooper, S.M.; Robinson, K.; Jain, A.; Cheng, Z.; Gaharwar, A.K. Self-Oscillating 3D Printed Hydrogel Shapes. *Adv. Mater. Technol.* 2021, *6*, 2100418. [CrossRef]
23. Kuss, M.A.; Wu, S.; Wang, Y.; Untrauer, J.B.; Li, W.; Lim, J.Y.; Duan, B. Prevascularization of 3D printed bone scaffolds by bioactive hydrogels and cell co-culture. *J. Biomed. Mater. Res. Part B Appl. Biomater.* 2018, *106*, 1788–1798. [CrossRef]
24. Liu, J.; Yan, C. 3D Printing of Scaffolds for Tissue Engineering. In *3D Printing*; InTech: London, UK, 2018; Volume 11, p. 13, ISBN 0000957720.
25. Auras, R.; Lim, L.-T.; Selke, S.E.M.; Tsuji, H. (Eds.) *Poly(lactic acid): Synthesis, Structures, Properties, Processing, and Application*; John Wiley & Sons, Inc.: Hoboken, NJ, USA, 2010; ISBN 9780470293669.
26. Ghalia, M.A.; Dahman, Y. Biodegradable poly(lactic acid)-based scaffolds: Synthesis and biomedical applications. *J. Polym. Res.* 2017, *24*, 74. [CrossRef]
27. Elmowafy, E.M.; Tiboni, M.; Soliman, M.E. Biocompatibility, biodegradation and biomedical applications of poly(lactic acid)/poly(lactic-co-glycolic acid) micro and nanoparticles. *J. Pharm. Investig.* 2019, *49*, 347–380. [CrossRef]
28. Jung, F.; Braune, S. Thrombogenicity and hemocompatibility of biomaterials. *Biointerphases* 2016, *11*, 029601. [CrossRef]
29. Esposito Corcione, C.; Gervaso, F.; Scalera, F.; Padmanabhan, S.K.; Madaghiele, M.; Montagna, F.; Sannino, A.; Licciulli, A.; Maffezzoli, A. Highly loaded hydroxyapatite microsphere/PLA porous scaffolds obtained by fused deposition modelling. *Ceram. Int.* 2019, *45*, 2803–2810. [CrossRef]
30. Hu, X.; He, J.; Yong, X.; Lu, J.; Xiao, J.; Liao, Y.; Li, Q.; Xiong, C. Biodegradable poly (lactic acid-co-trimethylene carbonate)/chitosan microsphere scaffold with shape-memory effect for bone tissue engineering. *Colloids Surf. B Biointerfaces* 2020, *195*, 111218. [CrossRef] [PubMed]
31. Zhu, N.; Li, M.G.; Cooper, D.; Chen, X.B. Development of novel hybrid poly(l-lactide)/chitosan scaffolds using the rapid freeze prototyping technique. *Biofabrication* 2011, *3*, 034105. [CrossRef]
32. Zhao, L.; Liu, Y.; Zhang, R.; He, H.; Jin, T.; Zhang, J. Unique morphology in polylactide/graphene oxide nanocomposites. *J. Macromol. Sci. Part B Phys.* 2015, *54*, 45–57. [CrossRef]
33. Murizan, N.I.S.; Mustafa, N.S.; Ngadiman, N.H.A.; Mohd Yusof, N.; Idris, A. Review on Nanocrystalline Cellulose in Bone Tissue Engineering Applications. *Polymers* 2020, *12*, 2818. [CrossRef]
34. Averianov, I.; Stepanova, M.; Solomakha, O.; Gofman, I.; Serdobintsev, M.; Blum, N.; Kaftuirev, A.; Baulin, I.; Nashchekina, J.; Lavrentieva, A.; et al. 3D-Printed composite scaffolds based on poly(ε-caprolactone) filled with poly(glutamic acid)-modified cellulose nanocrystals for improved bone tissue regeneration. *J. Biomed. Mater. Res. Part B Appl. Biomater.* 2022, *110*, 2422–2437. [CrossRef]
35. Stepanova, M.; Korzhikova-Vlakh, E. Modification of Cellulose Micro- and Nanomaterials to Improve Properties of Aliphatic Polyesters/Cellulose Composites: A Review. *Polymers* 2022, *14*, 1477. [CrossRef]
36. Wang, H.; Leeuwenburgh, S.C.G.; Li, Y.; Jansen, J.A. The Use of Micro- and Nanospheres as Functional Components for Bone Tissue Regeneration. *Tissue Eng. Part B. Rev.* 2012, *18*, 24–39. [CrossRef]
37. Gupta, V.; Khan, Y.; Berkland, C.J.; Laurencin, C.T.; Detamore, M.S. Microsphere-Based Scaffolds in Regenerative Engineering. *Annu. Rev. Biomed. Eng.* 2017, *19*, 135–161. [CrossRef]
38. Jiang, T.; Nukavarapu, S.P.; Deng, M.; Jabbarzadeh, E.; Kofron, M.D.; Doty, S.B.; Abdel-Fattah, W.I.; Laurencin, C.T. Chitosan-poly(lactide-co-glycolide) microsphere-based scaffolds for bone tissue engineering: In vitro degradation and in vivo bone regeneration studies. *Acta Biomater.* 2010, *6*, 3457–3470. [CrossRef] [PubMed]
39. Lee, B.K.; Yun, Y.; Park, K. PLA micro- and nano-particles. *Adv. Drug Deliv. Rev.* 2016, *107*, 176–191. [CrossRef]
40. Luciani, A.; Coccoli, V.; Orsi, S.; Ambrosio, L.; Netti, P.A. PCL microspheres based functional scaffolds by bottom-up approach with predefined microstructural properties and release profiles. *Biomaterials* 2008, *29*, 4800–4807. [CrossRef] [PubMed]
41. Nukavarapu, S.P.; Kumbar, S.G.; Brown, J.L.; Krogman, N.R.; Weikel, A.L.; Hindenlang, M.D.; Nair, L.S.; Allcock, H.R.; Laurencin, C.T. Polyphosphazene/Nano-Hydroxyapatite Composite Microsphere Scaffolds for Bone Tissue Engineering. *Biomacromolecules* 2008, *9*, 1818–1825. [CrossRef] [PubMed]
42. Duan, B.; Wang, M.; Zhou, W.Y.; Cheung, W.L.; Li, Z.Y.; Lu, W.W. Three-dimensional nanocomposite scaffolds fabricated via selective laser sintering for bone tissue engineering. *Acta Biomater.* 2010, *6*, 4495–4505. [CrossRef] [PubMed]
43. Masutani, K.; Kimura, Y. Chapter 1. PLA Synthesis. From the Monomer to the Polymer. In *Poly(lactic acid) Science and Technology: Processing, Properties, Additives and Applications*; RSC Publishing: Oxford, UK, 2014; pp. 1–36.
44. Pepelanova, I.; Kruppa, K.; Scheper, T.; Lavrentieva, A. Gelatin-Methacryloyl (GelMA) Hydrogels with Defined Degree of Functionalization as a Versatile Toolkit for 3D Cell Culture and Extrusion Bioprinting. *Bioengineering* 2018, *5*, 55. [CrossRef]

45. Kritchenkov, I.S.; Zhukovsky, D.D.; Mohamed, A.; Korzhikov-Vlakh, V.A.; Tennikova, T.B.; Lavrentieva, A.; Scheper, T.; Pavlovskiy, V.V.; Porsev, V.V.; Evarestov, R.A.; et al. Functionalized Pt(II) and Ir(III) NIR Emitters and Their Covalent Conjugates with Polymer-Based Nanocarriers. *Bioconjug. Chem.* **2020**, *31*, 1327–1343. [CrossRef]
46. Korzhikov-Vlakh, V.; Averianov, I.; Sinitsyna, E.; Nashchekina, Y.; Polyakov, D.; Guryanov, I.; Lavrentieva, A.; Raddatz, L.; Korzhikova-Vlakh, E.; Scheper, T.; et al. Novel pathway for efficient covalent modification of polyester materials of different design to prepare biomimetic surfaces. *Polymers* **2018**, *10*, 1299. [CrossRef]
47. Mandal, A.; Chakrabarty, D. Isolation of nanocellulose from waste sugarcane bagasse (SCB) and its characterization. *Carbohydr. Polym.* **2011**, *86*, 1291–1299. [CrossRef]
48. Wulandari, W.T.; Rochliadi, A.; Arcana, I.M. Nanocellulose prepared by acid hydrolysis of isolated cellulose from sugarcane bagasse. *IOP Conf. Ser. Mater. Sci. Eng.* **2016**, *107*, 012045. [CrossRef]
49. Le Troedec, M.; Sedan, D.; Peyratout, C.; Bonnet, J.P.; Smith, A.; Guinebretiere, R.; Gloaguen, V.; Krausz, P. Influence of various chemical treatments on the composition and structure of hemp fibres. *Compos. Part A Appl. Sci. Manuf.* **2008**, *39*, 514–522. [CrossRef]
50. Li, W.; Yue, J.; Liu, S. Preparation of nanocrystalline cellulose via ultrasound and its reinforcement capability for poly(vinyl alcohol) composites. *Ultrason. Sonochem.* **2012**, *19*, 479–485. [CrossRef] [PubMed]
51. Liang, J.; Dijkstra, P.J.; Poot, A.A.; Grijpma, D.W. Hybrid Hydrogels Based on Methacrylate-Functionalized Gelatin (GelMA) and Synthetic Polymers. *Biomed. Mater. Devices* **2022**. [CrossRef]
52. Lavrentieva, A.; Fleischhammer, T.; Enders, A.; Pirmahboub, H.; Bahnemann, J.; Pepelanova, I. Fabrication of Stiffness Gradients of GelMA Hydrogels Using a 3D Printed Micromixer. *Macromol. Biosci.* **2020**, *20*, 2000107. [CrossRef]
53. Hersel, U.; Dahmen, C.; Kessler, H. RGD modified polymers: Biomaterials for stimulated cell adhesion and beyond. *Biomaterials* **2003**, *24*, 4385–4415. [CrossRef]
54. Bellis, S.L. Advantages of RGD peptides for directing cell association with biomaterials. *Biomaterials* **2011**, *32*, 4205–4210. [CrossRef]

Disclaimer/Publisher's Note: The statements, opinions and data contained in all publications are solely those of the individual author(s) and contributor(s) and not of MDPI and/or the editor(s). MDPI and/or the editor(s) disclaim responsibility for any injury to people or property resulting from any ideas, methods, instructions or products referred to in the content.

Article

3D-Printed PLA Medical Devices: Physicochemical Changes and Biological Response after Sterilisation Treatments

Sara Pérez-Davila [1,2,*], Laura González-Rodríguez [1,2], Raquel Lama [3], Miriam López-Álvarez [1,2], Ana Leite Oliveira [4], Julia Serra [1,2], Beatriz Novoa [3], Antonio Figueras [3] and Pío González [1,2]

1. CINTECX, Universidade de Vigo, Grupo de Novos Materiais, 36310 Vigo, Spain
2. Galicia Sur Health Research Institute (IIS Galicia Sur), SERGAS-UVIGO, 36213 Vigo, Spain
3. Institute of Marine Research (IIM), National Research Council (CSIC), 36208 Vigo, Spain
4. Universidade Católica Portuguesa, CBQF—Centro de Biotecnologia e Química Fina, Laboratório Associado, Escola Superior de Biotecnologia, Rua Diogo de Botelho, 1327, 4169-005 Porto, Portugal
* Correspondence: saperez@uvigo.es

Citation: Pérez-Davila, S.; González-Rodríguez, L.; Lama, R.; López-Álvarez, M.; Oliveira, A.L.; Serra, J.; Novoa, B.; Figueras, A.; González, P. 3D-Printed PLA Medical Devices: Physicochemical Changes and Biological Response after Sterilisation Treatments. *Polymers* 2022, 14, 4117. https://doi.org/10.3390/polym14194117

Academic Editors: Beata Kaczmarek and Marcin Wekwejt

Received: 1 September 2022
Accepted: 27 September 2022
Published: 1 October 2022

Publisher's Note: MDPI stays neutral with regard to jurisdictional claims in published maps and institutional affiliations.

Copyright: © 2022 by the authors. Licensee MDPI, Basel, Switzerland. This article is an open access article distributed under the terms and conditions of the Creative Commons Attribution (CC BY) license (https://creativecommons.org/licenses/by/4.0/).

Abstract: Polylactic acid (PLA) has become one of the most commonly used polymers in medical devices given its biocompatible, biodegradable and bioabsorbable properties. In addition, due to PLA's thermoplastic behaviour, these medical devices are now obtained using 3D printing technologies. Once obtained, the 3D-printed PLA devices undergo different sterilisation procedures, which are essential to prevent infections. This work was an in-depth study of the physicochemical changes caused by novel and conventional sterilisation techniques on 3D-printed PLA and their impact on the biological response in terms of toxicity. The 3D-printed PLA physicochemical (XPS, FTIR, DSC, XRD) and mechanical properties as well as the hydrophilic degree were evaluated after sterilisation using saturated steam (SS), low temperature steam with formaldehyde (LTSF), gamma irradiation (GR), hydrogen peroxide gas plasma (HPGP) and CO_2 under critical conditions (SCCO). The biological response was tested in vitro (fibroblasts NCTC-929) and in vivo (embryos and larvae wild-type zebrafish *Danio rerio*). The results indicated that after GR sterilisation, PLA preserved the O:C ratio and the semi-crystalline structure. Significant changes in the polymer surface were found after HPGP, LTSF and SS sterilisations, with a decrease in the O:C ratio. Moreover, the FTIR, DSC and XRD analysis revealed PLA crystallisation after SS sterilisation, with a 52.9% increase in the crystallinity index. This structural change was also reflected in the mechanical properties and wettability. An increase in crystallinity was also observed after SCCO and LTSF sterilisations, although to a lesser extent. Despite these changes, the biological evaluation revealed that none of the techniques were shown to promote the release of toxic compounds or PLA modifications with toxicity effects. GR sterilisation was concluded as the least reactive technique with good perspectives in the biological response, not only at the level of toxicity but at all levels, since the 3D-printed PLA remained almost unaltered.

Keywords: polylactic acid (PLA); 3D printing; medical devices; sterilisation; physicochemical changes; biological response; zebrafish model

1. Introduction

Polylactic acid (PLA) is an aliphatic polyester produced as a racemic mixture of D and L lactide from nontoxic renewable sources, such as corn and sugarcane, with valuable properties for the biomedical field [1,2]. This polymer stands out for its behaviour in contact with biological media, as it gradually degrades into innocuous lactic acid or carbon dioxide and water and is metabolised intracellularly or excreted in urine and breath over time [1,3]. In addition to this immunologically inert response, PLA does not produce toxic or carcinogenic effects in local tissues, it is completely reabsorbed, and its production is relatively cost-efficient as compared to other traditional biodegradable polymers [4]. Given its biocompatible, biodegradable and bioabsorbable properties, PLA has become one of the most commonly used polymers in clinics with numerous applications including medical implants, porous scaffolds, sutures, cell carriers, drug delivery systems and a myriad of other fabrications [1,2,5].

Recently, with the emergence in the biomedical field of fused deposition modelling (FDM), one of the most common 3D printing techniques, interest in PLA has risen dramatically because of its favourable thermoplastic properties [6]. It can be heated to its melting point, cooled and reheated again without significant degradation. Together with FDM technology, this enables the rapid manufacture of customised structures and the fabrication of platforms for an extensive variety of applications both in research and in surgical practices, such as patient-specific implants, surgical guides (cranial and maxillofacial surgery) and surgical tools [5–8].

Once produced, the next critical step in the manufacturing process of these 3D-printed PLA-based medical devices is sterilisation, which is essential to prevent possible complications such as infections or rejections once in contact with the human body [9]. In this regard, despite the fact that PLA has been the focus of multiple pre-clinical and clinical trials, to what extent the different sterilisation techniques accurately affect its properties, and therefore its clinical functionality, continues to be the subject of debate, particularly now with the development of 3D printing [6]. It is clearly stated in the literature that conventional sterilisation techniques can cause physicochemical modifications in polymer-based medical devices that could limit their use in clinical applications [10]. Saturated steam is mostly not recommended for thermolabile and hydrolytically sensitive polymers—such as PLA—because of the high-temperature water vapour, which affects the structure of the sample prints. According to several authors [11,12], gamma radiation can induce chain scission or crosslinking reactions in polymers. Sterilisation by gas plasma with hydrogen peroxide is the recommended method for some authors, although it has an inferior penetrating capability compared to other methods [13]. On the other hand, low temperature steam formaldehyde sterilisation is widely used in European countries for the sterilisation of thermolabile medical equipment, but the residual levels of formaldehyde stipulated in the regulations must always be respected as it is known to be toxic and carcinogenic [14]. The same occurs with residues from ethylene oxide sterilisation as they may cause toxicity and induce a chemical reaction with the polymer matrix [10,11,13]. For these reasons, this methodology has been progressively prohibited by several hospitals in the EU and the USA [15]. The need to find new and effective sterilisation alternatives has brought the supercritical carbon dioxide methodology to the fore. This has emerged as a green and sustainable technology that requires moderate temperatures, avoiding physical and chemical damage in thermolabile and hydrolytically sensitive materials such as PLA and its derivatives [6,16]. It has already been proven to completely inactivate a wide variety of organisms [17] and even pores [18], however it is still a developing technology not yet optimised for every material, as is the case with PLA.

There is, therefore, a need to further study the extent to which sterilisation techniques affect PLA-based materials, preventing significant changes in their physicochemical, mechanical and biocompatibility properties that might give rise to adverse responses in the body or compromise bodily functions [12]. It must also be taken into account that the choice of sterilisation technique will be highly conditioned by the ease and cost-effectiveness of their implementation in the production routine. In fact, given the aforementioned evidence, it is expected that printed PLA, as a thermolabile and hydrolytically sensitive polymer, will be affected by conventional sterilization techniques such as saturated steam, which could in turn have effects on its biological performance.

Thus, the purpose of this paper was to study in detail the effect of the main sterilisation processes mentioned, both the current and emerging ones, on 3D-printed PLA samples in terms of their physicochemical (XPS, FTIR, DSC, XRD) and mechanical properties as well as their influence on the PLA wettability degree. Moreover, the biological response was also evaluated to determine whether the potential changes brought about by the different techniques significantly affect PLA behaviour in terms of toxicity when tested in vitro and in vivo. For the latter, the zebrafish model (*Danio rerio*) was used due to its advantages, as compared to conventional animal models, due to its close homology with the human genome including its immunogenic responses.

2. Materials and Methods

2.1. PLA Precursor Material

A 1.75 mm polylactic acid (PLA) filament 3D850, designed from the biodegradable resin formulation Ingeo™ PLA 3D850 by NatureWorks, was purchased from Smart Materials 3D, Jaén, Spain. In comparison to standard PLA, this improved biodegradable filament presents an elevated rate of crystallisation and very low thermal contraction, which makes printing more rapid and accurate without the risk of sample deformation. The main properties are summarised in Table 1.

Table 1. Technical data of PLA 3D850 filament taken from the Smart Materials 3D website: www.smartmaterials3d.com/pla-3d850 (2 August 2022).

Polylactic Acid 3D850	Value
Material density	1.24 g/cm^3
Tensile yield strength	65.5 MPa
Flexural strength	126 MPa
Flexural modulus	4357 MPa
Heat distortion temperature	144 °C
Extrusion temperature	190–230 °C

2.2. Obtaining 3D-Printed PLA

Two sets of PLA samples were first designed using the SolidWorks 2016 software (Dassault Systemes SolidWorksCorp., Waltham, MA, USA), with one of discs measuring 1 mm in height and 5 mm in diameter, and the second consisting of well-shaped samples replicating the wells of 6-well microplates, measuring 34 mm in height and 16 mm in diameter. See Figure 1 for images from the software showing the two corresponding sets of samples. The digital data were then saved as STL files to subsequently generate corresponding sets of G-code for 3D printing using the Simplify3D software (Simplify3DSoftware, Cincinnati, OH, USA).

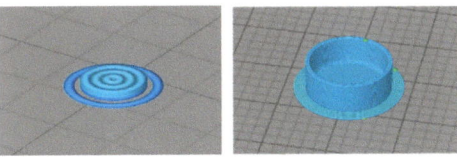

Figure 1. Software images showing the design of the PLA discs and the well-shaped samples.

A dual extruder 3D printer (BCN3D+, 3D RepRapBCN, Barcelona, Spain) based on fused filament deposition modelling (FDM) technology was used to print both sets, and the printing temperature was maintained at between 190–230 °C. A concentric infill pattern was chosen in order to create a uniform structure and avoid gaps. The main 3D printing parameters are summarised in Table 2.

Table 2. Main 3D printer parameters used.

Dual Extruder BCN3D+ Printer	Parameter Value
Nozzle	0.4 mm
Nozzle temperature	190–230 °C
Bed temperature	45 °C
Infill density	100%
Infill pattern	Concentric
Speed	60 mm/s
Layer height	0.2 mm

2.3. Sterilisation Techniques

PLA samples printed as discs were subjected to five different sterilisation treatments. Untreated 3D-printed material (PLA_C) was also considered and analysed as control samples. The specific sterilisation treatments evaluated are described below:

- Saturated steam (SS), carried out in an autoclave (Selecta Presoclave II 75, Cham, Switzerland) at Grupo de Novos Materiais (Vigo, Spain) operating at 121 °C and 2 bar for 20 min. The samples were sterilised in a surgical paper package.
- Low temperature steam with 2% formaldehyde (LTSF), carried out in a Matachana steriliser (Matachana 130 LF, Matachana Group, Barcelona, Spain) at Povisa Hospital (Vigo, Spain). This steriliser complied with EN 14180:2014 and used a mixture of steam and 2% formaldehyde in thermodynamic equilibrium. Sterilisation was performed at 78 °C, with a standard duration of 153 min at full load and the samples were sterilised in a surgical paper package.
- Gamma irradiation (GR), performed by Aragogamma S.L. (Barcelona, Spain) using a ^{60}Co source irradiator at a dose level of between 25 and 35 kGy at room temperature, as per ISO 13485:2018.
- Hydrogen peroxide gas plasma (HPGP), performed in a Sterrad NX steriliser (Advanced Sterilization Products, Irvine, CA, USA) at Hospital Universitario Lucus Augusti (Lugo, Spain). This was based on 59% aqueous hydrogen peroxide, which was concentrated to about 95% through removal of water from the peroxide solution before its evaporation and transfer to the chamber. PLA discs were conditioned in Tyvek packages compatible with the sterilisation process. The temperature during the sterilisation cycle was kept at between 45 °C and 55 °C in a standard cycle.
- CO_2 under critical conditions (SCCO) was carried out at the Biomaterials and Biomedical Technology lab (CBQF, Porto, Portugal). PLA discs were packed in sealed permeable plastic cartridges and placed in the reactor, to which hydrogen peroxide was added as an additive (300 ppm) to make sterilisation more effective. The optimised operating parameters for sterilisation were 40 °C temperature and 240 bar pressure with constant agitation of 600 rpm. Once pressurisation occurred, CO_2 under critical conditions acted for 4 h.

The PLA discs/well-shaped samples subjected to the five sterilisation processes were named as follows: PLA_{SS}, PLA_{LTSF}, PLA_{GR}, PLA_{HPGP} and PLA_{SCCO}.

2.4. Physicochemical Characterisation

The elemental compositional analysis of PLA_C, PLA_{SS}, PLA_{LTSF}, PLA_{GR}, PLA_{HPGP} and PLA_{SCCO} discs was determined in detail by X-ray photoelectron spectroscopy (XPS) using a Thermo Scientific K-Alpha ESCA instrument (Waltham, MA, USA) equipped with an Al Kα monochromatised X-ray source radiation at 1486.6 eV (CACTI, UVigo, Vigo, Spain). Photoelectrons were collected from a take-off angle of 90° relative to the sample surface and the measurement was taken in a Constant Analyser Energy mode (CAE) with a 100 eV pass energy for survey spectra and 20 eV pass energy for high resolution spectra. Charge referencing was performed by setting the lower binding energy C1s photo peak at the 285.0 eV C1s hydrocarbon peak. The surface elemental composition was determined using the standard Scofield photoemission cross sections.

Chemical modifications in the polymeric chains of the PLA discs induced by the different sterilisation treatments were identified by Fourier Transform Infrared Spectroscopy (FTIR) using a Thermo Scientific Nicolet 6700 spectrometer (Waltham, MA, USA) with a DTGS KBr detector (CACTI, UVigo). Spectra were collected in the 400 to 4000 cm^{-1} wavelength range by averaging 34 scans and with a resolution of 4 cm^{-1}.

To complement this information regarding the PLA chemical modifications, a thermal characterisation was also performed using a simultaneous TGA-DSC Setsys Evolution 1750 thermal analyser (Setaram, NJ, USA) in the CACTI, UVigo. The PLA discs were first subjected to a heating–cooling ramp (from 20 to 250 °C/min at 5 °C/min) with a constant supply of nitrogen. Secondly, a third ramp from 20 to 900 °C with a heating rate of 10 °C/min was applied with a constant supply of air. The information from DSC curves was used to determine the glass transition temperature (Tg) and melting point temperature (Tm), as well as the exothermal response relating to cold crystallisation (Tcc), which was obtained from the first heating cycle. The crystallinity index (Xc) was calculated according to Savaris et al. [19] based on the cold crystallisation enthalpy (ΔHcc), melting enthalpy (ΔHm) of the first heating calculated by peak fitting algorithms of the DSC curves and melting enthalpy of theoretically 100% crystalline PLA (ΔHm° = 93.7 J/g), in accordance with Equation (1):

$$Xc\ (\%) = \frac{\Delta Hm - \Delta Hcc}{\Delta Hm°} \times 100 \quad (1)$$

The PLA crystalline structure was evaluated by X-ray diffraction (XRD) in an X'Pert Pro Panalytical diffractometer (Malvern Panalytical, Malvern, UK) with monochromatic Cu-Kα radiation (λ = 1.5406 Å) and with a 2θ range of 4–100° (40 kV, 30 mA, 0.013° step size) (CACTI, UVigo).

Contact angle measurements were performed in a Pocket goniometer (Fibro System AB, Stockholm, Sweden) at the Grupo Novos Materiais (UVigo) to evaluate the hydrophilic degree of the 3D-printed PLA discs. A sessile drop of ultrapure water was dispensed on each sample at room temperature and analysed using the linked software. The reported values corresponded to the average of ten measurements of each of the five replicates per treatment ± standard deviation. Images were taken immediately after the drop was deposited on the surface of the sample.

Finally, the mechanical properties were analysed using a nanoindenter XP (MTS Inc., Huntsville, AL, USA), in CACTI, UVigo. Hardness and Young's Modulus values were measured using a 100 nm radius triangular pyramid indenter tip (Berkovich-type indenter) with the CSM (Continuous Stiffness Measurement) mode to perform dynamic measurements as a function of depth and XP head. A large number of indentations (30) were programmed, and the average of the valid results was calculated ± standard deviation.

Untreated 3D-printed PLA discs (PLA$_C$) were also subjected to the above-mentioned physicochemical characterisation as control to be able to identify the changes caused by the various sterilisation techniques.

2.5. Biological Response In Vitro: Cytotoxicity Assay

To evaluate the cytotoxicity of the potential release of (1) small particles from the 3D-printed PLA after the different sterilisations (as a thermosensitive polymer) and of (2) potential traces of the toxic additives required in some of the sterilisation methods, a solvent extraction test was performed. It was carried out following the indications of UNE-EN-ISO 10993-5:2009 with the cell line NCTC clone 929 (ECACC 88102702) from mouse fibroblasts. The cells were incubated throughout the experiment at 37 °C in a humidified atmosphere with 5% of CO_2 and with the cell growth medium DMEM (Lonza, Basilea, Switzerland), supplemented with 10% of foetal bovine serum (HyClone Laboratories LLC, Logan, UT,

USA) and 1% of a combination of penicillin, streptomycin and amphotericin B (Lonza, Basilea, Switzerland).

To prepare the extracts, PLA_{SS}, PLA_{LTSF}, PLA_{GR}, PLA_{HPGP} and PLA_{SCCO} discs were first placed in individual falcon tubes with the culture medium DMEM and kept at 37 °C for 24 h with 60 rpm agitation together with the controls. Then, different concentrations (100%, 50%, 30%, 10% and 0%) were prepared by diluting the initial extracts with fresh culture medium. A 6.4 g/L phenol solution was used as a positive control, while the negative control was the culture medium itself. The ratio between the material (PLA) and the volume of growth medium was 3 cm^2 of material per ml of medium (ISO 10993-12).

A suspension of 1×10^5 cells/m, in the same growth medium described above, were seeded in a 96-well microplate at a volume of 100 μL per well. After 72 h of incubation, a sub-confluent layer was formed, and the cell medium was replaced by the previously prepared extracts from 24 h before. Four replicates per concentration were incubated with the cells for 24 h. After that time, the cellular viability was quantified using the MTS Cell Proliferation Assay Kit (Abcam, Cambridge, UK). This colorimetric assay is based on the MTS tetrazolium compound, exclusively reduced by viable cells to generate a coloured formazan dye that is soluble in the culture medium. A volume of 10 μL of MTS reactive was added to each well. After 45 min of incubation, the reagent was renewed by fresh medium and the absorbance of the resulting solutions was read at a wavelength of 490 nm in a microplate spectrophotometer (Bio-Rad, Hercules, CA, USA). This test was repeated 3 times and the results were expressed as the percentage of viability compared to the negative control ± standard error.

2.6. Biological Response In Vivo: Acute Toxicity Test in Zebrafish

To evaluate in direct contact the in vivo toxicity caused by the physicochemical changes and/or toxic residues detected on the 3D-printed PLA after the sterilisation methods, a zebrafish model in direct contact was carried out and the corresponding 3D-printed well-shaped PLA samples were used. Embryos and larvae of wild-type zebrafish (*Danio rerio*, AB strain) were obtained from the experimental facilities at the Institute of Marine Research (IIM-CSIC, Vigo, Spain), where the animals were cultured and maintained following established protocols [20,21]. Adult zebrafish were maintained in a recirculating water system on a 12:12 h light–dark cycle and the water composition was maintained at pH 7.0 and 28–29 °C. The zebrafish were fed twice a day with commercial food (Nutrafin Max Tropical Fish Flakes) and once a day with live Artemia. Breeding crosses were conducted with a female/male ratio of 3:2. After the adult fish spawned, the embryos were obtained according to the protocols outlined in The Zebrafish Book [21] and maintained at 28 °C in an incubator (INE-500; Memmert, Schwabach, Germany) in zebrafish water during the whole experiment. All of the procedures in the experiment were reviewed and approved by the CSIC National Committee of Bioethics under approval number ES360570202001/17/FUN.01/INM06/BNG.

To evaluate the developmental toxicity of the zebrafish embryos, well-shaped PLA samples were placed on 6-well plates to prevent water loss. The wild type zebrafish fertilised embryos were collected and deposited in direct contact within these PLA samples (18 embryos per PLA well in triplicate for each condition) until the end of the experiment. First, the percentage of viable embryos was calculated at 24 hpf (hours post-fertilisation). Viable embryos were those that have live larvae inside and dead embryos became opaque, and this was also confirmed by the absence of a heartbeat under a Nikon SMZ800 microscope (Nikon, Tokyo, Japan). Once the non-viable embryos were removed from the well-shaped samples, the experiment continued to calculate the hatching rate at 24, 48 and 72 exposure hours by visual inspection and the survival larvae determined after 178 h (7 days). Mortality was assessed every 24 h by visual inspection discarding dead larvae, those that did not swim and those that did not have a heartbeat when viewed under a Nikon SMZ800 microscope (Nikon, Japan). Ten zebrafish larvae were randomly selected for each treatment at 3 dpf (days post-fertilisation), properly rinsed with zebrafish water and anesthetised with tricaine (0.05%) (Sigma-Aldrich, Merck Group, Darmstadt, Germany) to detect the main deformations under a Nikon AZ100 microscope (Nikon, Japan). Different measures, such as body length and yolk sac diameter, were also calculated with the aid of the ImageJ programme. In addition, the heart rate of embryos was counted at 3 days. Their heartbeat rates were measured from video recordings for 10 s at room temperature in set conditions. The data were analysed using a design macro in the Image J programme, which provided beats per minute. This allowed for the analysis of the heart rate of 3-dpf larvae. For all experiments, a blank control treatment without PLA was added in addition to the untreated PLA control material to validate the process. Each treatment was tested in triplicate and repeated twice, and all the results compared the mean values of the measurements after the treatments with respect to the control ± standard error.

2.7. Statistical Analysis

All biological data were analysed using GraphPad Prism 8 (GraphPad Software Inc., San Diego, CA, USA) and the results were represented graphically as the mean ± standard error of means (mean ± SEM). The nonparametric Mann–Whitney U test was used to determine the statistical differences between the control and the different sterilisation treatments. Statistical significance was determined to be * ($p \leq 0.05$) at the 95% confidence level. In the zebrafish assay significant differences were obtained using the Mann–Whitney U test and displayed as *** ($0.0001 < p < 0.001$), ** ($0.001 < p < 0.01$) or * ($0.01 < p < 0.05$). Kaplan–Meier survival curves were analysed with a log–rank (Mantel–Cox) test.

3. Results and Discussion

3.1. Physicochemical Characterisation

The elemental composition of the surface of the PLA printed discs after sterilisation by the different techniques, PLA_{SS}, PLA_{LTSF}, PLA_{GR}, PLA_{HPGP} and PLA_{SCCO}, was first evaluated by XPS and presented as an atomic percentage (Table 3). The surface layer of the PLA printed disc without any sterilisation treatment (PLA_C) was also subjected to the analysis and, as can be seen in Table 3, it was composed of a major contribution in C followed by O, in a ratio O:C of 0.50, with a minor contribution of Na. The PLA-tested discs where the elemental composition was more similar to the control were the PLA_{SCCO} and PLA_{GR}, with an O:C ratio of 0.48 and 0.46, respectively. Conversely, the discs of PLA_{LTSF}, PLA_{HPGP} and PLA_{SS} presented a relevant decrease in the O:C ratio to values in the range 0.35–0.27, which meant a lower oxygen content. Moreover, with these latter treatments, PLA_{LTSF}, PLA_{HPGP} and PLA_{SS}, a higher contribution of minor elements such as Si, Cl, S, I, Na and N was found compared to the two previous techniques. In relation to this, it is well known that in autoclave sterilisation, due to the high humidity conditions during the process, the transference of elements from the chamber to the sample surface is common. With respect to the small amounts of unexpected components, detected in the rest of the techniques, they can also be due to contaminations produced in the manufacture of the material and its subsequent handling (S, Cl, Na, N, C and O) or even the use of plastic gloves (Si, Na, S and Cl), with Si being the second most common contaminant in this technique (https://xpslibrary.com/contamination-2/) (25 July 2022). Finally, it is important to note that the differences found in the O:C ratios can be used as a guide to determine the degree of functionalisation of the polymer surface. Thus, the most significant change was detected in the autoclaved PLA discs (PLA_{SS}) with a sudden increase in C at the expense of a decrease in the percentage of O, compared to the control (PLA_C). On the other hand, the HPGP treatment led to a greater "oxidation" of the surface than the formaldehyde treatment (PLA_{LTSF}). However, sterilisation by supercritical CO_2 (PLA_{SCCO}) and gamma radiation (PLA_{GR}) caused only very slight changes at this level.

Table 3. Elemental composition of untreated PLA and after sterilisation treatments obtained from XPS analysis (at. %).

Samples	C	O	Si	S	N	Na	I	Cl	O/C
PLA$_C$	66.24	33.42	-	-	-	0.34	-	-	0.50
PLA$_{GR}$	68.65	31.35	-	-	-	-	-	-	0.46
PLA$_{SCCO}$	66.52	31.82	-	-	1.24	0.42	-	-	0.48
PLA$_{LTSF}$	73.78	21.20	3.34	-	1.07	0.61	-	-	0.29
PLA$_{HPGP}$	70.92	24.73	1.85	-	1.11	0.93	-	0.47	0.35
PLA$_{SS}$	74.94	20.00	1.23	0.18	2.53	0.92	0.19	-	0.27

XPS high resolution spectra for C1s were taken to make an in-depth evaluation of the changes to the polymer by assigning energy binding peaks. Three binding energy assignments were obtained for all the tested PLA discs: C-H, C-C bonds at 285 eV, C-OH, C-O-C bonds around 287 eV and COOH, O=C-O bonds around 289 eV. However, differences were clearly detected. By way of example, Figure 2 shows the high-resolution spectra of the printed PLA control (PLA$_C$) and PLA sterilised with saturated steam (PLA$_{SS}$) where a significant change to the polymer surface after this latter treatment was clear. Thus, the C1s binding assignments obtained for PLA$_C$ showed the three well-defined peaks with the assignments of C-H, C-C bonds being detected at higher intensity (285 eV) and the other two (at 287 and 289 eV) in slightly lower intensities attributed to C-OH, C-O-C and COOH, O=C-O, respectively. When the assignments obtained for the PLA$_{SS}$ disc were observed, differences were clear with far less intense peaks at the binding energies at 287 eV and 289 eV and increased intensity (counts/s) for the assignment C-H, C-C at the binding energy 285 eV.

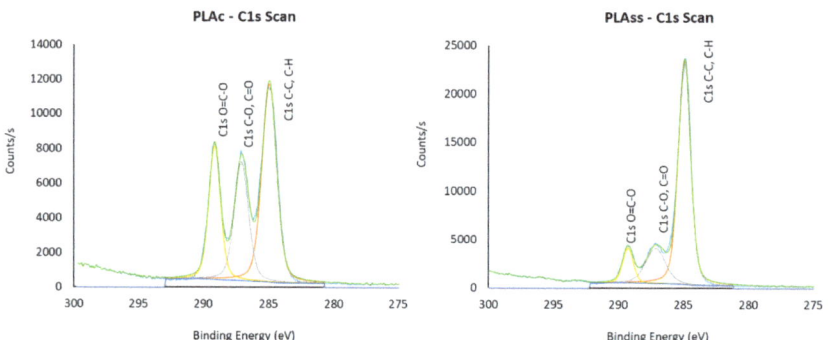

Figure 2. XPS high resolution spectra for C1s of printed PLA control (PLA$_C$) and PLA sterilised with saturated steam (PLA$_{SS}$).

The C1s binding energy assignments with relative percentages for all the printed discs (shown in Table 4) indicated variations in the binding energy ranging from 287.10 eV for PLA$_C$ to 287.20 eV for PLA$_{HPGP}$ and PLA$_{SS}$ and from 289.18 eV (PLA$_C$) to 289.41 eV for PLA$_{LTSF}$. The two sterilisation treatments that delivered more similar C1s binding assignments with respect to the control PLA$_C$, which presented C-H, C-C in 47.12 rel.%, C-OH, C-O-C in 27.35 rel.% and COOH, O=C-O in 25.54 rel.%, were PLA$_{SCCO}$, firstly, followed by PLA$_{GR}$. Meanwhile, PLA$_{SS}$ and PLA$_{LTSF}$ presented C-H, C-C in around 72 rel.%, C-OH, C-O-C in 17–18 rel.% and COOH, O=C-O in 9–10 rel.%. Both techniques, together with HPGP, caused a marked decrease in O and related bonds on the surface of the PLA.

Table 4. Carbon binding energy (BE) assignments with relative percentage (rel.%) from high resolution XPS spectra of untreated PLA and PLA after sterilisation treatments.

	C-H, C-C		C-OH, C-O-C		COOH, O=C-O	
	BE	Rel.%	BE	Rel.%	BE	Rel.%
PLA_C	285	47.12	287.10	27.35	289.18	25.54
PLA_{GR}	285	54.45	287.12	23.95	289.20	21.61
PLA_{SCCO}	285	49.62	287.13	27.55	289.20	22.83
PLA_{LTSF}	285	72.64	287.18	17.32	289.41	10.04
PLA_{HPGP}	285	67.36	287.20	18.75	289.35	13.89
PLA_{SS}	285	72.06	287.20	18.29	289.36	9.66

The chemical changes in terms of vibrational modes were evaluated by FTIR spectroscopy to identify potential modifications in the characteristic absorption bands of PLA when subjected to the different sterilisation treatments. Figure 3 presents the FTIR spectra in the 400–2000 cm^{-1} wavenumber range, and the main absorption bands are identified. First, the absorption bands detected for the 3D-printed untreated disc, PLA_C, were evaluated and a correspondence with those of neat PLA fibres [22] was observed. This included a strong peak at 1747 cm^{-1} attributed to the stretching mode of carbonyl group (υ C=O) and a weaker one at 1266 cm^{-1} corresponding to the bending vibration of the same group (δ C=O). Second, major bands were recorded in the 1050–1200 cm^{-1} range, with three peaks at 1180, 1127 and 1080 cm^{-1} assigned to υ C-O vibrations. Bands in the 1300–1500 cm^{-1} range were also detected, and this was attributed to the deformational vibrations of C-H in the CH_3 groups, with the asymmetric bending mode at 1452 cm^{-1} and the symmetric deformation at the doublet 1381–1360 cm^{-1} [23]. Other bands at 1043 and 868 cm^{-1}, respectively, corresponding to δ O-H and υ C-C vibrations, were also identified.

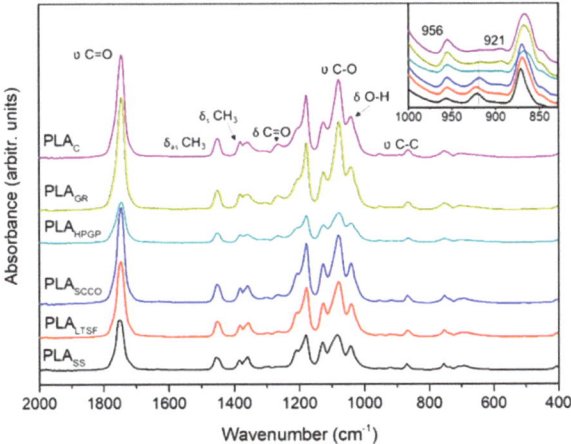

Figure 3. FTIR spectra of the PLA discs subjected to the different sterilisation treatments with absorption bands identified: PLA_{GR}, PLA_{SCCO}, PLA_{LTSF}, PLA_{SS} and PLA_{HPGP} together with the untreated PLA disc as control, PLA_C. Amplification of the 400–2000 cm^{-1} wavenumber range in the inset with the two main peaks identified.

The comparison between the FTIR spectra before (PLA_C) and after the different sterilisation processes did not exhibit major differences, coinciding with the control PLA of 90–100% in the total wavenumber, 400–4000 cm^{-1}. Indeed, the PLA_{GR} sample was the most similar to the control with a 99.7% fit, followed by PLA_{SCCO} at 97.3%, with the least similar to the control being the one sterilised by steam heat, PLA_{SS}, 89.6%. Minor changes related to small chemical alterations were observed as slight shifts in some characteristic peaks. On the general PLA$_C$ spectrum, the characteristic peaks of the υ C=O and δas CH$_3$ bands were observed at 1747 and 1452 cm^{-1}. Thus, for the PLA$_{SCCO}$, PLA$_{LTSF}$ and PLA$_{SS}$ samples, an inversion of the intensities of the components of doublet 1380 and 1365 cm^{-1} and an increase in the signal at 1210 cm^{-1} were also observed and already reported [23]. These minor chemical alterations were clearly confirmed when the 960 to 830 cm^{-1} range was amplified and evaluated in depth (inset at Figure 3). According to the literature, this band is sensitive to the degree of crystallinity [24], where the crystalline α-phase corresponds to bands at 921 and 872 cm^{-1} and the amorphous phase with a band at 956 cm^{-1} [11,24]. The PLA$_{GR}$ and PLA$_{HPGP}$ spectra at the inset clearly presented only two peaks: at 872 cm^{-1} (crystalline α-phase) and 956 cm^{-1} (amorphous phase), exactly the same as with PLA$_C$. In addition to these two peaks, in PLA$_{SCCO}$, PLA$_{LTSF}$ and PLA$_{SS}$, the crystalline α-phase peak at 921 cm^{-1} also appeared with higher intensity at the saturated steam samples, with PLA$_{SS}$ overcoming its amorphous band at 956 cm^{-1}. The absence of the crystallisation peak at 921 cm^{-1} for the untreated PLA was in accordance with other authors [11,25]. Its presence for PLA$_{SCCO}$, PLA$_{LTSF}$ and PLA$_{SS}$ suggests the crystallisation of the polymer to a greater or lesser extent for these latter three methods, as during crystallisation the 956 cm^{-1} band area decreased in a synchronised way with the appearance of the new band at 921 cm^{-1}, characteristic of R crystals [26]. This tendency was confirmed by measuring the areas of both bands (956 cm^{-1}/921 cm^{-1}) with Magicplot software and a ratio of 921:956 was obtained for each PLA spectrum, with the obtained values for PLA$_C$, PLA$_{GR}$ and PLA$_{HPGP}$ being 0.19, 0.16 and 0.21, respectively (no contribution of the crystalline peak at 921 cm^{-1}). These values were slightly above 1 for PLA$_{LTSF}$ and PLA$_{SCCO}$ (1.13 and 1.14 respectively), which means an increased contribution of the crystalline peak to reach the same level as the amorphous one, and the highest value was for PLA$_{SS}$ at 3.13 (clear higher contribution of the crystalline band in relation to the amorphous one). When revising the literature, Zhao and colleagues also describe the appearance of the band at 921 cm^{-1}, characteristic of the crystalline α-phase at the same time as the decrease in intensity in the band at 956 cm^{-1} in PLA for disposable medical devices [10]. It is established that during sterilisation the presence of a high concentration of water, hydrogen peroxide or other small molecules can plasticise a thin layer on the surface of the polymer, facilitating crystallisation [25].

Next, the effect of the sterilisation on the thermal stability of the polymer was studied in detail using differential scanning calorimetry (DSC) in terms of how it influenced the degree of crystallinity. The first heat DSC thermograms for the different PLA samples are presented in Figure 4. The PLA$_C$ thermogram showed the three distinct transitions typical of semi-crystalline thermoplastics: (1) heat flux at the glass transition temperature (Tg), (2) an exotherm associated with cold crystallisation (Tcc) and (3) a melting endotherm (Tm). A small exothermic band was observed immediately before melting occurred, associated with a pre-melt recrystallisation. Despite finding small differences in the values of the peaks of the three transitions, the thermograms obtained for the PLA$_{GR}$, PLA$_{HPGP}$ and PLA$_{LTSF}$ samples presented the same semi-crystalline behaviour. This was not the case for the samples that were sterilised using autoclave and supercritical CO$_2$ (PLA$_{SCCO}$ and PLA$_{SS}$), which presented the greatest changes. For the PLA$_{SCCO}$ sample, the glass transition and cold crystallisation almost disappeared, which again suggests PLA crystallisation. In the same way, for PLA$_{SS}$, these peaks were no longer observed, with only the melting peak (Tm) appearing, which is characteristic of crystalline polymers. These results supported the findings obtained from the FTIR analysis.

To confirm this, the corresponding crystallinity indices were calculated and are presented in percentage form (X$_C$ %) in Table 5, together with peaks for key transitions. The first transition detected for PLA$_C$ was a change in the specific heat of the material at the glass transition region at 63.2 °C immediately followed by another transition at 100.7 °C, which corresponded to cold crystallisation. This process was associated with the self-nucleation of the crystalline phases, where molecular chains that were preciously locked into position in the amorphous regions now had enough molecular mobility to reorganise into a more ordered and lower energy state crystalline phase. The final transition observed was a large endothermic peak at 175.5 °C associated with the heat of fusion for melting [25]. The index of crystallinity obtained was 18.5%. The peaks obtained in the key transitions were in accordance with other authors [1], with low glass transition temperature (Tg) values in the 60–65 °C range and a melting temperature (Tm) of 173–178 °C for PLA. When this untreated PLA_C was compared to the other samples, PLA_{GR} and PLA_{HPGP} were both observed to present a lower glass transition (Tg) in comparison to the control, PLA_C. This could indicate some degradation of the amorphous regions of the polymer via hydrolysis or chain scission in these two types of sterilisation [27]. The degree of crystallinity obtained for these two samples was 15.1 and 17.4%, respectively, which was relatively close to the untreated PLA_C (18.5%). As expected after seeing the DSC thermograms in Figure 4, the PLA_{SS} sample only presented a melting endotherm transition at 176.8 °C and the highest degree of crystallinity with a value of 39.3%. During sterilisation with saturated steam, the crystallisation temperature of PLA was exceeded, crystallising the amorphous sample [10,19]. The same occurred for the PLA_{SCCO} and PLA_{LTSF} samples, although to a lesser extent than for autoclaving.

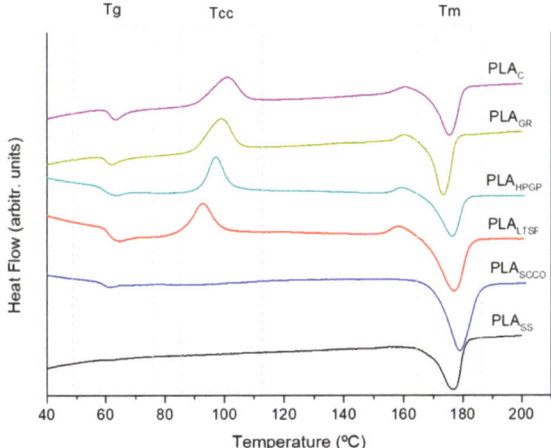

Figure 4. DSC thermograms for PLA discs subjected to the different sterilisation treatments: PLA_{GR}, PLA_{SCCO}, PLA_{LTSF}, PLA_{SS} and PLA_{HPGP} together with the untreated PLA disc as control, PLA_C.

Table 5. Glass transition temperatures (Tg), cold crystallisation (Tcc), crystalline melting (Tm) and crystallinity index of PLA subjected to the different sterilisation treatments: PLA_{GR}, PLA_{SCCO}, PLA_{LTSF}, PLA_{SS} and PLA_{HPGP} together with the untreated control, PLA_C.

	Tg (°C)	Tcc (°C)	Tm (°C)	Xc (%)
PLA_C	63.2	100.7	175.5	18.5
PLA_{GR}	61.9	98.6	173.3	15.1
PLA_{HPGP}	59.1	97.1	176.4	17.4
PLA_{LTSF}	66.6	92.9	176.9	29.0
PLA_{SCCO}	61.1	99.4	178.9	32.6
PLA_{SS}	-	-	176.8	39.3

In accordance with our results, other authors [19] stated that gamma radiation can promote only slight changes in crystallinity by inducing ionisation reactions in the polymeric chains. At the same time, and again in line with our work, hydrogen peroxide gas plasma was proven to not chemically or morphologically affect PLA microfibers in electrospun scaffolds [23]. Moreover, the disappearance of a measurable glass transition (Tg), a cold crystallisation (Tcc) temperature and an increase in crystallinity in the DSC thermograms after PLA autoclave sterilisation has also previously been published [19]. This behaviour for PLA_{SS} was also observed for the PLA_{SCCO} samples (Figure 4), however previous works have established that supercritical CO_2 sterilisation treatments did not induce changes in the calorimetric properties of macroscopic poly(L-lactic acid) porous scaffolds (on crystallinity and Tm) [18].

In order to supplement these results by identifying the PLA crystalline planes after the sterilisation treatments, the 3D-printed PLA samples were evaluated by XRD. Figure 5 shows the XRD diffraction patterns obtained for the most representative samples PLA_C, PLA_{GR} (both at the inset) and PLA_{SS}. The spectra obtained for PLA_C was of a semi-crystalline polymeric material. The same behaviour was observed after sterilisation of the sample by gamma radiation. However, the XRD profile of the saturated steam sample, PLA_{SS}, exhibited two sharp diffraction peaks, a very intense one at $2\theta = 16.6°$ and another at 19°, respectively attributed to (200)/(110) and (203) lattice planes. Weaker peaks were also detected at 12.4°, 14.7° and 22.3°, corresponding to the (004)/(103), (010) and (015) crystal planes, respectively. These peaks were very similar to those obtained by Zhao et al. [10] for PLA sterilised using saturated steam. In relation to the main crystalline form, it has been shown that PLA exhibits two different crystalline phases termed as α and α'. The latter is described as the disordered form of the stable α phase. According to Jalali and coworkers [28], for crystallisation processes at temperatures below 100 °C the main crystalline form present is the α' phase, whereas above 120 °C it is the α phase, and within the 100–120 °C range a mixture of both crystalline phases is present. The same authors associated an exothermic peak at the DSC prior to the single melting temperature with the $\alpha'-\alpha$ solid-state transition (100–120 °C), and a single melting of the α phase for samples crystallised above 120 °C. In our study, the crystallisation process referred to saturated steam sterilisation at 121 °C for 20 min and the DSC thermogram for PLA_{SS} (Figure 4) did not present any exothermic peak below the melting temperature. Taking all this into account, the α phase may have been the main crystalline form present.

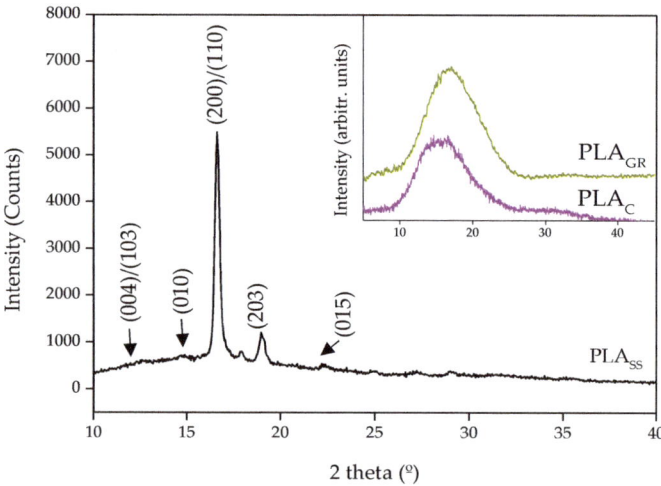

Figure 5. XRD diffraction pattern of untreated PLA_{SS} with the identification of the characteristic peaks and PLA_C, and PLA_{GR} diffraction pattern at the inset.

The structural changes to the 3D-printed PLA samples caused by certain sterilisation methods may affect the mechanical properties of the biomaterial. The hardness and Young Modulus values were measured in a nanoindenter and are both presented in Figure 6. The results obtained for hardness of the control sample PLA_C presented a value of 0.27 ± 0.01 GPa, which was in agreement with other authors [29] with 0.23 ± 0.03 GPa. The sterilisation of PLA by gamma radiation did not affect the hardness of the polymer, which remained practically constant, which also occurred for PLA_{SCCO}, PLA_{HPGP} and PLA_{LTSF}. An intense increment was clearly detected when PLA was sterilised by autoclaving, PLA_{SS} (0.34 ± 0.03 GPa). This increase in hardness was in concordance with the increment in the crystallinity degree of the PLA [30]. In relation to the Young Modulus, the result obtained was 4.7 ± 0.1 GPa for PLA_C, again in accordance with the literature [29] with 4.6 ± 0.4 GPa. A slight decrease was detected, and therefore in the PLA stiffness, after sterilisation by gamma radiation, supercritical CO_2 and hydrogen peroxide gas plasma, and there was a steeper decrease with formaldehyde (PLA_{LTSF}). Conversely, sterilisation by saturated steam promoted an increase in the value of the stiffness in PLA_{SS}. It was, therefore, in this sample (PLA_{SS}) where the physicochemical and structural changes seemed to be relevant enough to promote certain modifications at the mechanical behavioural level.

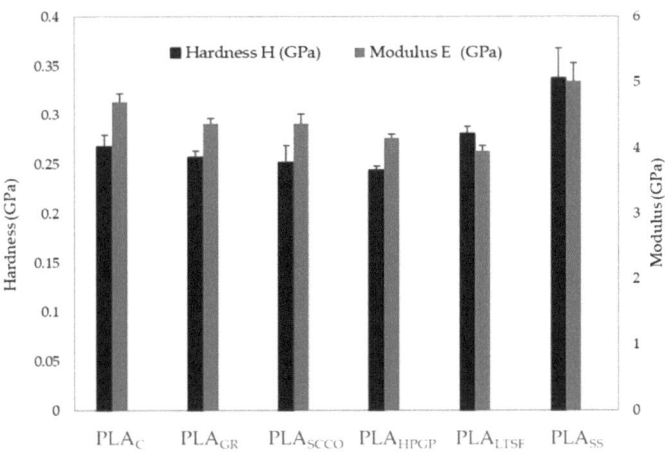

Figure 6. Modulus and hardness of PLA discs subjected to the different sterilisation treatments: PLA_{GR}, PLA_{SCCO}, PLA_{LTSF}, PLA_{SS} and PLA_{HPGP} together with the untreated PLA disc as control, PLA_C. Results are presented as mean ± standard deviation.

Finally, once the changes at the compositional, structural and mechanical level were evaluated, the wettability was also analysed, which is relevant to predicting how a biomaterial will behave within a biological environment. The contact angle measurements presented in Figure 7 confirmed the expected hydrophilic behaviour (<90°) of the untreated PLAC with a mean value of 72.1°. The PLAC contact angle obtained was in agreement with that measured by Savaris and colleagues (75.1°) [19]. When subjected to the sterilisation processes, slight modifications in the contact angle were evidenced, with the highest value being 79.3° for PLA_{SS}, which was therefore the least hydrophilic sample in terms of the mean value. This latter sterilisation caused a decrease in the amount of oxygen at the printed PLA surface (Table 3). Hydrophilicity is generally related to oxygen concentration and the presence of polar species such as carbonyl, carboxyl and hydroxyl groups [31]. A slight decrease in wettability was then expected for PLA_{SS} sterilisation. Significant variability was detected for the other samples, PLA_{SCCO}, PLA_{HPGP}, PLA_{GR} and PLA_{LTSF}, with the lowest angle measured being in the latter (61.9°). According to the amount of oxygen in the surface of the PLA_{LTSF}, the expected decrease in wettability did not occur, and the sterilisation even favoured it. More exhaustive studies on this specific technique could be performed, as the wettability is influenced by several factors, not only oxygen. When performing our experiment, the contact angle was measured at the smooth part of the

3D-printed PLA disc (the one in contact with the hot bed), as the side with the semi-circular pattern gave very unstable values.

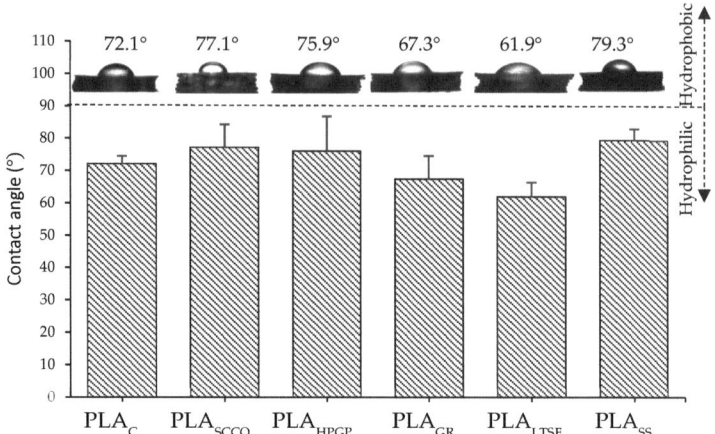

Figure 7. Water contact angle for PLA discs subjected to the different sterilisation treatments: PLA_{GR}, PLA_{SCCO}, PLA_{LTSF}, PLA_{SS} and PLA_{HPGP} together with the untreated PLA disc as control, PLA_C. Results are represented as mean ± standard deviation.

In summary, the analyses carried out showed that the physicochemical properties of PLA can be affected to a greater or lesser extent depending on the sterilisation technique used. Compositionally, gamma radiation did not contribute with external minor elements, providing a similar O:C ratio with respect to the control sample PLA_C. Conversely, autoclave sterilisation presented a relevant decrease in the O:C ratio and the highest contribution of minor elements (Si, S, N, Na and I). Moreover, due to the high temperatures and pressure in the PLA_{SS} process, some degree of crystallisation took place, demonstrated by the results of the analyses carried out using FTIR, XRD and DSC techniques, as indicated by other authors [13,32]. This change in crystallinity was also observed in the PLA_{SCCO} and PLA_{LTSF} samples, although to a lesser extent. Other techniques, such as gamma radiation, maintained the original semi-crystalline structure of the unsterilised material (PLA_C). In terms of hardness and Young Modulus, PLA_{GR}, PLA_{SCCO}, PLA_{HPGP} and PLA_{LTSF} presented minor changes when compared to the PLA_C. However, saturated steam sterilisation promoted an increment in the hardness properties of the PLA_{SS}, in concordance with the increment in its crystallinity degree. Changes to the wettability revealed the expected hydrophilic behaviour for all the tested samples with, again, the highest mean value of the contact angle being that for PLA_{SS}, which implies the lowest hydrophilic degree. Saturated steam sterilisation applied to 3D-printed PLA samples was then clearly proven to promote changes to the physicochemical properties of the polymer by increasing its crystallisation and, to a lesser extent, the super critical CO_2 technique, which is still experimental and surely needs the parameters to be optimised for these types of polymers. On the contrary, gamma radiation was clearly the least reactive with the biomaterial causing barely any alterations to it. To transfer these results into clinical practice, it is important to take into account that some methods are easier or more viable to implement than others. Furthermore, it is of great relevance to assess the extent to which the changes promoted in the biomaterial by the different sterilisation methods significantly affect its biological response. This aspect was tested first in vitro and later in vivo using the zebrafish model.

3.2. Biological Response: Cytotoxicity In Vitro and Acute Toxicity in Zebrafish Model

The in vitro evaluation of the potential cytotoxicity caused by the physicochemical changes detected in the 3D-printed PLA discs after the different sterilisation methods is presented in Figure 8. Given the absence of sterilisation in the control PLA, this sample could not be tested, and the sample extracts tested were from PLA_{GR}, PLA_{SCCO}, PLA_{LTSF}, PLA_{HPGP} and PLA_{SS} in a wide concentration percentage from 0 to 100%. All the experiments were compared with a negative control, which was the same culture medium without a sample extract, and a positive control, represented by phenol. A total of 100% of the extract (black) for all the PLA samples tested presented a lower mean value of cell viability than the negative control, however these differences were not found to be statistically significant for any of them, even for 100% of PLA_{SS}. The only PLA sample where the mean cell viability values for the different extracts were higher than the negative control were 50%, 30% and 10% of PLA_{GR} extract. However, when considering the standard deviation, the values were again not found to be statistically significant. Under all the test conditions, all values were above 80% of cell viability and therefore non-cytotoxic, being above the cytotoxicity limit of 70% established by the standard (UNE-EN-ISO 10993-5:2009). The cytotoxicity found at the 100%, 50% and 30% extracts of the positive control (phenol) validated the experiment. An absence of cytotoxicity in the PLA_{GR}, PLA_{SCCO}, PLA_{LTSF}, PLA_{HPGP} and PLA_{SS} extracts towards fibroblasts NCTC clone 929 was demonstrated.

Once the absence of cytotoxicity in vitro had been proven, the 3D-printed PLA well-shaped samples of the sterilisation processes that triggered the most significant physicochemical changes (PLA_{SS}) and the least significant (PLA_{GR}), together with untreated PLA (PLA_C), were evaluated in direct contact with the zebrafish model for the in vivo evaluation of potential acute toxicity. Figure 9 shows an optical image of the 3D-printed PLA well-shaped samples sterilised by the five techniques and the control, where a clear change in the colour of the PLA was observed. The material turned from transparent to white in the PLA_{SS} samples, and to a lesser extent with PLA_{SCCO}. According to the literature and in agreement with our results, Weir et al. [32] clearly stated that when PLA crystallises it becomes opaque. Moreover, a certain deformation was clearly observed for PLA_{SS}, with such a deformation after the autoclaving process in 3D-printed parts having previously been reported by several authors [13].

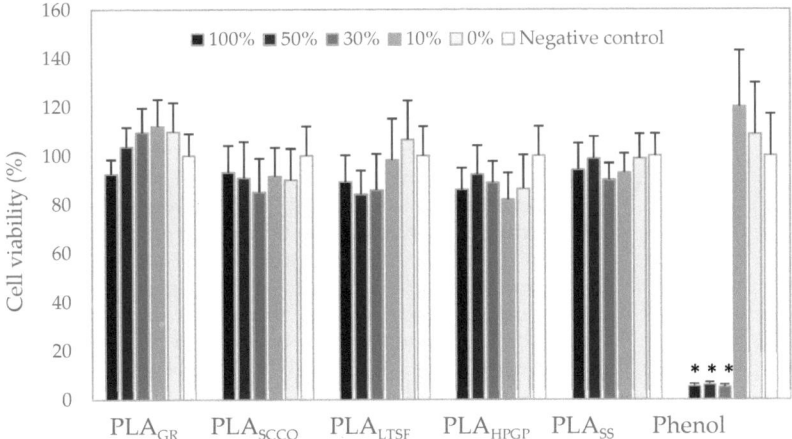

Figure 8. Cell viability detected in mouse fibroblast cells (NCTC clone 929) in the presence of different extracts of PLA_{GR}, PLA_{SCCO}, PLA_{LTSF}, PLA_{HPGP} and PLA_{SS} discs, compared to the positive control (phenol) and negative control (DMEM). Results are expressed as percentage compared to the negative control ± standard error. Statistical significance was determined to * ($p \leq 0.05$).

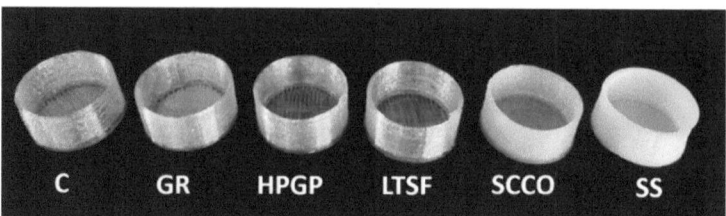

Figure 9. Image of 3D-printed PLA well samples before (untreated control C) and after five sterilisation processes: (GR) gamma radiation, (HPGP) hydrogen peroxide gas plasma, (LTSF) low temperature steam formaldehyde, (SCCO) supercritical CO_2 and (SS) saturated steam.

Figure 10 displays the development and growth of zebrafish embryos on the 3D-printed PLA samples PLA_{SS}, PLA_{GR} and PLA_C. The figures show viable embryos (%) after 24 h (A); hatching rate (%) at 24, 48 and 72 hpf (B); survival rate of larvae after 7 dpf (C) and larvae at 3 dpf on the PLA_C and PLA_{SS} well samples (D). As already discussed, for the acute toxicity experiments in the zebrafish model, a blank control treatment without PLA was added in addition to the untreated PLA control material to validate them. Given that similar results with non-significant differences were observed for the two controls, the results presented in Figure 10, and at the evaluation of the main morphological structures of 72 hpf larvae at Figure 11, are represented only against the untreated PLA_C. These similar results in both controls proved, in the first place, that PLA_C itself had no effect on the hatching or survival rates. Secondly, viable embryos were counted after 24 hpf (Figure 10A) and the embryos exposed to both PLA_{GR} and PLA_{SS} showed a percentage of viability similar to the control. The hatching rates of embryos exposed to the materials at 24, 48 and 72 h were also compared (Figure 10B). Following their normal development, the embryos began to hatch on the second day (48 h), with no significant differences between the two treatments compared to the control. At 72 h, all viable embryos had fully hatched. The rate was not delayed for the embryos exposed to PLA samples sterilised by gamma radiation or saturated steam. In the case of the larvae, the mortality rate was estimated in a continuing observation period every 24 h to acquire the survival percentage (Figure 10C). For the two sterilisation processes tested, the mortality rate did not present significant differences from the control.

An assessment of the toxic effects on embryonic developmental stages was also carried out, and the main morphological structures of the 72 hpf larvae zebrafish models were inspected under a microscope. Figure 11 represents the images of the normal body shape morphology of 72 hpf larvae after exposure to PLA_C (A), PLA_{GR} (B) and PLA_{SS} (C) together with body length (D), yolk sac diameter (E) and heart rate of 72 hpf larvae (F). Images B and C show that the body shape of the larvae at 3 dpf was completely normal after direct exposure to the PLA_{GR} and PLA_{SS} samples, respectively, and compared to the control group. None of the most common morphological defects, which would imply some degree of cytotoxicity, were observed, such as pericardial oedema, yolk sac oedema, ocular oedema, spine curvature, head malformation, shortened body stature, lowered yolk consumption and underdeveloped brain and eye [33]. To verify these visual inspection results, measurements of the body length (Figure 11 bottom left) of the larvae were taken, presenting a mean of 3.8 mm in the control sample PLA_C. This value was very close to that described by Nüsslein et al. of 3.5 mm at 3 dpf, validating our PLA_C control [20]. For the PLA_{GR} and PLA_{SS} samples, the values were very close to the control with no significant differences, indicating that there was no delay in larval growth. The same was repeated for the yolk sac diameter, with no significant differences compared to the control. The yolk sac was completely normal in all three conditions, since no oedema or increase in size was observed that would indicate a lowered yolk consumption. Finally, the heart rate (Figure 11 bottom right) can be used as an indicator of the state of cardiac function of the embryo, which is an important parameter for studying toxicity. Due to the transparency of the embryos, the heart development, heart rate and deformations can be seen at a simple resolution [33]. The heartbeat ratios of the embryos exposed to PLA sterilised by gamma radiation and autoclaving were counted and presented no significant differences between them or as compared to the control (130 beats/minute). Moreover, the results were in agreement with the values of other studies at 133.08 beats/minute [34]. These results revealed that direct exposure to the PLA sterilised using these techniques did not bring about any changes in heart rate such as brachycardia or tachycardia. There was no toxic influence on the zebrafish hearts.

In summary, the in vivo evaluation proved that direct contact exposure to both the 3D-printed PLA_{GR} and the PLA_{SS} well-shaped samples did not induce embryonic developmental abnormalities and was not toxic to developing zebrafish embryos or early larvae stages. Other authors have used the zebrafish model to study the toxicity of different printing methods on zebrafish embryos, but with other materials (ABS) [35] and only cleaning the parts according to the manufacturing specifications [36]. As indicated by Zhu and colleagues, a limited number of research works have been completed on the toxicity of 3D materials to cells, tissues and model organisms. For this reason, they evaluate the toxicity of different polymers used in the 3D printing processes, with rapid methods to test for toxicity, including zebrafish assays. Their results for PLA_C, sample extracts and direct contact of PLA with embryos were also non-toxic, however they did not indicate any sterilisation method [37,38].

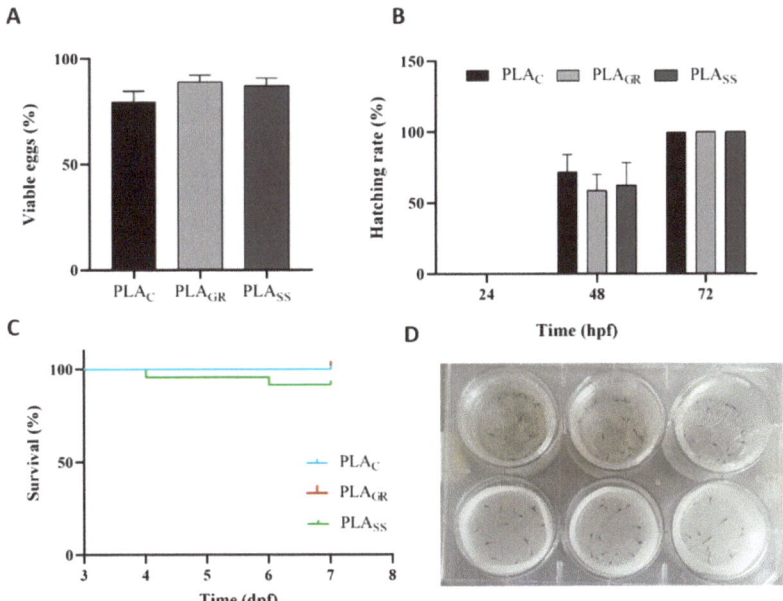

Figure 10. Viable embryos (%) after 24 h after exposure to PLA_{GR}, PLA_{SS} and PLA_C (**A**); hatching rate (%) at 24, 48 and 72 hpf (**B**); survival rate of larvae after 7 dpf (days post-fecundation) (**C**); larvae at 3 dpf on the PLA_C (above) and PLA_{SS} (below) well samples (**D**). Results are represented as the mean ± standard error.

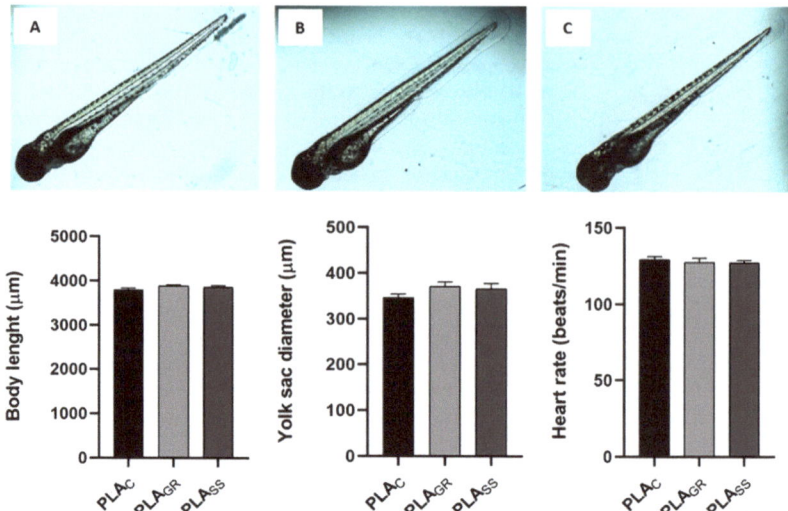

Figure 11. (**A–C**): Images of normal body shape morphology of 72 hpf larvae after exposure to PLA$_C$, PLA$_{GR}$ and PLA$_{SS}$. Body length, yolk sac diameter and heart rate of 72 hpf larvae after exposure to PLA$_C$, PLA$_{GR}$ and PLA$_{SS}$. Results are represented as the mean ± standard error.

4. Conclusions

The present work demonstrated that 3D-printed PLA remained practically unaltered when sterilised using gamma irradiation in terms of physicochemical and elemental composition, O:C ratio and distribution of binding assignments, maintaining also the original semi-crystalline structure of the unsterilised material. Conversely, saturated steam sterilisation was proven to generate significant physicochemical changes to the 3D-printed PLA, beginning with a 54% reduction in the O:C ratio caused by an intense decrease in O and related bonds. Moreover, an increase in the crystallinity degree of the 3D-printed PLA was demonstrated in the form of α-phase R crystals, confirmed by thermogravimetry, with an FTIR band ratio of 921:955 cm^{-1} of 3.13, which implied a clear contribution of the band sensitive to the degree of crystallinity (921 cm^{-1}), a 52.9% increase in the index of crystallinity and the diffraction pattern. These changes, promoted by saturated steam sterilisation, slightly affected the mechanical properties and hydrophilic degree of the 3D-printed PLA. The three remaining sterilisation techniques evaluated (hydrogen peroxide gas plasma, low temperature steam with formaldehyde and CO_2 under supercritical conditions) were situated at the intermediate point in terms of physicochemical changes. In fact, a decrease in the O:C ratio was also found after sterilisation using hydrogen peroxide gas plasma and low temperature steam together with a change in crystallinity after sterilisation using CO_2 under critical conditions and low temperature steam with formaldehyde. Finally, the present work also demonstrated that despite all these changes the biological response of the 3D-printed PLA was not affected in terms of toxicity for any of the tested sterilisations, either in vitro or in the zebrafish animal model. However, GR sterilisation was clearly concluded as the least reactive technique with good perspectives in terms of the biological response, not only at the level of toxicity but also in terms of biofunctionality.

Author Contributions: Conceptualization and methodology, S.P.-D., L.G.-R., R.L., P.G. and J.S.; Writing—Original Draft Preparation, S.P.-D. and M.L.-Á.; Writing—Review, S.P.-D., R.L., M.L.-Á., A.L.O., P.G. and J.S.; Editing, S.P.-D.; Supervision, J.S., B.N., A.F. and P.G. All authors have read and agreed to the published version of the manuscript.

Funding: This research was funded by the BLUEBIOLAB project (POCTEP INTERREG Spain-Portugal), the BLUEHUMAN project (EAPA_151/2016, Atlantic Area 2016), the BIOHEAT project (PID 2020-115415RB-100, Spanish Ministry of Science and Innovation) and the Xunta de Galicia ED431C 2021/49 Programme for the consolidation and structuring of competitive research units (GRC). Pérez-Davila, S. and Lama, R. are grateful for funding support from the Xunta de Galicia's pre-doctoral grants (ED481A 2019/314 and IN606A-2017/011, respectively). This work was also supported by National Funds from the FCT (Fundação para a Ciência e a Tecnologia) through the UIDB/50016/2020 project. Funding for open access charge: University of Vigo/CISUG is acknowledged.

Institutional Review Board Statement: Not applicable.

Informed Consent Statement: Not applicable.

Acknowledgments: The technical staff from CACTI (University of Vigo) and from the IIM-CSIC (Judit Castro, Lucía Sánchez and the aquarium staff) are gratefully acknowledged.

Conflicts of Interest: The authors declare no conflict of interest.

References

1. Cuiffo, M.A.; Snyder, J.; Elliott, A.M.; Romero, N.; Kannan, S.; Halada, G.P. Impact of the Fused Deposition (FDM) Printing Process on Polylactic Acid (PLA) Chemistry and Structure. *Appl. Sci.* **2017**, *7*, 579. [CrossRef]
2. Singhvi, M.S.; Zinjarde, S.S.; Gokhale, D.V. Polylactic Acid: Synthesis and Biomedical Applications. *J. Appl. Microbiol.* **2019**, *127*, 1612–1626. [CrossRef] [PubMed]
3. da Silva, D.; Kaduri, M.; Poley, M.; Adir, O.; Krinsky, N.; Shainsky-Roitman, J.; Schroeder, A. Biocompatibility, Biodegradation and Excretion of Polylactic Acid (PLA) in Medical Implants and Theranostic Systems. *Chem. Eng. J.* **2018**, *340*, 9–14. [CrossRef] [PubMed]
4. Marcatto, V.A.; Sant'Ana Pegorin, G.; Barbosa, G.F.; Herculano, R.D.; Guerra, N.B. 3D Printed-polylactic Acid Scaffolds Coated with Natural Rubber Latex for. Pdf. *J. Appl. Polym. Sci.* **2022**, *139*, 51728. [CrossRef]
5. Tyler, B.; Gullotti, D.; Mangraviti, A.; Utsuki, T.; Brem, H. Polylactic Acid (PLA) Controlled Delivery Carriers for Biomedical Applications. *Adv. Drug Deliv. Rev.* **2016**, *107*, 163–175. [CrossRef]
6. Pérez Davila, S.; González Rodríguez, L.; Chiussi, S.; Serra, J.; González, P. How to Sterilize Polylactic Acid Based Medical Devices? *Polymers* **2021**, *13*, 2115. [CrossRef]
7. Hamad, K.; Kaseem, M.; Yang, H.W.; Deri, F.; Ko, Y.G. Properties and Medical Applications of Polylactic Acid: A Review. *Express Polym. Lett.* **2015**, *9*, 435–455. [CrossRef]
8. Serra, T.; Planell, J.A.; Navarro, M. High-Resolution PLA-Based Composite Scaffolds via 3-D Printing Technology. *Acta Biomater.* **2013**, *9*, 5521–5530. [CrossRef]
9. Lerouge, S. Introduction to Sterilization: Definitions and Challenges. In *Sterilisation of Biomaterials and Medical Devices*; Lerouge, S., Simmons, A., Eds.; Woodhead Publishing Limited: Cambridge, MA, USA, 2012; pp. 1–19.
10. Zhao, Y.; Zhu, B.; Wang, Y.; Liu, C.; Shen, C. Effect of Different Sterilization Methods on the Properties of Commercial Biodegradable Polyesters for Single-Use, Disposable Medical Devices. *Mater. Sci. Eng. C* **2019**, *105*, 110041. [CrossRef]
11. Valente, T.A.M.; Silva, D.M.; Gomes, P.S.; Fernandes, M.H.; Santos, J.D.; Sencadas, V. Effect of Sterilization Methods on Electrospun Poly(Lactic Acid) (PLA) Fiber Alignment for Biomedical Applications. *ACS Appl. Mater. Interfaces* **2016**, *8*, 3241–3249. [CrossRef]
12. Qiu, Q.Q.; Sun, W.Q.; Connor, J. *Sterilization of Biomaterials of Synthetic and Biological Origin*; Elsevier Ltd.: Amsterdam, The Netherlands, 2017; Volume 4, ISBN 9780081006924.
13. Aguado-Maestro, I.; De Frutos-Serna, M.; González-Nava, A.; Merino-De Santos, A.B.; García-Alonso, M. Are the Common Sterilization Methods Completely Effective for Our In-House 3D Printed Biomodels and Surgical Guides? *Injury* **2021**, *52*, 1341–1345. [CrossRef] [PubMed]
14. Kanemitsu, K.; Kunishima, H.; Saga, T.; Harigae, H.; Imasaka, T.; Hirayama, Y.; Kaku, M. Residual Formaldehyde on Plastic Materials and Medical Equipment Following Low-Temperature Steam and Formaldehyde Sterilization. *J. Hosp. Infect.* **2005**, *59*, 361–364. [CrossRef] [PubMed]
15. Ribeiro, N.; Soares, G.C.; Santos-Rosales, V.; Concheiro, A.; Alvarez-Lorenzo, C.; García-González, C.A.; Oliveira, A.L. A New Era for Sterilization Based on Supercritical CO2 Technology. *J. Biomed. Mater. Res. Part B Appl. Biomater.* **2020**, *108*, 399–428. [CrossRef]
16. Soares, G.C.; Learmonth, D.A.; Vallejo, M.C.; Davila, S.P.; González, P.; Sousa, R.A.; Oliveira, A.L. Supercritical CO_2 Technology: The next Standard Sterilization Technique? *Mater. Sci. Eng. C* **2019**, *99*, 520–540. [CrossRef] [PubMed]
17. Dillow, A.K.; Dehghani, F.; Hrkach, J.S.; Foster, N.R.; Langer, R. Bacterial Inactivation by Using Near- and Supercritical Carbon Dioxide. *Proc. Natl. Acad. Sci. USA* **1999**, *96*, 10344–10348. [CrossRef]
18. Lanzalaco, S.; Campora, S.; Brucato, V.; Carfì Pavia, F.; Di Leonardo, E.R.; Ghersi, G.; Scialdone, O.; Galia, A. Sterilization of Macroscopic Poly(l-Lactic Acid) Porous Scaffolds with Dense Carbon Dioxide: Investigation of the Spatial Penetration of the Treatment and of Its Effect on the Properties of the Matrix. *J. Supercrit. Fluids* **2016**, *111*, 83–90. [CrossRef]

19. Savaris, M.; dos Santos, V.; Brandalise, R.N. Influence of Different Sterilization Processes on the Properties of Commercial Poly(Lactic Acid). *Mater. Sci. Eng. C* **2016**, *69*, 661–667. [CrossRef]
20. Nüsslein-Volhard, C.; Dahm, R. *Zebrafish. A Practical Approach.*; Oxford University Press: Oxford, UK, 2002.
21. Westerfield, M. *The Zebrafish Book. A Guide for the Laboratory Use of Zebrafish (Danio Rerio)*, 4th ed.; University of Oregon Press: Eugene, OR, USA, 2000.
22. Torres-Giner, S.; Gimeno-Alcañiz, J.V.; Ocio, M.J.; Lagaron, J.M. Optimization of Electrospun Polylactide-Based Ultrathin Fibers for Osteoconductibe Bone Scaffolds. *J. Appl. Polym. Sci.* **2011**, *122*, 914–925. [CrossRef]
23. Rainer, A.; Centola, M.; Spadaccio, C.; Gherardi, G.; Genovese, J.A.; Licoccia, S.; Trombetta, M. Comparative Study of Different Techniques for the Sterilization of Poly-L-Lactide Electrospun Microfibers: Effectiveness vs. Material Degradation. *Int. J. Artif. Organs* **2010**, *33*, 76–85. [CrossRef]
24. Zhang, H.; Shao, C.; Kong, W.; Wang, Y.; Cao, W.; Liu, C.; Shen, C. Memory Effect on the Crystallization Behavior of Poly(Lactic Acid) Probed by Infrared Spectroscopy. *Eur. Polym. J.* **2017**, *91*, 376–385. [CrossRef]
25. Peniston, S.J.; Choi, S.J. Effect of Sterilization on the Physicochemical Properties of Molded Poly(L-Lactic Acid). *J. Biomed. Mater. Res. Part B Appl. Biomater.* **2007**, *80*, 67–77. [CrossRef] [PubMed]
26. Meaurio, E.; López-Rodríguez, N.; Sarasua, J.R. Infrared Spectrum of Poly(L-Lactide): Application to Crystallinity Studies. *Macromolecules* **2006**, *39*, 9291–9301. [CrossRef]
27. Savaris, M.; Braga, G.L.; Dos Santos, V.; Carvalho, G.A.; Falavigna, A.; MacHado, D.C.; Viezzer, C.; Brandalise, R.N. Biocompatibility Assessment of Poly(Lactic Acid) Films after Sterilization with Ethylene Oxide in Histological Study in Vivo with Wistar Rats and Cellular Adhesion of Fibroblasts in Vitro. *Int. J. Polym. Sci.* **2017**, *2017*, 7158650. [CrossRef]
28. Jalali, A.; Huneault, M.A.; Elkoun, S. Effect of Thermal History on Nucleation and Crystallization of Poly(Lactic Acid). *J. Mater. Sci.* **2016**, *51*, 7768–7779. [CrossRef]
29. Wright-Charlesworth, D.D.; Miller, D.M.; Miskioglu, I.; King, J.A. Nanoindentation of Injection Molded PLA and Self-Reinforced Composite PLA after in Vitro Conditioning for Three Months. *J. Biomed. Mater. Res. Part A* **2005**, *74*, 388–396. [CrossRef]
30. Gong, M.; Zhao, Q.; Dai, L.; Li, Y.; Jiang, T. Fabrication of Polylactic Acid/Hydroxyapatite/Graphene Oxide Composite and Their Thermal Stability, Hydrophobic and Mechanical Properties. *J. Asian Ceram. Soc.* **2017**, *5*, 160–168. [CrossRef]
31. Izdebska-Podsiadły, J.; Dörsam, E. Storage stability of the oxygen plasma modified PLA film. *Bull. Mater. Sci.* **2021**, *44*, 79. [CrossRef]
32. Weir, N.A.; Buchanan, F.J.; Orr, J.F.; Farrar, D.F.; Boyd, A. Processing, Annealing and Sterilisation of Poly-L-Lactide. *Biomaterials* **2004**, *25*, 3939–3949. [CrossRef]
33. Manjunatha, B.; Park, S.H.; Kim, K.; Kundapur, R.R.; Lee, S.J. In Vivo Toxicity Evaluation of Pristine Graphene in Developing Zebrafish (Danio Rerio) Embryos. *Environ. Sci. Pollut. Res.* **2018**, *25*, 12821–12829. [CrossRef]
34. Wang, Z.G.; Zhou, R.; Jiang, D.; Song, J.E.; Xu, Q.; Si, J.; Chen, Y.P.; Zhou, X.; Gan, L.; Li, J.Z.; et al. Toxicity of Graphene Quantum Dots in Zebrafish Embryo. *Biomed. Environ. Sci.* **2015**, *28*, 341–351. [CrossRef]
35. MacDonald, N.P.; Zhu, F.; Hall, C.J.; Reboud, J.; Crosier, P.S.; Patton, E.E.; Wlodkowic, D.; Cooper, J.M. Assessment of Biocompatibility of 3D Printed Photopolymers Using Zebrafish Embryo Toxicity Assays. *Lab Chip* **2016**, *16*, 291–297. [CrossRef] [PubMed]
36. Oskui, S.M.; Diamante, G.; Liao, C.; Shi, W.; Gan, J.; Schlenk, D.; Grover, W.H. Assessing and Reducing the Toxicity of 3D-Printed Parts. *Environ. Sci. Technol. Lett.* **2016**, *3*, 1–6. [CrossRef]
37. Zhu, F.; Friedrich, T.; Nugegoda, D.; Kaslin, J.; Wlodkowic, D. Assessment of the Biocompatibility of Three-Dimensional-Printed Polymers Using Multispecies Toxicity Tests. *Biomicrofluidics* **2015**, *9*, 061103. [CrossRef] [PubMed]
38. Zhu, F.; Skommer, J.; Friedrich, T.; Kaslin, J.; Wlodkowic, D. 3D Printed Polymers Toxicity Profiling: A Caution for Biodevice Applications. *Micro+ Nano Mater. Devices Syst.* **2015**, *9668*, 96680Z. [CrossRef]

Article

Development of Indicator Film Based on Cassava Starch–Chitosan Incorporated with Red Dragon Fruit Peel Anthocyanins–Gambier Catechins to Detect Banana Ripeness

Valentia Rossely Santoso [1], Rianita Pramitasari [1,*] and Daru Seto Bagus Anugrah [2]

[1] Food Technology Study Program, Faculty of Biotechnology, Atma Jaya Catholic University of Indonesia, BSD Campus, Tangerang 15345, Indonesia; valentiarossely@gmail.com
[2] Biotechnology Study Program, Faculty of Biotechnology, Atma Jaya Catholic University of Indonesia, BSD Campus, Tangerang 15345, Indonesia; daru.seto@atmajaya.ac.id
* Correspondence: rianita.pramitasari@atmajaya.ac.id or rianitta@gmail.com

Citation: Santoso, V.R.; Pramitasari, R.; Anugrah, D.S.B. Development of Indicator Film Based on Cassava Starch–Chitosan Incorporated with Red Dragon Fruit Peel Anthocyanins–Gambier Catechins to Detect Banana Ripeness. *Polymers* **2023**, *15*, 3609. https://doi.org/10.3390/polym15173609

Academic Editors: Marcin Wekwejt and Beata Kaczmarek

Received: 12 August 2023
Revised: 27 August 2023
Accepted: 29 August 2023
Published: 31 August 2023

Copyright: © 2023 by the authors. Licensee MDPI, Basel, Switzerland. This article is an open access article distributed under the terms and conditions of the Creative Commons Attribution (CC BY) license (https:// creativecommons.org/licenses/by/ 4.0/).

Abstract: Banana ripeness is generally determined based on physical attributes, such as skin color; however, it is considered subjective because it depends on individual factors and lighting conditions. In addition, improper handling can cause mechanical damage to the fruit. Intelligent packaging in the form of indicator film incorporated with anthocyanins from red dragon fruit peel has been applied for shrimp freshness detection; however, this film has low color stability during storage, necessitating the addition of gambier catechins as a co-pigment to increase anthocyanin stability. Nevertheless, the characteristics of films that contain gambier catechins and their applications to bananas have not been studied yet; therefore, this study aims to develop and characterize indicator films that were incorporated with red dragon fruit peel anthocyanins and gambier catechins to detect banana ripeness. In this study, the indicator films were made via solvent casting. The films were characterized for their structural, mechanical, and physicochemical properties, and then applied to banana packaging. The results show that the film incorporated with anthocyanins and catechins in a ratio of 1:40 (w/w) resulted in better color stability, mechanical properties, light and water vapor barrier ability, and antioxidant activity. The application of the indicator films to banana packaging resulted in a change in color on the third day of storage. It can be concluded that these films could potentially be used as an indicator to monitor banana ripeness.

Keywords: red dragon fruit peel anthocyanins; gambier catechins; indicator film; co-pigmentation; banana ripeness

1. Introduction

Bananas are Indonesia's leading horticultural product. In 2022, Indonesia produced 9,245,427 tons of bananas, which were then marketed both globally and domestically [1]. The distribution of bananas for retail marketing is carried out when the banana reaches a certain level of maturity according to a reference color chart, with a score of 3 indicating a banana that is green–yellow, a score of 4 indicating a banana that is green and more yellow, and a score of 5 indicating a banana that is yellow [2]. The maturity level of a banana is generally determined by comparing the color of the fruit skin with a reference color chart, but this method is considered to be very subjective and inconsistent because it depends on operator skill and lighting [3]. In addition, improper handling of the fruit when determining ripeness can cause mechanical damage [4].

Intelligent packaging is a type of food packaging that can detect and provide information about the quality of food ingredients in such aspects as freshness, ripeness, gas production, and the presence of microbial contamination in the ingredient [5]. Intelligent packaging is generally equipped with sensor technology, enabling the collection of information without damaging or opening the package. Sensors that are commonly used in intelligent packaging are color indicators, which are incorporated into polymer-based films and are sensitive to pH, allowing for a color change when there is a change in the pH of the product [6].

Anthocyanins are water-soluble natural pigments that originate from red-, blue-, purple-, and black-colored materials, such as red dragon fruit [7], butterfly pea flower [8], purple sweet potato [9], and black rice [10]. These pigments can change color based on changes in the pH of their environment. This characteristic has led to the widespread development of anthocyanins as color indicators incorporated into films made from materials such as whey protein isolate nanofibers [11], chitosan [12], alginate [13], gelatin [14], and pectin–chitosan [10], primarily to monitor the freshness of meat and seafood. Additionally, anthocyanins incorporated into films can also be utilized to determine the maturity levels of fruits, including bananas. Bananas are climacteric fruits, where the respiration process will continue even beyond the harvest process. The fruit ripening process causes an increase in the pH, which can be used as an indicator of the level of fruit maturity [5].

Research on indicator films based on cassava–chitosan starch with anthocyanins from red dragon fruit peel has been carried out to detect freshness in shrimp; however, it produces less stable colors during storage [15]. Anthocyanin-based indicator films are susceptible to light and temperature. Light and temperature cause discoloration on the films even when they are not used for food monitoring [16]. One method that can be applied to increase the color stability of anthocyanins is co-pigmentation. Co-pigmentation of anthocyanins using gambier catechins (*Uncaria gambier*) has been carried out and proven to minimize the degradation of black rice anthocyanins in isotonic drinks [17]; however, improvements in the color stability of anthocyanin-based indicator films with gambier catechin co-pigmentation have never been carried out.

In order to fill this research gap, this study aimed to develop and characterize indicator films based on a cassava starch–chitosan complex and anthocyanins of red dragon fruit (*Hylocereus polyrhizus*) peel co-pigmented with gambier catechins as indicators to monitor banana ripeness. The success of the film's development can be understood based on the film's ability to detect the maturity level of bananas as climacteric fruits that produce volatile compounds that increase the pH in the environment inside the package.

2. Materials and Methods

2.1. Materials

Red dragon fruit, cavendish banana (*Musa acuminata* L.), and tapioca flour (Rose Brand) were obtained from AEON Mall BSD City, Tangerang, Indonesia. Chitosan with a deacetylation degree of 83.3% and molecular weight of 70 kDa was obtained from Xi'an Gawen, China. Ethanol, glacial acetic acid, and 2,2-diphenyl-1-picrylhydrazyl (DPPH) were obtained from Smart-Lab, Tangerang, Indonesia. Glycerol was obtained from Merck KGaA, Darmstadt Germany. The other ingredients included gambier catechins and distilled water. All chemicals used in this research were analytical grade.

2.2. Methods

2.2.1. Anthocyanin Extraction from Red Dragon Fruit Peel

The red dragon fruit was washed, and the flesh was removed. Then, the red dragon fruit peel was mashed using a food processor (Philips 2061, Shanghai, China) until it became mushy. Once smooth, 100 g of red dragon fruit peel pulp were weighed and placed into 500 mL of a 70% ethanol solution. Then, the mixture was stirred using a stir bar until the mixture became evenly distributed. The mixture was then macerated in a chiller at 5 °C for 24 h. After maceration, the extract was filtered using Whatman No. 1 paper. The filtrate obtained was then concentrated using a rotary evaporator (R-300 Buchi, Flawil, Switzerland) at 55 °C, 60 rpm speed, and 130 mbar pressure [18].

2.2.2. Analysis of the Anthocyanin Levels of Red Dragon Fruit Peel

Every 100 µL of the concentrated anthocyanin extract was mixed into 1.9 mL of a potassium chloride buffer solution (pH 1) and a sodium acetate buffer solution (pH 4.5) in a test tube. Next, the solution was homogenized using a vortex (Thermo Fisher, Waltham, MA, USA) and incubated for 15 min. After that, the absorbance of the solution was measured at wavelengths of 536 and 700 nm using a UV-Vis spectrophotometer (Thermo Scientific, Waltham, MA, USA). The blank used in this measurement was distilled water. The total anthocyanin level was calculated using the following equation [19]:

$$\text{Anthocyanin level (mg/L)} = \frac{A \times MW \times DF \times 1000}{\varepsilon \times L} \quad (1)$$

where A is the absorbance value, with $A = (A_{536} - A_{700})$ pH 1 $- (A_{536} - A_{700})$ pH 4.5; MW is the molecular weight of anthocyanin (449.2 g/mol); DF is the dilution factor; ε is the molar attenuation coefficient of anthocyanin (26,900 L/mol·cm); and L is the width of the cuvette (1 cm).

2.2.3. Addition of Gambier Catechins

Gambier catechins were added to the anthocyanin extract of the red dragon fruit peel following four discrete formulations, as seen in Table 1 [20].

Table 1. Color extract formulation.

Sample	Anthocyanins:Gambier Catechins (w/w)
SCh–A	1:0
SCh–AC20	1:20
SCh–AC28	1:28
SCh–AC40	1:40

The color extract was prepared by dissolving gambier catechin powder in concentrated anthocyanin extract, which was then homogenized using a magnetic hotplate stirrer (Heidolph MR Hei-Standard, Schwabach, Germany) for 30 min. The solution was then filtered using Whatman No. 1 paper to obtain the filtrate.

2.2.4. Color Change Analysis of the Color Extracts

A total of 13 cuvettes were prepared. Then, 1 mL of color extract was placed into each cuvette and 1 mL of pH 1–13 buffer was added to each cuvette, so that a final volume of 2 mL was obtained. The solution was then incubated for 30 min, and then the color change was observed and photographed [21].

2.2.5. Indicator Film Preparation

The indicator films were prepared by first preparing a 1% (w/v) chitosan solution involving the mixing of 1 g of chitosan into 100 mL of 1% (v/v) acetic acid solution. The solution was homogenized using a magnetic hotplate stirrer (Thermo Scientific, Waltham, MA, USA), with stirring at 300 rpm for 24 h. Next, a 2% (w/v) cassava starch solution was prepared by mixing 2 g of cassava starch powder into 100 mL of distilled water. The mixture was then gelatinized using a magnetic hotplate stirrer (Thermo Scientific, Waltham, MA, USA) until it reached a temperature of 80 °C.

The chitosan solution was then mixed with the gelatinized cassava starch solution in a ratio of 1:1. After mixing, the solution was then homogenized using a homogenizer (Heidolph, Schwabach, Germany) at 530 rpm for 10 min. Then, 85% glycerol was added to the solution, and homogenization was performed using a homogenizer (IKA T10 ultra-turrax, Potsdam, Germany) at 14,500 rpm for 5 min (Table 2). The color extract was then added to the film solution and stirred using a stir bar until it became smooth [22].

Table 2. Composition of indicator film solution per print unit (20 mL).

Composition	Amount (mL)
1% (w/v) chitosan solution	7.46
2% (w/v) cassava starch solution	7.46
85% glycerol	0.08
Color extract	5

2.2.6. Indicator Film Casting

Each variation of the film solution was poured into a plastic Petri dish 6 cm in diameter at 20 mL. Next, the solution was dried in a fume cupboard for 72 h. After the solution dried and formed a film, the film was removed from the Petri dish using a spatula and stored in a closed, dark container at a chiller temperature of 5 °C [23].

2.2.7. Characterization of Indicator Films' Surface Color

The surface colors of the four indicator film samples were analyzed using a colorimeter (Nh3, China) on the 1st and 14th day of storage at a temperature of 5 °C. After obtaining the L, a, and b values from the film samples, the total color difference (ΔE) was calculated using the following formula [21]:

$$\Delta E = \sqrt{(L_s - L)^2 + (a_s - a)^2 + (b_s - b)^2} \tag{2}$$

where L is lightness, a is red (+) and green (−), and b is yellow (+) and blue (−).

2.2.8. Analysis of Indicator Film Sensitivity to Various pH Conditions

The films' sensitivity to pH and their color-changing ability were analyzed. The indicator film was cut to a size of 1×1 cm and then dipped in each buffer solution with a pH value of 1–13 for 2 min, respectively [21].

2.2.9. Characterization of Indicator Film Structure

The surface morphology of the films was observed using scanning electron microscopy (SEM) (Hitachi SU3500, Tokyo, Japan) with an electric voltage of 10 kV and high vacuum conditions (10^{-3}). An FTIR analysis in the range of 500–4000 cm^{-1} (Bruker-Tensor II, Borken, Germany) was performed to observe the functional groups on the films [24].

2.2.10. Characterization of Mechanical Properties of the Films

The film thickness was measured using a micrometer with an accuracy of 0.01 mm. The mechanical properties were then analyzed using a texture analyzer (Agrosta Texturometer, Serqueux, France). The film samples were cut to a size of 4×1 cm and affixed to the towing lever. A tensile test was then carried out with an initial grip of 1 cm and a cross-head speed of 0.8 mm/s. The tensile strength and elongation at break, as measurements of film strength and flexibility, were determined using the following formula [21]:

$$Tensile\ strength\ (\text{MPa}) = \frac{F}{A_0} \tag{3}$$

$$Elongation\ at\ break\ (\%) = \frac{\delta}{L_0} \times 100 \tag{4}$$

where F is the force applied to the film sample, A_0 is the area of the film sample, δ is the increase in length, and L_0 is the initial length of the film sample.

2.2.11. Physical Characterization of Films

Analysis of Transparency and Transmittance of the Indicator Films

The films were cut to a size of 4 × 1 cm and affixed to the clear side of the cuvette. The light transmittance of the films could be measured using a UV-Vis spectrophotometer (Shimadzu UV-2450, Tokyo, Japan) by measuring the transmittance at a wavelength of 200–800 nm. After determining the percentage of light transmittance, the transparency of the film was calculated using the following formula [21]:

$$Transparency = \frac{\log T600}{D} \qquad (5)$$

where $T600$ is the percent transmittance of light at a wavelength of 600 nm and D is the thickness of the film sample (mm).

Analysis of Moisture Content and Solubility of the Indicator Films

Analyses of the film water content and solubility were carried out based on the method of Alizadeh-Sani et al. [21]. The films were cut to a size of 2 × 2 cm. The moisture content of the film samples was then measured by weighing the film samples before drying ($W1$) and after drying ($W2$) using an oven (Memmert UN110, Greifensee, Switzerland) at 105 °C for 24 h using the following formula [21]:

$$Moisture\ content\ (\%) = \frac{W1 - W2}{W1} \times 100 \qquad (6)$$

For the film solubility analysis, 2 × 2 cm films were dried in an oven (Memmert UN110, Greifensee, Switzerland) at 105 °C for 24 h and then weighed as $W1$. The dried film samples were immersed in 25 mL of distilled water involving shaking using an orbital shaker (GFL 3017, Greifensee, Germany) at 100 rpm for 24 h. The film samples were then dried again using an oven (Memmert UN110, Greifensee, Switzerland) at 105 °C for 24 h and then weighed as $W2$. After weighing, the solubility of the film samples was calculated using the following formula [21]:

$$Water\ solubility\ (\%) = \frac{W1 - W2}{W1} \times 100 \qquad (7)$$

Analysis of Water Vapor Transmission Rate (WVTR)

The WVTR value was calculated by cutting a circular film sample 4 cm in diameter. The cut film was then used as a lid for a container containing 5 g of silica gel (0% RT), which was placed in a desiccator filled with distilled water (99% RH). The silica gel was weighed and replaced periodically every 24 h for 3 days. From the results, a graph of the mass of the container against time was obtained and a linear equation was developed using the formula y = mx + c. Then, the WVTR value was calculated using the following formula [21]:

$$WVTR = \frac{m}{A \times t} \qquad (8)$$

The WVTR value obtained is in units of $gm^{-2} \cdot 24\ h^{-1}$. The value of m is obtained from the linear equation, A is the film area (m^2), and t is the time (hours).

2.2.12. Analysis of Antioxidant Activity

As much as 50 mg of each film sample was weighed and dissolved in 5 mL of distilled water and homogenized using a vortex (Thermo Fisher, Waltham, MA, USA). The solution was then shaken using an orbital shaker (GFL 3017, Burgwedel, Germany) at 200 rpm for 24 h. The extract from the film was then taken and used for testing using the radical scavenging 2,2-diphenyl-1-picrylhydrazyl (DPPH) method. The test solution was prepared by dissolving 2.5 mg of DPPH in 100 mL of 99.7% methanol, with distilled water as a blank. A total of 4 µL of extract from the film was mixed with 196 µL of DPPH solution, then the mixture was homogenized using a vortex (Thermo Fisher, Waltham, MA, USA). The absorbance was measured at a wavelength of 490 nm using a microplate reader (NanoQuant TECAN, Hombrechtikon, Switzerland). The DPPH activity was then calculated using the following formula [25]:

$$Antioxidant\ activity\ (\%) = \frac{A_{DPPH} - A_{film}}{A_{DPPH}} \times 100 \qquad (9)$$

A_{DPPH} is the absorbance of the blank DPPH in the form of a mixture of DPPH solution and distilled water, while A_{film} is the absorbance of the DPPH film sample.

Monitoring the Ripeness of Bananas

Each indicator film sample was cut to a size of 1.5 × 1.5 cm and then affixed to PVC cling wrap, which was part of the lid of the container. The film was not allowed to come into direct contact with the banana sample. The prepared samples were stored at room temperature for 7 days. The color changes of the films were observed and photographed every day [26].

Statistical Analysis

The experiments in film-making and film characterization were repeated three times. The data obtained were then statistically processed using the IBM SPSS Statistics 25 (IBM Corp., Armonk, NY, USA) application. First of all, a normality test was performed on the data using the Shapiro-Wilk test. If the data were normally distributed, then the data were tested for homogeneity. If the data were normally distributed and homogeneous, then the test was followed by a one-way analysis of variance (ANOVA). The real differences obtained were analyzed using Duncan's test at $\alpha = 0.05$. If the data were otherwise not normally distributed and/or not homogeneous, then the data were processed using the Kruskal–Wallis test, followed by the stepwise step-down test [27].

3. Results

3.1. Anthocyanin Level of Red Dragon Fruit Peel Extract

The anthocyanin extract of red dragon fruit peel obtained was purplish-red in color (Figure 1). The extraction of anthocyanin by the maceration method using 70% ethanol resulted in an anthocyanin level of 0.122 ± 0.01 mg/g.

Figure 1. Anthocyanin extract of red dragon fruit peel.

3.2. Changes in Extract Color under Various pH Conditions

The color extract was divided into two categories, namely, anthocyanin extract from red dragon fruit peel without the addition of gambier catechin and anthocyanin extract from red dragon fruit peel with the addition of gambier catechin. In the color extract without the addition of gambier catechin, the color changed to purple–yellow at a pH of 12 and to yellow at a pH of 13. In the color extract with the addition of gambier catechin, color changes appeared at pH 1 to purple and at pH 13 to yellowish-brown (Figure 2).

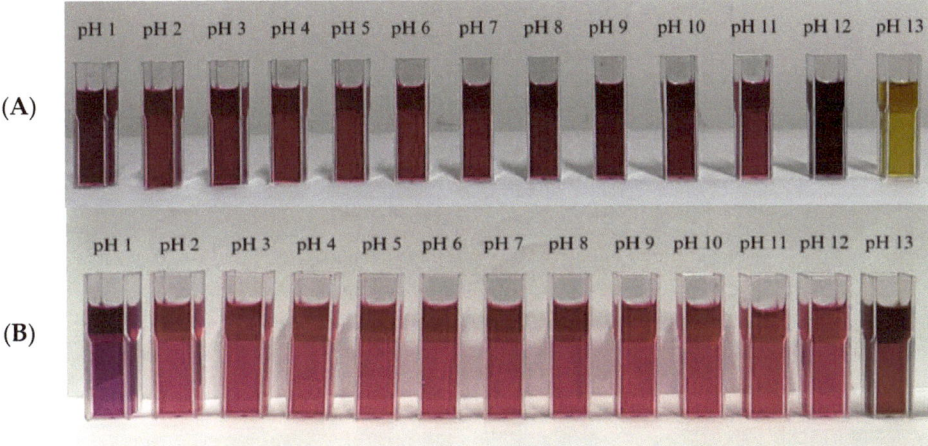

Figure 2. Color change test of the color extract (**A**) before the addition of gambier catechin and (**B**) after the addition of gambier catechin.

3.3. Indicator Films

The final results of the film-making and the condition of each film after 14 days of storage can be seen in Figure 3. During the 14 days of storage, it was clear that the SCh–A and SCh–AC20 films had changed in color from pink to red–yellow, whereas the SCh–A, SCh–AC28, and SCh–AC40 films had only a slightly yellowish tint to the outside of the films.

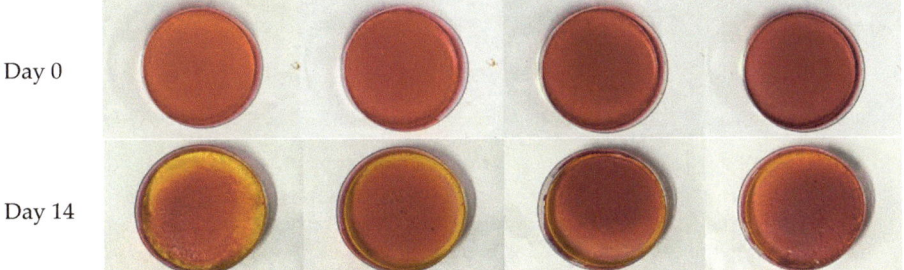

Figure 3. The color changes of the indicator films (SCh–A (cassava starch–chitosan, glycerol, anthocyanin), SCh–AC20 (cassava starch–chitosan, glycerol, anthocyanin–catechin 1:20), SCh–AC28 (cassava starch–chitosan, glycerol, anthocyanin–catechin 1:28), and SCh–AC40 (cassava starch–chitosan, glycerol, anthocyanin–catechin 1:40) after 14 days of storage.

3.4. Surface Colors of the Indicator Films

After 14 days of storage, the SCh–A film showed increased L^*, a^*, and b^* values. The SCh–AC20 film showed increased L^* and b^* values, but the a^* value decreased. The

SCh–AC28 film showed decreased L*, a*, and b* values. Meanwhile, the SCh–AC40 film showed decreased L* and a* values, but the b* value increased. This indicates that the color of the film grew darker and yellower, and the redness decreased toward green. Based on the L*, a*, and b* values, the SCh–AC40 film had the lowest color change. This is indicated by the smaller ΔE value of the SCh–AC40 film compared to the other film samples (Table 3).

Table 3. The results of the surface color analysis of the indicator films using a colorimeter.

Sample	1st Day			14th Day			
	L*	a*	b*	L*	a*	b*	ΔE
SCh–A	32.50 ± 1.10 [a]	35.62 ± 3.51 [a]	14.45 ± 3.25 [a]	34.87 ± 1.65 [bc]	38.59 ± 1.10 [b]	19.36 ± 2.14 [b]	8.09 ± 5.29 [a]
SCh–AC20	34.13 ± 2.16 [a]	41.16 ± 6.73 [a]	12.25 ± 3.02 [a]	36.51 ± 0.86 [c]	39.51 ± 0.90 [b]	19.25 ± 0.92 [b]	7.83 ± 1.90 [a]
SCh–AC28	34.99 ± 0.65 [a]	39.92 ± 4.14 [a]	15.1 ± 1.65 [a]	32.83 ± 0.20 [ab]	37.29 ± 0.12 [b]	14.72 ± 0.09 [a]	4.43 ± 3.34 [a]
SCh–AC40	31.67 ± 0.16 [a]	33.32 ± 0.28 [a]	11.07 ± 0.21 [a]	31.41 ± 0.22 [a]	31.09 ± 0.63 [a]	11.46 ± 0.42 [a]	2.33 ± 0.57 [a]

All data are presented as means ± standard deviations. Different letters in each column indicate a real difference; $\alpha \leq 0.05$.

3.5. Indicator Film Sensitivity to Various pH Conditions

The sensitivity of an indicator film to various pH conditions was tested only in SCh–A40 because of the film's smallest ΔE value. The results of the evaluation of the SCh–AC40 indicator film that was dipped in pH 1–13 buffer can be seen in Figure 4. It was discovered that the indicator film had the ability to change color according to the pH conditions from red at pH 1–12 to yellow at pH 13.

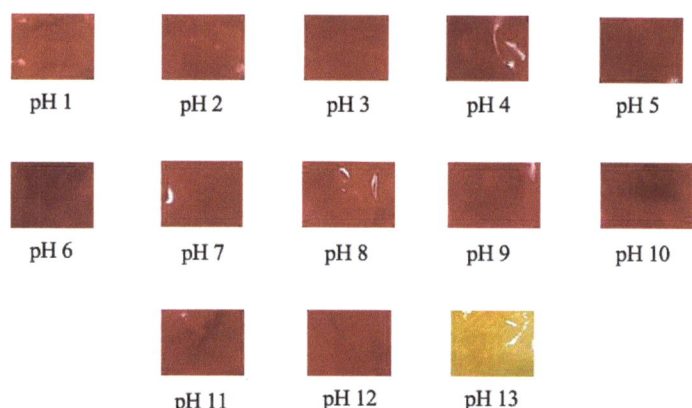

Figure 4. The SCh-AC40 indicator film's sensitivity to various pH conditions.

3.6. Structural Properties of the Indicator Films

The morphology of the indicator films observed using SEM can be seen in Figure 5. The SCh–AC40 film showed very clear gray and white aggregates. As the amount of gambier catechins added increased, the number of aggregates captured by SEM increased as well.

The results of the FTIR curve analysis can be seen in Figure 6. The spectrum produced by each film was similar, with differences found only in the absorption at 500–1000 cm^{-1}, which was caused by differences in the stretching of C–O due to the addition of the gambier catechins.

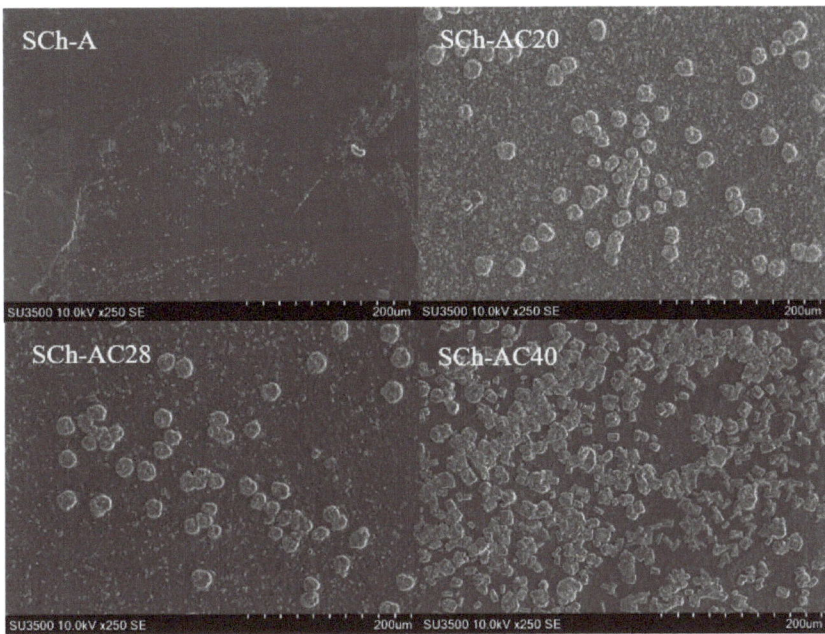

Figure 5. Indicator film morphology.

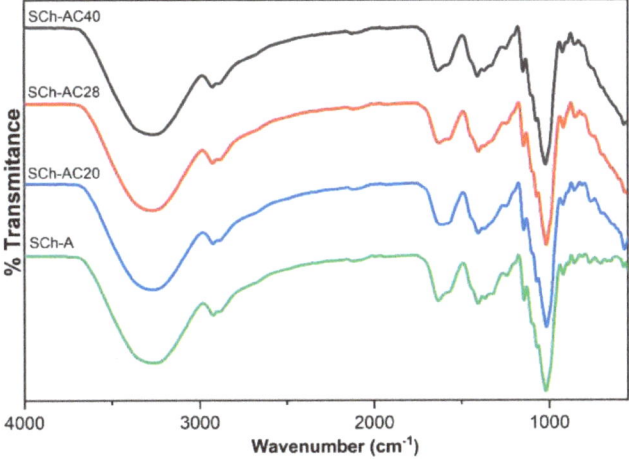

Figure 6. FTIR results of the indicator films.

3.7. Thickness and Mechanical Properties of the Indicator Films

The SCh–AC20 and SCh–AC28 films showed the highest thickness values. Based on the mechanical properties results, the SCh–AC40 film had the highest tensile strength and elongation at break values, while the SCh–AC20 film had the lowest tensile strength and elongation at break values (Table 4).

3.8. Physical Properties of the Indicator Films

3.8.1. Transmittance and Transparency of the Indicator Films

The light transmittance value of each film increased at wavelengths of 380 nm and 570 nm; however, it experienced a decrease at a wavelength of 450 nm (Figure 7a).

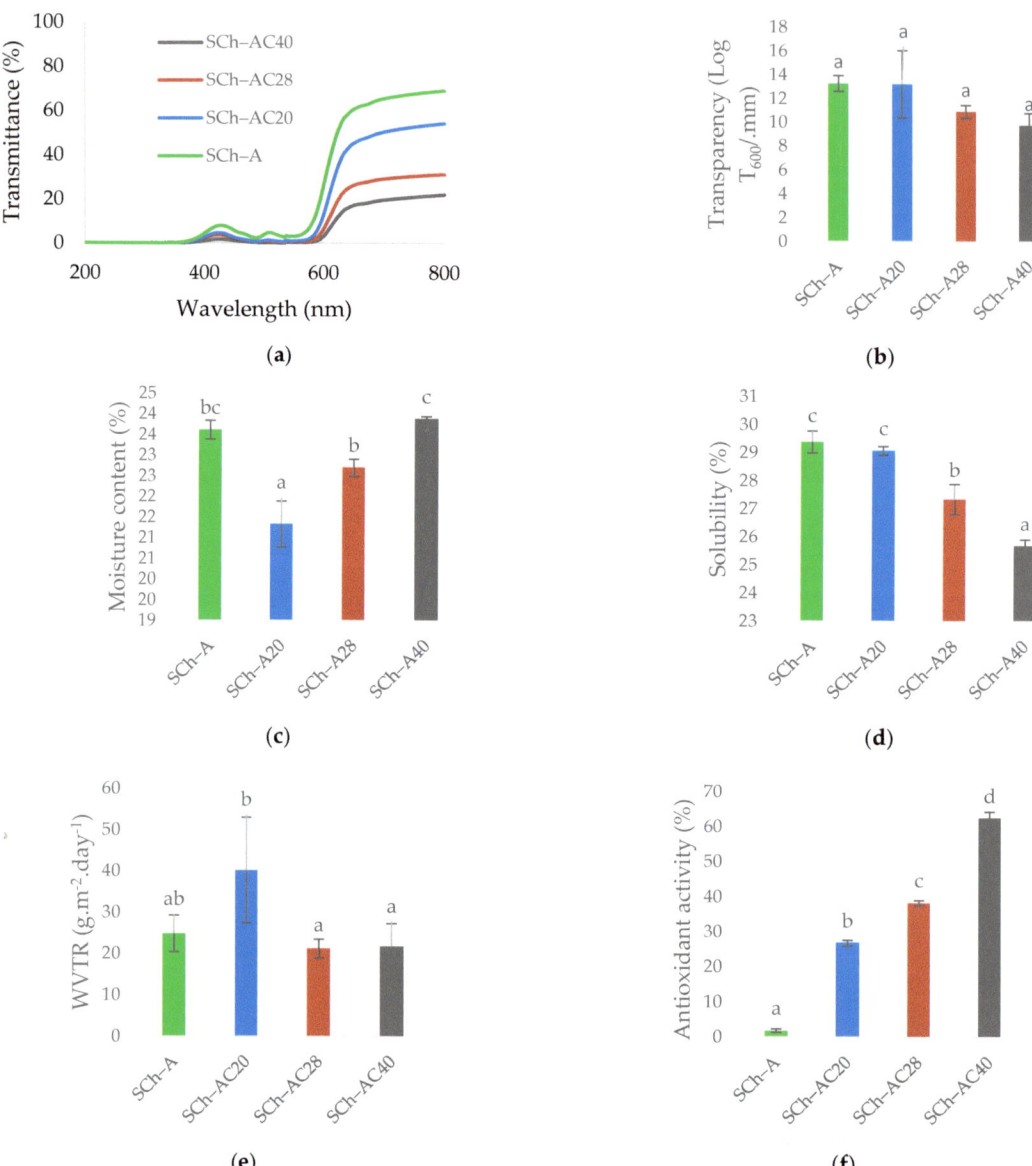

Figure 7. Physicochemical properties of SCh-A, SCh-A20, SCh-A28, and SCh-A40 indicator films through (**a**) transmittance; (**b**) transparency; (**c**) moisture content; (**d**) solubility; (**e**) water vapor transmission rate (WVTR); and (**f**) antioxidant activity. Bars represent means ± standard deviations. Different letters indicate a real difference; $\alpha \leq 0.05$.

The results for the transparency of the indicator films can be seen in Figure 7b. The transparency values obtained for the SCh–A, SCh–AC20, SCh–AC28, and SCh–AC40 films were not significantly different.

Table 4. Thickness and mechanical properties of the indicator film.

Films	Thickness (mm)	Tensile Strength (MPa)	Elongation at Break (%)
SCh–A	0.13 ± 0.00 [a]	0.19 ± 0.02 [b]	2.22 ± 0.35 [b]
SCh–AC20	0.15 ± 0.01 [b]	0.12 ± 0.03 [a]	1.29 ± 0.01 [a]
SCh–AC28	0.15 ± 0.00 [b]	0.12 ± 0.01 [a]	1.42 ± 0.24 [a]
SCh–AC40	0.14 ± 0.00 [ab]	0.23 ± 0.02 [b]	4.88 ± 0.27 [c]

All data are presented as means ± standard deviations. Different letters in each column indicate a real difference; $\alpha \leq 0.05$.

3.8.2. Moisture Content and Solubility of the Indicator Films

The SCh–AC40 film had the highest water content, while the SCh–AC20 film had the lowest. As for film solubility, the SCh–A film had the highest value, followed by the SCh–AC20, SCh–AC28, and SCh–AC40 values (Figure 7c,d).

3.8.3. Water Vapor Transmission Rate (WVTR)

The SCh–AC40 film had the lowest WVTR value, while the SCh–AC20 film had the highest (Figure 7e).

3.9. Antioxidant Activity of the Indicator Films

The SCh–AC40 film had the highest antioxidant activity value, followed by the SCh–AC28, SCh–AC20, and SCh–A films (Figure 7f).

4. Banana Ripening Monitoring

The results of the application of the SCh–A indicator film in monitoring banana ripeness during room temperature storage (25 °C) can be seen in Figure 8. The SCh–A indicator film showed a color change from red to pink–yellow beginning on the 3rd day of storage and another color change to almost entirely yellow on the 6th day of storage.

The results of the application of the SCh–AC20 indicator film in monitoring the ripeness of a banana during room temperature storage (25 °C) can be seen in Figure 9. The SCh–AC20 indicator film showed a change in film color from red to pink beginning on the 3rd day of storage and another change into red–yellow on day 5; however, until day 7 of storage, the color of the film did not change completely to yellow.

The results of the application of the SCh–AC28 indicator film in monitoring the ripeness of a banana during room temperature storage (25 °C) can be seen in Figure 10. The SCh–AC28 indicator film showed a color change from red to faded red on the 3rd day of storage and another change into red–yellowish on the 4th day, but until the 7th day of storage, the film discoloration only occurred on a portion of the film surface.

The results of the application of the SCh–AC40 indicator film in monitoring the ripeness of a banana during room temperature storage (25 °C) can be seen in Figure 11. The SCh–AC40 indicator film showed a color change from red to red–yellow on the 3rd day of storage, and on the 7th day of storage, the film color changed almost completely to yellow.

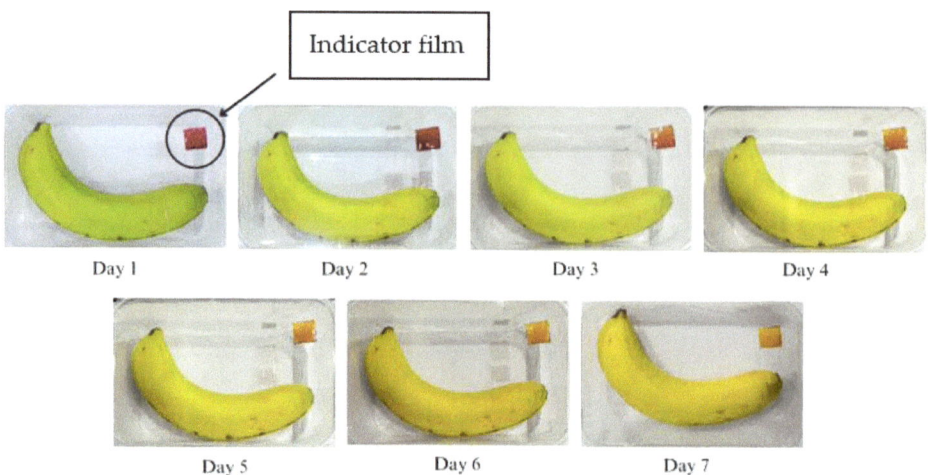

Figure 8. Changes in color of the SCh–A indicator film on banana packaging.

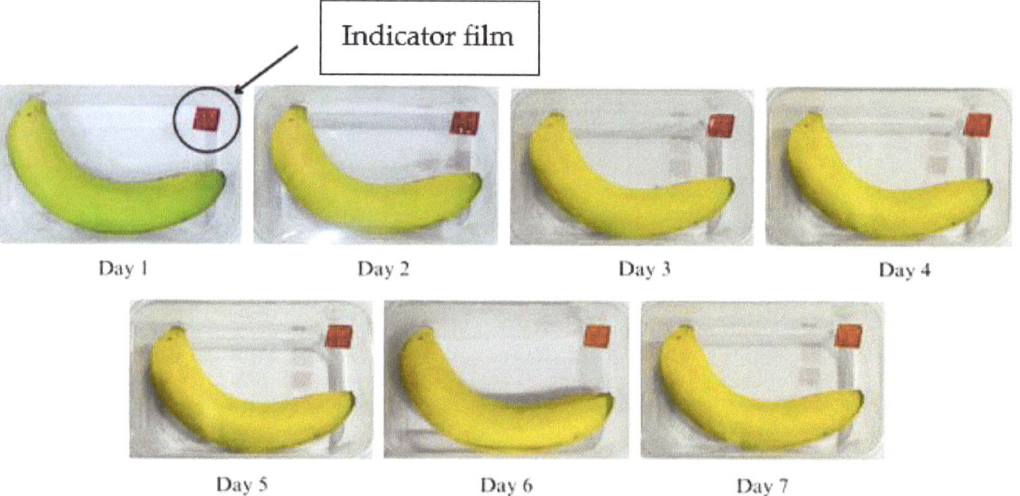

Figure 9. Changes in color of the SCh–AC20 indicator film on banana packaging.

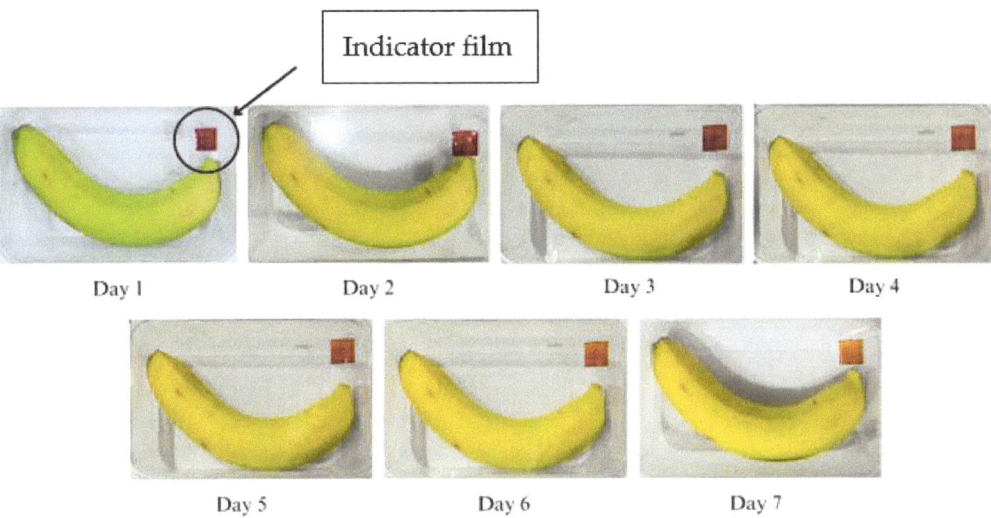

Figure 10. Changes in color of the SCh–AC28 indicator film on banana packaging.

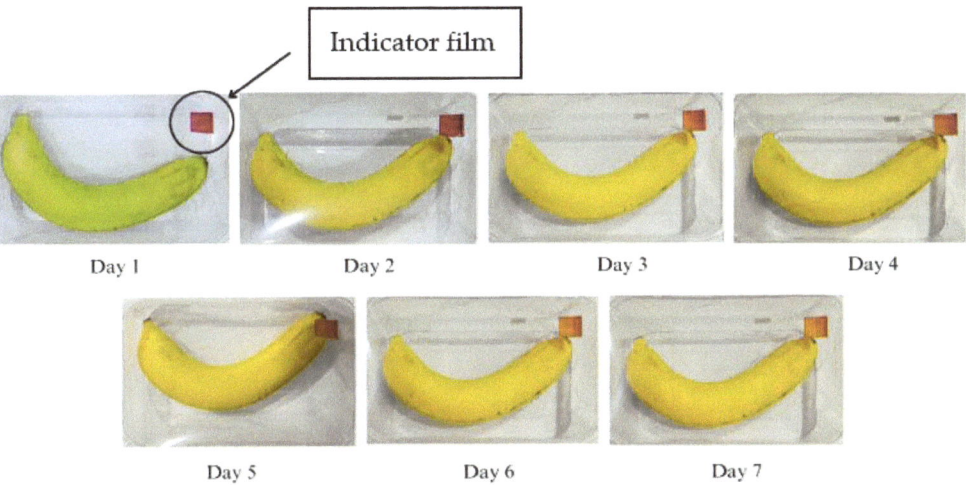

Figure 11. Changes in color of the SCh–AC40 indicator film for banana packaging.

5. Discussion

In this study, anthocyanin was contained in red dragon fruit peel at 0.122 ± 0.01 mg/g. Other research similarly discovered anthocyanin in red dragon fruit peel at a level of 0.105 mg/g [15]. The concentration method was applied using a rotary evaporator at 55 °C for the extraction of anthocyanins because anthocyanins are thermolabile compounds. This method was selected in order to prevent anthocyanin damage due to high temperatures, in which case the temperature at which this method was implemented is considered to be the optimal condition for anthocyanin extraction [15].

The anthocyanin extracts obtained were then used as color extracts, and it was determined that two of the color extracts used had the ability to change color depending on the pH conditions. When the color extract without the addition of gambier catechins was used, a significant color change was seen at pH 12 into purple and at pH 13 into yellow. This is in accordance with the research conducted by Apriliyanti et al. [28], which found that at pH 2–10, red dragon fruit peel extract has a reddish-purplish color, which changes into yellow at pH 12–13. In general, anthocyanins will form a pink color at pH 4–5, purple at pH 6–7, and yellow at pH values above 10 [16]. In the color extract added with gambier catechins, a significant color change was obtained at pH 1 to purple. Then, another color change occurred at pH 13 to purple–yellow. The difference in the color series produced by the two types of color extracts was due to the anthocyanin co-pigmentation process rendered by the addition of gambier catechins. Co-pigmentation is a phenomenon of the formation of molecular or complex associations by anthocyanins and co-pigment compounds. The effect of co-pigmentation is manifested as an increase in the color intensity and a change in the color tone of the anthocyanin solution [29]. This phenomenon is known as the hyperchromatic effect (ΔA), which refers to an increase in the intensity of the anthocyanin color, which results in an increase in absorbance and a bathochromic effect ($\Delta \lambda max$), i.e., a shift in the maximum absorbance wavelength (λmax) towards a higher value [30].

Analyses of the surface colors of the indicator films using a colorimeter were performed on the 1st and 14th days after casting. Throughout the storage period, the SCh–A film showed increased L*, a*, and b* values. This indicates a change in the color of the film into brighter red-yellowish. The SCh–AC20 film showed increased L* and b* values, but the a* value decreased. This indicates that the color of the film turned lighter and yellower, but the redness was reduced. The SCh–AC28 film showed decreased L*, a*, and b* values, indicating that the color of the film grew darker, and that the reddish and yellowish colors of the film had decreased toward green and blue. The SCh–AC40 film showed decreased L* and a* values, but the b* value increased. This indicates that the color of the film became darker and yellower, while the redness decreased toward green. Based on the L*, a*, and b* values, the SCh–A film obtained the highest total color difference (ΔE), followed by SCh–AC20, SCh–AC28, and SCh–AC40. These results indicate that the most visible discoloration occurred on the SCh–A film, with decreases proportionate to the amount of gambier catechins added. The addition of gambier catechins as a co-pigment can form intramolecular bonds with anthocyanins, which increases the stability [31,32]. Anthocyanin molecules in the form of electron-poor flavylium cations can associate with catechin compounds as electron-rich co-pigments, so that the hydration reaction that changes the anthocyanin structure from colored flavylium cations to colorless chalcones can be prevented [33].

The results of the sensitivity analysis of the SCh–AC40 indicator film after dipping in pH 1–13 buffer showed a color change that was similar to the change that occurred to the color extract. The color change varied slightly, but it was clearly visible at pH 13 as a yellow color. Pramitasari et al. [15] previously found that an indicator film made from the anthocyanin extract of red dragon fruit peel experienced a color change to red–yellow at pH 12 and turned completely yellow at pH 13. This finding bears a similarity with the color change experienced by the SCh–AC40 film, which was analyzed for its sensitivity; therefore, it was concluded that dye based on the anthocyanin of red dragon fruit peel has the ability to change color and has the potential to be used as an indicator film [14].

The results of the SEM morphological analysis showed that the SCh–AC40 film had a heterogeneous structure, as indicated by the presence of highly visible gray and white aggregates. The higher concentration of gambier catechins added increased the amount of sedimentation of insoluble components in the film matrix captured by SEM. This is because gambier catechins are insoluble in cold water and form crystals in dry conditions [34]. The crystalline form of this catechin compound is an aggregate in the film matrix, which induces the formation of a heterogeneous film structure [35]. The results of the FTIR analysis show a similarity with the results of the research by González et al. [24], where the curve shows peaks at wavelengths of 3268 cm^{-1} (O–H stretching), 2926 cm^{-1} (C–H stretching), and 1633 cm^{-1} (O–H bending from water bonds). The band at 1411 cm^{-1} corresponds to CH$_2$ bending, the band at 1148 cm^{-1} corresponds to CO deformation, and the band at 1074 cm^{-1} corresponds to COH bending. In addition, the 924 cm^{-1} band is associated with the skeletal mode vibration of α-1,4-glycosidic, namely, the COC bond. The addition of gambier catechins did not cause any new bands, but it caused changes in the absorption area between 1000 and 500 cm^{-1}. This may be due to changes in the matrix vibration in CO caused by the addition of gambier catechins [24].

The indicator film thickness analysis results show that the SCh–A film was the thinnest of all the films. This finding is in line with the research by Santoso et al. [34], who reported that the addition of gambier catechins increases the thickness of the film as an added material is used. Gambier catechins, which can form crystals in dry conditions, cause the total solids in the film suspension to increase, so that the thickness of the cast film increases [34]. Based on the results of the analysis of the mechanical properties, the SCh–A film obtained higher TS and EAB values than SCh–AC20 and SCh–AC28. This indicates that the SCh–AC20 and SCh–AC28 films tended to be more brittle and stiffer. As explained by Santoso et al., the addition of gambier catechins can increase the thickness of the film, thereby reducing its tensile strength [34]. In addition, the conformation of the binding between the starch molecular chains and the formation of the double helix structure can hinder the relative displacement between the chains, which results in a decrease in the elongation at break value [36]. The SCh–AC40 film obtained the highest tensile strength and elongation at break values, indicating that it possessed the strongest and most elastic film properties. Chen et al. [36] reported that polyphenols in tea in a certain concentration can increase the film's tensile and resistance-to-stress properties without reducing its elongation ability. As catechins are natural resins suitable for use as reinforcing agents, they can improve the mechanical properties of the film [37]. In addition, gambier catechins have active hydroxyl groups (-OH); therefore, the higher the concentration of gambier catechins added, the higher the number of OH groups in the film matrix. The OH group can bind to starch molecules and interfere with the rolling and binding between starch molecules, giving it a significant role in increasing the mobility of the film matrix polymer chains, as well as the elasticity and elongation at break value of the film [34,36].

The results of the analysis of film transparency are supported by the results of light transmittance. It was found that the transmittance value of the SCh–AC40 film formed the gentlest curve of all the films, indicating that the SCh–AC40 film had the best ability to inhibit visible light at a wavelength of 380–700 nm. As for transparency, no significant difference was found between the treatments; however, it was found that the transparency value decreased as the concentration of gambier catechins increased, which is in line with the transmittance curve that is formed [29]. In addition, a higher number of solids in the film matrix, as well as greater thickness of the film could prevent light from penetrating the film surface better [28].

Based on the analysis of water content and film solubility, it was found that the SCh–A film had higher water content and solubility than the SCh–AC20, SCh–AC28, and SCh–AC40 films. The addition of gambier catechins could reduce the value of the water content and solubility of the film, which is similar to the findings of Gao et al. [38] that the water content of a film decreases as polyphenols from tea are added to the film. This is because the addition of gambier catechins increases the amount of polymer and the viscosity that forms the film matrix. As the polymer making up the film matrix increases, the total solids also increase, resulting in lower film water content [28]. It should be noted, however, that to a certain extent, the addition of catechins in large quantities can cause a resurgence in the water content of the film. When catechins interact with anthocyanins or other polymers, there will be free catechins, which causes an increase in the availability of OH groups. Based on the results of the analysis of film solubility, the higher the concentration of catechins, the lower the solubility value of the film [34]. This is because gambier catechins are difficult to dissolve in cold water.

According to the WVTR analysis results, the SCh–AC40 film had the lowest WVTR value of all the films, indicating that the SCh–AC40 film had the best ability as a water vapor barrier. This is in accordance with Ku et al.'s finding that the addition of catechins in the film matrix can reduce the WVTR value of the film. This could be because the addition of catechins causes changes in matrix interactions and modifications in the structure of the film formed [39].

From the antioxidant activity test results, it was found that the SCh–A film had the lowest antioxidant activity of all the films. In the SCh–A film, the antioxidant activity comes from chitosan and anthocyanin compounds. Anthocyanins, which are polyphenolic compounds, can act as antioxidants, while chitosan can prevent the initial stage of oxidation through the interaction between the free NH_2 group in the C_2 position of chitosan, which binds to the H^+ group from the film solution to form NH_3^+. This compound will then interact with the DPPH free radical to form a stable molecule [40]. Chen et al. previously reported that the addition of tea polyphenols, which is comparable to the gambier catechins in the case of this research, increased the antioxidant activity of the film [36]. According to research by Anggraini et al. [41], gambier catechins contain many polyphenolic compounds belonging to the flavonoid group, which act as good antioxidant compounds. Catechins can donate hydrogen atoms or their single electrons to radical compounds, which causes oxidation–reduction reactions to occur, so that radical compounds that initially had unpaired electrons now have their electrons paired and thus become non-radical [42].

The SCh–A, SCh–AC20, SCh–AC28, and SCh–AC40 films all showed color changes from red to dull red or red–yellow on the 3rd day of storage. This is in line with the research of Ardiyansyah et al., which reported an increase in pH on the 4th day of storage, indicating that the bananas were starting to ripen. The ripening process of bananas is marked by increases in the texture value, pH, and CO_2 gas produced and by a decrease in the vitamin C level in the fruit. The color change occurring on the indicator films is due to respiration or metabolism that takes place during the fruit ripening process, as well as the activity of enzymes and microorganisms during the fruit decomposition process. This process causes an increase in the pH of the environment to become alkaline, resulting in a change in the color of the indicator film to yellow [43]. In addition, this can be caused by the presence of other compounds, namely, betalain color pigments, which consist of betacyanin, a red–purple pigment, and betaxanthin, a yellow–orange pigment. Betalain pigments are stable under acidic pH conditions. As the pH increases, the structure of betalain can be degraded into colorless cyclo-DOPA 5-O-(malonyl)-β-glucoside and yellow betaxanthin [44].

6. Conclusions

The addition of gambier catechins to the anthocyanin color extract of red dragon fruit peel could successfully increase the color stability of the film without reducing the film's ability to change under certain pH conditions. The SCh–AC40 indicator film produced the best color stability, mechanical properties, light and water vapor barrier ability, and antioxidant activity. When applied to banana packaging, the SCh–AC40 indicator film showed a color change that was easily observable during storage, supported by changes in the environmental pH toward an alkaline condition. Based on the results obtained, the SCh–AC40 indicator film has the most potential to be developed as smart packaging for banana ripeness detection.

Author Contributions: Conceptualization, R.P. and D.S.B.A.; methodology, R.P., D.S.B.A. and V.R.S.; software, V.R.S.; validation, R.P., D.S.B.A. and V.R.S.; formal analysis, V.R.S., R.P. and D.S.B.A.; investigation, V.R.S., D.S.B.A. and R.P.; resources, R.P., D.S.B.A. and V.R.S.; data curation, R.P., D.S.B.A. and V.R.S.; writing—original draft preparation, V.R.S. and R.P.; writing—review and editing, R.P. and D.S.B.A.; visualization, V.R.S., R.P. and D.S.B.A.; supervision, R.P. and D.S.B.A.; project administration, R.P.; funding acquisition, R.P. and D.S.B.A. All authors have read and agreed to the published version of the manuscript.

Funding: This research was funded by the Penelitian Dasar Kompetitif Nasional (PDKN) grant from the Directorate General of Higher Education, Research, and Technology, Ministry of Education, Culture, Research, and Technology of the Republic of Indonesia, grant number SP DIPA-023.17.1690523/2023, 1415/LL3/AL.04/2023, 0388.13/III/LPPM.PM10.01-HD/5/2023.

Institutional Review Board Statement: Not applicable.

Data Availability Statement: The data presented in this study are available on request from the corresponding author.

Acknowledgments: The authors acknowledge the facilities and scientific and technical support from Advanced Characterization Laboratories Serpong, National research and innovation agency through E-Layanan Sains, Badan Riset dan Inovasi Nasional.

Conflicts of Interest: The authors declare no conflict of interest. The funder had no role in the design of the study; in the collection, analyses, or interpretation of data; in the writing of the manuscript; or in the decision to publish the results.

References

1. Badan Pusat Statistik: Produksi Tanaman Buah-Buahan 2022. Available online: https://www.bps.go.id/indicator/55/62/1/produksi-tanaman-buah-buahan.html (accessed on 27 August 2023).
2. Sastrahidayat, I.R.; Jauhary, S. *Introduction Study of Cavendish Banana and Its Pests and Diseases*; Brawijaya University Press: Malang, Indonesia, 2015.
3. Pramono, E.K. Measuring the maturity level of cavendish bananas based on the reflectance of led light. *J. Agric. Postharvest Res.* **2021**, *17*, 88–94.
4. Firouz, M.S.; Mohi-Alden, K.; Omid, M. A critical review on intelligent and active packaging in the food industry: Research and development. *Food Res. Int.* **2021**, *141*, 110113. [CrossRef]
5. Nature, A.U.; Rathi, P.; Beshai, H.; Sarabha, G.K.; Deen, M.J. Fruit quality monitoring with smart packaging. *Sensors* **2021**, *21*, 1509.
6. Dong, H.; Ling, Z.; Zhang, X.; Zhang, X.; Ramaswamy, S.; Xu, F. Smart colorimetric sensing films with high mechanical strength and hydrophobic properties for visual monitoring of shrimp and pork freshness. *Sens. Actuators B Chem.* **2020**, *309*, 127752. [CrossRef]
7. Khoo, H.E.; He, X.; Tang, Y.; Li, Z.; Li, C.; Zeng, Y.; Tang, J.; Sun, J. Betacyanins and anthocyanins in pulp and peel of red pitaya (*Hylocereus polyrhizus* cv. Jindu), inhibition of oxidative stress, lipid reducing, and cytotoxic effects. *Front. Nutr.* **2022**, *9*, 894438. [CrossRef]
8. Netravati; Gomez, S.; Pathrose, B.; Raj, M.; Joseph, M.P.; Kuruvila, B. Comparative evaluation of anthocyanin pigment yield and its attributes from butterfly pea (*Clitorea ternatea* L.) flowers as prospective food colorant using different extraction methods. *Future Foods* **2022**, *6*, 100199. [CrossRef]
9. Ekaputra, T.; Pramitasari, R. Evaluation of physicochemical properties of anthocyanin extracts and powders from purple sweet potato (*Ipomoea batatas* L.). *Food Res.* **2020**, *4*, 2020–2029. [CrossRef] [PubMed]
10. Zeng, F.; Ye, Y.; Liu, J.; Fei, P. Intelligent pH indicator composite film based on pectin/chitosan incorporated with black rice anthocyanins for meat freshness monitoring. *Food Chem. X* **2023**, *17*, 100531. [CrossRef]
11. Han, B.; Chen, P.; Guo, J.; Yu, H.; Zhong, S.; Li, D.; Liu, C.; Feng, Z.; Jiang, B. A novel intelligent indicator film: Preparation, characterization, and application. *Molecules* **2023**, *28*, 3384. [CrossRef]
12. Anugrah, D.S.B.; Delarosa, G.; Wangker, P.; Pramitasari, R.; Subali, D. Utilising n-glutaryl chitosan-based film with butterfly pea flower anthocyanin as a freshness indicator of chicken breast. *Packag. Technol. Sci.* **2023**, *36*, 681–697. [CrossRef]
13. Anugrah, D.S.B.A.; Darmalim, L.V.; Sinanu, J.D.; Pramitasari, R.; Subali, D.; Prasetyanto, E.A.; Cao, X.T. Development of alginate-based film incorporated with anthocyanins of red cabbage and zinc oxide nanoparticles as freshness indicator for prawns. *Int. J. Biol. Macromol.* **2023**, *251*, 126203. [CrossRef] [PubMed]
14. Azlim, N.A.; Mohammadi Nafchi, A.; Oladzadabbasabadi, N.; Ariffin, F.; Ghalambor, P.; Jafarzadeh, S.; Al-Hassan, A.A. Fabrication and characterization of a pH-sensitive intelligent film incorporating dragon fruit skin extract. *Food Sci. Nutr.* **2022**, *10*, 597–608. [CrossRef] [PubMed]
15. Pramitasari, R.; Gunawicahya, L.N.; Anugrah, D.S.B. Development of an indicator film based on cassava starch–chitosan incorporated with red dragon fruit peel anthocyanin extract. *Polymers* **2022**, *14*, 4142. [CrossRef] [PubMed]
16. Zhao, L.; Liu, Y.; Zhao, L.; Wang, Y. Anthocyanin-based pH-sensitive smart packaging films for monitoring food freshness. *J. Agric. Food Res.* **2022**, *9*, 100340. [CrossRef]
17. Pramitasari, R.; Marcel; Lestari, D. Co-pigmentation with catechin derived from Indonesian gambier increases the stability of black rice anthocyanin in isotonic sports drinks during one-month storage in 4 °C. *Food Res.* **2022**, *6*, 118–123. [CrossRef] [PubMed]
18. Rawdkuen, S.; Fasha, A.; Benjakul, S.; Kaewprachu, P. Application of anthocyanin as a color indicator in gelatin films. *Food Biosci.* **2020**, *36*, 100603. [CrossRef]
19. Jampani, C.; Raghavarao, K.S.M.S. Process integration for purification and concentration of red cabbage (*Brassica Oleracea* L.) anthocyanins. *Sep. Purif. Technol.* **2015**, *141*, 10–16. [CrossRef]
20. Tan, C.; Celli, G.B.; Selig, M.J.; Abbaspourrad, A. Catechin modulates the copigmentation and encapsulation of anthocyanins in polyelectrolyte complexes (PECs) for natural colorant stabilization. *Food Chem.* **2018**, *264*, 342–349. [CrossRef] [PubMed]

21. Alizadeh-Sani, M.; Tavassoli, M.; Mohammadian, E.; Ehsani, A.; Khaniki, G.J.; Priyadarshi, R.; Rhim, J.W. pH-responsive color indicator films based on methylcellulose/chitosan nanofiber and barberry anthocyanins for real-time monitoring of meat freshness. *Int. J. Biol. Macromol.* **2021**, *166*, 741–750. [CrossRef]
22. Shao, Y.; Wu, C.; Wu, T.; Li, Y.; Chen, S.; Yuan, C.; Hu, Y. Eugenol-chitosan nanoemulsions by ultrasound-mediated emulsification: Formulation, characterization and antimicrobial activity. *Carbohydr. Polym.* **2018**, *193*, 144–152. [CrossRef]
23. Bilgic, S.; Söğüt, E.; Seydim, A.C. Chitosan and starch based intelligent films with anthocyanins from eggplant to monitor pH variations. *Turk. JAF Sci.Technol.* **2019**, *7*, 61–66. [CrossRef]
24. Gonzalez, C.M.O.; Schelegueda, L.I.; Ruiz-Henestrosa, V.M.P.; Campos, C.A.; Basanta, M.F.; Gerschenson, L.N. Cassava starch films with anthocyanins and betalains from agroindustrial by-products: Their use for intelligent label development. *Foods* **2022**, *11*, 3361. [CrossRef] [PubMed]
25. López-Fandiño, R.; Otte, J.; Van Camp, J. Physiological, chemical and technological aspects of milk-protein-derived peptides with antihypertensive and ace-Inhibitory activity. *Int. Dairy J.* **2006**, *16*, 1277–1293. [CrossRef]
26. Iskandar, A.; Yuliasih, I.; Warsiki, E. Performance improvement of fruit ripeness smart label based on ammonium molybdate color indicators. *Indones. Food Sci. Technol. J.* **2020**, *3*, 48–57. [CrossRef]
27. Liu, W. Stepwise tests when test statistics are independent. *Aust. J. Stat.* **1997**, *39*, 169–177. [CrossRef]
28. Apriliyani, M.W.; Purwadi, P.; Manab, A.; Ikhwan, A.D. Characteristics of moisture content, swelling, opacity and transparency with the addition of chitosan as edible films/coating base on casein. *Adv. J. Food Sci. Technol.* **2020**, *18*, 9–14. [CrossRef]
29. Zhu, Y.; Chen, H.; Lou, L.; Chen, Y.; Ye, X.; Chen, J. Copigmentation effect of three phenolic acids on color and thermal stability of Chinese bayberry anthocyanins. *Food Sci. Nutr.* **2020**, *8*, 3234–3242. [CrossRef]
30. Gençdağ, E.; Özdemir, E.; Demirci, K.; Görgüç, A.; Yılmaz, F. Copigmentation and stabilization of anthocyanins using organic molecules and encapsulation techniques. *Curr. Plant Biol.* **2022**, *29*, 100238. [CrossRef]
31. Teixeira, N.; Cruz, L.; Brás, N.F.; Matthew, N.; Ramos, M.J.; de Freitas, V. Structural features of copigmentation of oenin with different polyphenol copigments. *J. Agric. Food Chem.* **2013**, *61*, 6942–6948. [CrossRef]
32. Trouillas, P.; Sancho-García, J.C.; De Freitas, V.; Gierschner, J.; Otyepka, M.; Dangles, O. Stabilizing and modulating color by copigmentation: Insights from theory and experiment. *Chem. Rev.* **2016**, *116*, 4937–4982. [CrossRef]
33. Escribano, T.; Santos Buelga, C. Anthocyanin copigmentation—Evaluation, mechanisms and implications for the color of red wines. *Curr. Org. Chem.* **2012**, *16*, 715–723. [CrossRef]
34. Santoso, B.; Marsega, A.; Priyanto, G.; Pambanyun, R. Improvement of physical, chemical and antibacterial characteristics of edible film based on *Canna edulis* Kerr. starch. *Agritech* **2017**, *36*, 378. [CrossRef]
35. Patil, P.; Pawar, S. Wound dressing applications of nano-biofilms. In *Biopolymer-Based Nano Films for Applications in Food Packaging and Wound Healing*; Elsevier: Baramati, India, 2021; pp. 247–268.
36. Chen, N.; Gao, H.-X.; Hey, Q.; Zeng, W.C. Potato starch-based film incorporated with tea polyphenols and its application in fruit packaging. *Polymers* **2023**, *15*, 588. [CrossRef] [PubMed]
37. Quader, F.; Khan, R.; Islam, M.A.; Saha, S.; Nazira Sharmin, K. Development and characterization of a biodegradable colored film based on starch and chitosan by using *Acacia catechu*. *J. Environ. Sci. Nat. Resour.* **2015**, *8*, 123–130. [CrossRef]
38. Gao, L.; Zhu, T.; Hey, F.; Ou, Z.; Xu, J.; Ren, L. Preparation and characterization of functional films based on chitosan and corn starch incorporated tea polyphenols. *Coatings* **2021**, *11*, 817. [CrossRef]
39. Ku, K.J.; Hong, Y.H.; Song, K.B. Mechanical properties of a *Gelidium corneum* edible film containing catechin and its application in sausages. *J. Food Sci.* **2008**, *73*, C217–C221. [CrossRef] [PubMed]
40. Zhang, X.; Liu, Y.; Yong, H.; Qin, Y.; Liu, J.; Liu, J. Development of multifunctional food packaging films based on chitosan, TiO_2 nanoparticles and anthocyanin-rich black plum peel extract. *Food Hydrocoll.* **2019**, *94*, 80–92. [CrossRef]
41. Anggraini, T.; Tai, A.; Yoshino, T.; Itani, T. Antioxidative activity and catechin content of four kinds of *Uncaria gambier* extracts from West Sumatra, Indonesia. *Afr. J. Biochem. Res.* **2011**, *5*, 33–38.
42. Fadhilah, Z.H.; Prime, F.; Syamsudin, R.A.M.R. Review: Telaah kandungan senyawa katekin dan epigalokatekin galat (EGCG) sebagai antioksidan pada berbagai jenis the. *J. Pharmascience* **2021**, *8*, 31–44. [CrossRef]
43. Ardiyansyah; Kurnianto, M.F.; Poerwanto, E.; Wahyono, A.; Apriliyanti, M.W.; Lestari, I.P. Monitoring of banana deteriorations using intelligent-packaging containing brazilian extract (*Caesalpina Sappan* L.). *IOP Conf. Ser. Earth Environ. Sci.* **2020**, *411*, 012043. [CrossRef]
44. Liu, D.; Zhang, C.; Pu, Y.; Chen, S.; Liu, L.; Cui, Z.; Zhong, Y. Recent advances in pH-responsive freshness indicators using natural food colorants to monitor food freshness. *Foods* **2022**, *11*, 1884. [CrossRef] [PubMed]

Disclaimer/Publisher's Note: The statements, opinions and data contained in all publications are solely those of the individual author(s) and contributor(s) and not of MDPI and/or the editor(s). MDPI and/or the editor(s) disclaim responsibility for any injury to people or property resulting from any ideas, methods, instructions or products referred to in the content.

Article

Criteria for Assessing Sustainability of Lignocellulosic Wastes: Applied to the Cellulose Nanofibril Packaging Production in the UK

Samantha Islam * and Jonathan M. Cullen

Department of Engineering, University of Cambridge, Trumpington Street, Cambridge CB2 1PZ, UK
* Correspondence: si313@cam.ac.uk

Abstract: Extensive use of petrochemical plastic packaging leads to the greenhouse gas emission and contamination to soil and oceans, posing major threats to the ecosystem. The packaging needs, hence, are shifting to bioplastics with natural degradability. Lignocellulose, the biomass from forest and agriculture, can produce cellulose nanofibrils (CNF), a biodegradable material with acceptable functional properties, that can make packaging among other products. Compared to primary sources, CNF extracted from lignocellulosic wastes reduces the feedstock cost without causing an extension to agriculture and associated emissions. Most of these low value feedstocks go to alternative applications, making their use in CNF packaging competitive. To transfer the waste materials from current practices to the packaging production, it is imperative to assess their sustainability, encompassing environmental and economic impacts along with the feedstock physical and chemical properties. A combined overview of these criteria is absent in the literature. This study consolidates thirteen attributes, delineating sustainability of lignocellulosic wastes for commercial CNF packaging production. These criteria data are gathered for the UK waste streams, and transformed into a quantitative matrix, evaluating the waste feedstock sustainability for CNF packaging production. The presented approach can be adopted to decision scenarios in bioplastics packaging conversion and waste management.

Keywords: biodegradable packaging; lignocellulose; cellulose nanofibrils; feedstock selection; sustainability assessment; waste management

Citation: Islam, S.; Cullen, J.M. Criteria for Assessing Sustainability of Lignocellulosic Wastes: Applied to the Cellulose Nanofibril Packaging Production in the UK. *Polymers* **2023**, *15*, 1336. https://doi.org/10.3390/polym15061336

Academic Editors: Beata Kaczmarek and Marcin Wekwejt

Received: 21 February 2023
Revised: 27 February 2023
Accepted: 28 February 2023
Published: 7 March 2023

Copyright: © 2023 by the authors. Licensee MDPI, Basel, Switzerland. This article is an open access article distributed under the terms and conditions of the Creative Commons Attribution (CC BY) license (https://creativecommons.org/licenses/by/4.0/).

1. Introduction

Plastics, the fossil-derived polymers, with strength, flexibility and durability, have wide range of applications, including packaging [1]. Packaging holds the largest global plastic market, presenting 36% of the overall demand in 2021 [2]. Most of the plastic packaging are single use and often end up in incineration or landfilling, causing major global greenhouse gas (GHG) emissions [1,3]. According to OECD [4], plastic life cycle globally accounts for 1.8 Gt CO_2-equivalent emissions in 2019, which is projected to grow to 4.3 Gt by 2060. When proper disposal does not take place, plastics often enter the terrestrial and marine environments, negatively impacting the ecosystems for thousands of years, due to being non-biodegradable [5,6].

These prevalent environmental impacts have led to the shift of packaging consumption towards bioplastics, derived from biological precursors (e.g., starch, cellulose, alginate, gelatin, collagen, proteins, chitosan, pectin) with natural biodegradability [3,7,8]. The global bioplastics production capacity standing at 2.42 mtonnes in 2021 is projected to grow to 7.59 mtonnes in 2026 [9]. Starch-blends derived from food crops (e.g., maize, sugarcane) dominate the commercial market of bioplastic feedstocks [10,11] but present a number of problems. Consumption of food crop feedstocks threatens food security, increasing both the market demand and price [12]. This also increases the use of land

and fertiliser with associated GHG emissions, negating the sought environmental benefits of bioplastics [13,14]. Moreover, material characteristics, e.g., poor mechanical and barrier properties of starch-based bioplastics make them an inferior alternative to their petrochemical counterparts [15,16].

Whereas, lignocellulose (LC), the biomass composites of cellulose, hemicellulose and lignin, deriving from forestry and agriculture, does not compete with food and are abundant in nature [17]. Some potential LC sources include: wood (softwood, hardwood), seed (cotton), bast (flax, hemp), leaf (sisal, brassica), stalk (wheat, barley) and grass/weed (miscanthus, Arabidopsis, bamboo) [18]. The cellulose fibres in LC are composed of microfibrils of 10–50 nm diameter, that in turn is comprised of elementary fibrils with a diameter of 3–5 nm, each of which consist of around 30–100 cellulose polymer chains (see Figure 1) [19]. Biosynthesis of LC can produce native nanocellulose materials: weblike cellulose nano fibrils (CNFs) and rodlike cellulose nanocrystals (CNCs) [17]. CNF gels, with larger surface areas, possess better film formation capability than CNC, and are therefore recommended for packaging applications [8,20].

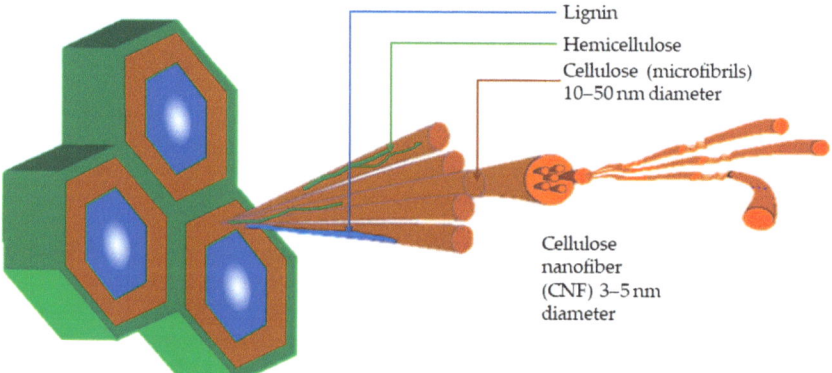

Figure 1. Hierarchical structure of cellulose fibres in LC cell wall.

Numerous studies discuss the use of CNF packaging films for: food, health care and various consumer goods [8,16,21]. These films are not only biodegradable and recyclable, but also demonstrate functionalities better or comparable to that of petrochemical polymers and other LC derivatives, e.g., regular paper [22,23]. CNF films demonstrate high mechanical strength, optical transmittance, thermal stability and gas (e.g., oxygen, air) barrier properties [24,25]. They also show better water vapor barrier properties than paper, though that remains somewhat lower compared to petrochemical plastics (e.g., polyolefins) [23]. This limits their application for packaging products with high moisture content (e.g., horticulture, fish, meat) and/or being stored at high relative humidity. However, this shortcoming could be overcome by various processes: incorporation of inorganic fillers (i.e., silver), chemical modification (i.e., plasma polymerization) and adsorption of other film matrix materials (e.g., chitosan) [26,27].

CNF packaging films are produced mainly in four generic steps (See Figure 2): (1) size reduction, e.g., chopping or grinding of LCs; (2) chemical/biological pre-treatment for removal of non-cellulosic compounds (e.g., lignin, hemicellulose) or modifying properties; (3) mechanical disintegration of cellulose; and (4) film preparation [17,28]. The CNF films can be either recycled or converted into compost, returning organic matter to the soil [29,30]. Ease of preparation, competitive properties and circular end-of-life treatments spur commercial interests in CNF packaging production [26,29]. Large-scale production ought to fulfil a major proportion of the global demand for flexible packaging that stood at 33.5 million metric tons in 2022 [31,32].

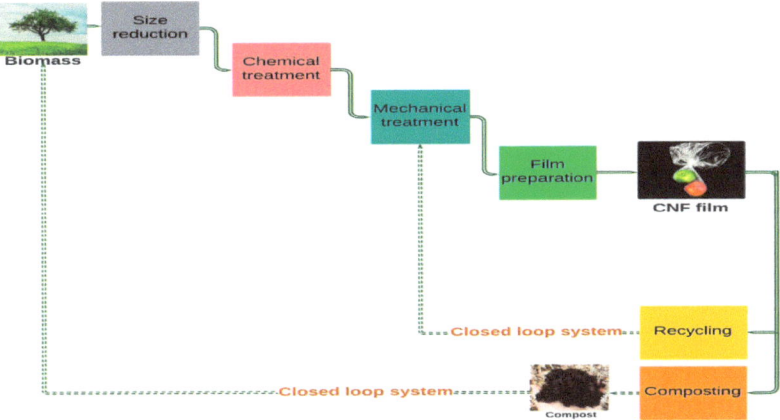

Figure 2. Generic CNF film processing from LC and end-of-life treatments.

Economic and environmental consequences are major obstacles for large-scale biodegradable packaging production [33,34]. Production of dedicated LC feedstock (i.e., purposefully cultivated for bioplastics) can lead to the land use changes as well as enhanced agricultural activities and fertilizer uses, causing a massive environmental burden [12,35,36]. Moreover, feedstock price, a major contributor to the LC processing cost, is higher for the dedicated biomass [34]. These present a need to identify more sustainable feedstock options for commercial CNF packaging production, providing environmental neutrality while maintaining the economic benefits [23].

Compared to dedicated LC, the use of lignocellulosic wastes, i.e., the leftover and eliminated substances of primary processes and applications, lowers the feedstock price and removes the need for land use changes, while producing CNFs with similar properties [37,38]. These wastes— comprising primary residues from forestry (e.g., bark, branches, stump) and agriculture (straw) as well as secondary wastes from municipality, businesses and industries (e.g., waste paper, saw dust, and waste food)— are collectively known as lignocellulosic waste and residue (LCW&R) (See Figure 3) [22,39]. CNFs can also be extracted from algae, bacteria and some animals (e.g., tunicates); however, this study focuses on the readily available, carbon neutral and low-cost feedstock alternative LCW&R [7,23,26].

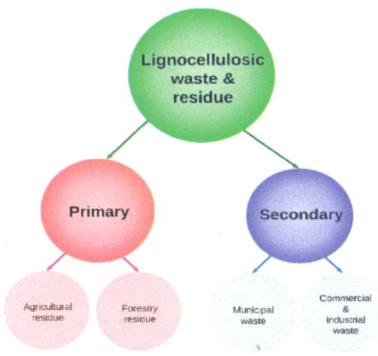

Figure 3. Typology of lignocellulosic wastes.

Many of the LCW&R options either have alternative uses, e.g., straw use in power generation, compost media, and animal bedding, or they go through different end-of-life treatments, e.g., incineration and landfilling of paper wastes [40,41]. Diversion of these materials from current uses and treatments to CNF packaging production requires an evaluation of their economic viability and emission mitigation efficacy [33,34,40]. In addition, it is also imperative to assess the feedstock technical characteristics, i.e., physical and chemical attributes that largely influence the properties and processing requirements for CNF-based packaging [42–44].

Existing studies in the literature discuss the impact of various feedstock criteria on CNF film properties, processing requirements and overall LC-based supply chains [20,24,42,45–47]. Shanmugam et al. [23] and Ang et al. [29] investigated how the mechanical properties, barrier and recycling performance of CNFs differ for the processed (i.e., dried) and virgin (i.e., never dried) LC. The impact of cell wall structure and composition (e.g., lignin, hemicellulose) of the LC on the CNF properties and process energy consumption were also examined by many authors [19,46,48–50]. Existing studies also relate the CNF production yield to the raw material carbohydrate composition [49,51], whereas for industrial LC processing, several studies [20,34,52,53] indicate the influence of feedstock physical properties (e.g., bulk density, durability) and price on the overall production cost. The impact of biomass supply chains on environment, soil and biodiversity were also widely analysed [35,40,54,55]. However, a study consolidating all the above criteria, defining the sustainability of LC wastes for large-scale CNF production, is still absent in the literature.

This study aims to coalesce the sustainability criteria, incorporating technical, economic, and environmental aspects of LCW&R for large-scale CNF packaging production. To this end, we adopted an iterative literature review and expert interviews, and identify thirteen relevant attributes. To demonstrate the use of this criteria pool, we collected the data on LCW&R streams in the UK and analysed how they perform across the given criteria. This helps in better understanding of their sustainability potential in their use for CNF packaging. The approach could be applied to various scenarios to support sustainable feedstock selection for bioplastic packaging and waste management decisions.

2. Materials and Methods

An iterative review of relevant academic and grey literature (e.g., reports, briefs and websites) followed by experts' interviews were conducted to identify the technical, economic and environmental criteria, defining the sustainability of LCW&R for the CNF packaging production. A total of thirteen criteria were identified, categorised as: positive/beneficial, whose higher values are desired; and negative/non-beneficial which were to be as low as possible. Availability (C1), physical composition (C2), cellulose content (C3), hemicellulose content (C4) and bulk density (C5) form the positive criteria, whereas the negative criteria are comprised of lignin content (C6), ash content (C7), cell wall thickness (C8), price of feedstock (C9), seasonal variability (C10), particle size (C11), environmental emission (C12), and soil and biodiversity impact (C13). These criteria are discussed in Section 3.

To demonstrate the waste feedstock sustainability evaluation based on these criteria, data for the current LCW&R streams (See Tables 1–3) were collated for the UK. While most of these criteria are objective and measurable that include C1, C3, C4, C5, C6, C7, C8, C9 and C11, others, i.e., C2, C10, C12, C13, are subjective. The quantitative values of the objective attributes are gathered from the existing sources while subjective criteria values are approximated using categorial scales shown in Table 4. The criteria values are then transformed into a coherent quantitative matrix, assessing the sustainability of different waste streams for CNF film production. The criteria values and the sustainability measuring matrix are shown in Section 3. Results. The criteria analysis for the UK waste streams are discussed below.

Table 1. LCW&R streams for primary agriculture residues.

	Wheat Straw	Barley Straw	Oat Straw	Oilseed Rape Straw
Animal bedding	Wheat straw → Animal bedding (F1)	Barley straw → Animal bedding (F6)	Oat straw → Animal bedding (F9)	—
Animal feed	Wheat straw → Animal feed (F2)	Barley straw → Animal feed (F7)	Oat straw → Animal feed (F10)	—
Heat & power	Wheat straw → Heat & power (F3)	—	—	Oilseed rape straw → Heat & power (F12)
Mushroom and carrot production	Wheat straw → Mushroom and carrot production (F4)	—	—	—
Soil incorporation	Wheat straw → Soil incorporation (F5)	Barley straw → Soil incorporation (F8)	Oat straw → Soil incorporation (F11)	Oilseed rape straw → Soil incorporation (F13)

Table 2. LCW&R streams for primary forest residues.

	Conifer Leftover	Broadleaf Leftover
Uncollected	Conifer leftover → Uncollected (F14)	Broadleaf leftover → Uncollected (F16)
Heat & power	Conifer leftover → Heat & power (F15)	Broadleaf leftover → Heat & power (F17)

Table 3. LCW&R streams for secondary municipal and industrial wastes.

	Paper and Cardboard Waste	Wood Waste	Organic Waste
Incineration with/out recovery	Paper and cardboard waste → Incineration with/out recovery (F18)	Wood waste → Incineration with/out recovery (F21)	Organic waste → Incineration with/out recovery (F25)
Recycling & reuse	Paper and cardboard waste → Recycling & reuse (F19)	Wood waste → Recycling & reuse (F22)	—
Backfilling	—	Wood waste → Backfilling (F23)	Organic waste → Backfilling (F26)
Landfilling	Paper and cardboard waste → Landfilling (F20)	Wood waste → Landfilling (F24)	Organic waste → Landfilling (F27)
Composting and anaerobic digestion	—	—	Organic waste → Composting and anaerobic digestion (F28)

Table 4. Scales used for quantitative estimation of subjective criteria.

Subjective Rating	Quantitative Rating
Physical composition	
Raw & homogenous	4
Raw & mixed	3
Raw & mixed to processed & mixed	2
Processed & mixed	1
Seasonal variability	
High	3
Medium	2
Low	1
Environmental emission	
Increase	1
Decrease or unchanged	0
Soil and biodiversity impact	
Yes	1
No	0

2.1. LCW&R Streams in the UK

In this paper the LCW&R streams refer to the flows of specific LC wastes to various end applications or treatments [56]. To indicate an LCW&R stream, we use the name of the waste material and their existing end use/treatment with a graphical rightward arrow ($\rightarrow$) in between, demonstrating the conversion direction (See Tables 1–3). Including end uses and treatments within the feedstock options helps to consider their differing environmental impacts as a criterion for feedstock selection. For example, diversion to CNF packaging from two waste streams—'Wheat straw $\rightarrow$ Animal bedding' and 'Wheat straw $\rightarrow$ Heat & power'—ought to result in different emission mitigations due to different end uses at present, though both comprise the same material (i.e., wheat straw). A total of 28 LCW&R streams were considered in this study and are denoted with alphanumeric code F1–F28 for the ease of the readers. The LCW&R streams in the UK under primary and secondary categories are discussed below.

2.1.1. Primary Agriculture Residues

The primary agriculture residues are comprised of crop stem, leaves, dead shoots and chaff; for simplification we use the term "straw" to generically denote the residues remaining after extracting grains. Straw is the second largest food supply chain waste and the cost of collection is relatively high [54]. In the UK, the major produced crops are: wheat, barley, oat and oilseed rape [40]. Residues from these crops have many alternative uses that compete and influence the uncertainty of their availability and price [57]. Even when they are not collected for specific applications, they house small insects and return nutrients to the soil [41].

Wheat straw, being less palatable, contributes a small fraction to 'animal feed' while the main uses are: 'animal bedding', 'heat and power', and 'mushroom and carrot production' [40,54,57]. Barley and oat straws, being highly nutritious, are mostly used in 'animal feed' with a small portion going to 'animal bedding', owing to higher price [54]. Oilseed rape straws are brittle and not ideal for 'animal bedding' but are increasingly being used for bioenergy, i.e., 'heat and power'. A proportion of all crop residues are left or chopped and ploughed back into the land, broadly considered here as 'soil incorporation'. Combining the four crop types with alternative uses, a total of 13 LCW&R streams were identified for the UK primary agriculture residues as shown in Table 1.

2.1.2. Primary Forest Residues

Forest residues are the mix of tree remains, i.e., bark, tops, branches, distorted wood, and in some cases stumps that are left after harvesting [40]. This biomass is expensive to collect and transport [41]. Moreover, extensive collection can cause soil erosion and risks to biodiversity [58].

Forest residues derive from two types of wood: hardwood that comes from broadleaved trees, such as oak, ash and beech; and softwood produced by coniferous trees, e.g., pine, fir, spruce and larch. Considering two material types (conifers, broadleaves) with two end applications (uncollected, heat and power), a total of four LCW&R streams were identified for the UK primary forest residues as shown in Table 2.

2.1.3. Secondary Municipal and Industrial Waste

The secondary LCW&R derive from the lignocellulosic municipal and industrial wastes. The three material groups presenting this waste category are: paper and cardboard, wood and organic. Unlike the primary residues of homogenous materials, secondary LC wastes are mostly comprised of processed and mixed material. Paper and cardboard waste includes paper and card packaging from businesses and households as well as sludges and rejects from the pulp and paper industries [59], whereas wooden packaging, saw dust, bark, chips and cuttings from these industries make up the wood waste. The key components of organic waste come from green and food wastes [59].

The waste treatment routes were identified from the 2018 UK national statistics [60]. The major paper and cardboard wastes go to 'recycling and reuse' which is followed by 'incineration with/out recovery' and 'landfilling'. Wood wastes also follow the same treatment routes although 'backfilling' is performed to some extent. A major portion of the organic wastes go to organic recycling, i.e., 'composting and anaerobic digestion'. The LCW&R streams for secondary municipal and industrial wastes are presented in Table 3.

2.2. LCW&R Streams Criteria Data Compilation

The positive and negative criteria data for the current LCW&R streams in the UK are gathered in Section 3. The data collection approach used for objective and subjective data are discussed below.

2.2.1. Objective Criteria Data

The quantitative values of objective criteria are gathered from the existing literature except for availability (C1) that is estimated based on both the literature and recent statistics as follows:

Availability of primary agriculture residues: The data on the UK straw availability are not reported, though the crop production data is publicly available [61]. To estimate the current amounts of dry crop residue in the UK, 2021 data on crop areas, yields, moisture content and harvest indices (i.e., the proportion of total dry crop biomass harvested as grain) were used [61–64]. The proportions of various straws' applications were then determined based on public datasets and the existing literature [40,54,65,66].

Availability of primary forest residues: Forest residues are not part of the UK national statistics. This study estimated the current dry wood residue biomass for the year 2021 from known forestry statistics [67,68] with the assumptions of harvest site area, wood density, moisture content and the ratio of harvest residues [40,41]. About 50% of the forest residues were considered "uncollected" to comply with the sustainable and good management practices, e.g., ensuring soil cover or adding organic fertilizers [41]. The only application identified for rest of the biomass (collected) was the production of "heat and power" through domestic and industrial combustion [41].

Availability of secondary municipal and industrial waste: To devise the secondary LCW&R streams, generation and treatments of non-hazardous municipal and industrial LC-based wastes in the UK 3454 considered. The latest dataset reporting this information derives from 2018 UK waste statistics [60]. No moisture content was assumed for paper and wood waste, though 82.5% moisture was considered for organic waste [69]. Waste statistics for later years were not used due to being incomplete, and not reflecting the standard waste management practices due to COVID-19.

2.2.2. Subjective Criteria Data

The subjective criteria values shown in Section 3 are defined by various terms based upon the literature and authors' perception. The four subjective criteria considered in this study are discussed below.

Physical composition: Four subjective ratings were used to define physical composition (C2) (See Section 3). The term 'raw and homogenous' is used for all the primary residues from forestry and agriculture. The other three ratings are used for the secondary waste streams: 'raw and mixed' for organic; 'processed and mixed' for paper; and an intermediate category between these two 'raw & mixed to processed & mixed' for wood wastes which were comprised of both processed and unprocessed materials.

Seasonal variability: Seasonal variability (C10), comprising three ratings (high, medium and low) defines three levels of uncertainty associated with the potential availability of the biomass (See Section 3).

Environmental emission: To gauge the change in environmental emission (C12) for feedstock diversion from current practices to CNF packaging, we used the EU Waste Hierarchy, i.e., an order of preference for waste management based on their environmental impact [70]. In this hierarchy, bioplastic production falls in the third step, i.e., reuse, recycling and composting [70]. The waste currently flowing to the treatments below the third step, i.e., energy recovery (i.e., combustion, incineration) and disposal (incineration, landfilling) are considered 'decrease' emission when diverted to bioplastic (i.e., CNF) production. The current practices that are likely to involve less processing and chemical use (e.g., feed and bedding material production) are considered 'increase' emissions when moved to CNF production [71]. All types of soil incorporation and composting are considered 'unchanged' emissions as the CNF end-of-life treatment can take the same route (See Section 3).

Soil and biodiversity impact: Soil and biodiversity impact (C13) only applies to the primary biomass extracted from nature. The primary residue, going to the soil, were considered 'yes' (i.e., having an impact on soil and biodiversity) for C13 when moved to CNF films production. The rest of the material streams were considered 'no' for C13 (See Section 3).

2.3. LCW&R Performance Matrix

Simple calculations are performed to devise the LCW&R performance matrix (See Section 3). All quantitative criteria values were converted to discrete numbers by taking the average if they are expressed as a range. The subjective attributes presented via qualitative data were approximated in discrete quantitative values using the categorial scales shown in Table 4. The data were then normalised to dimensionless indicators in a coherent scale of 0 to 1, using a technique described in the literature [72,73]. The values are presented via data bars in green and red colours for positive and negative criteria, respectively (See Section 3).

3. Results

This section presents the thirteen criteria of LCW&R comprising technical, economic and environmental aspects that collectively determine the feedstock sustainability for CNF packaging production. The criteria values were collated for the 28 LC-waste streams in the UK, as shown with the units of measurements in Tables 5 and 6. This was converted into a quantitative matrix in Figure 4, mapping sustainability performance of the waste streams along the criteria between 0 to 1.

3.1. Sustainability Criteria

The criteria, evaluating the sustainability of using LCW&R from their current practices to the CNF packaging production, are discussed below.

3.1.1. Availability (C1)

Feedstock availability refers to the maximum amount of LCW&R at hand for the CNF packaging production [34,40,74,75]. Knowing the material quantity flowing to various applications/processes at a given time helps to identify which LCW&R stream diversion can achieve economies of scale in the packaging production. Lack of consideration of the availability criteria may cause overstretch or underutilization of the waste material [76]. In the UK (Table 5), F19, i.e., paper and cardboard waste flowing to the recycling operations, presents the overall highest availability, although wheat straws from livestock bedding (F1) enables the maximum feedstock accessibility if primary residues are concerned.

3.1.2. Physical Composition (C2)

This criterion indicates whether an LC stream is comprised of raw, processed, homogenous or mixed materials, determining its requirements for handling or processing operations and the resulting bioplastic quality [77]. Refined biomass is different by chemical composition and processing history than its raw counterpart, and therefore results in CNF films differing in properties, processability or performance [28,29,46]. For example, recycled pulp (i.e., dried once), contrasting to virgin pulp (i.e., never-dried), produces CNF films with reduced tensile strength and swelling capacity, thereby reducing recyclability [28,48,78]. Whereas mixed wastes, e.g., food and garden waste in MSW, possessing heterogenous compositions may cause high costs, requiring more flexible and complex processing in the biorefineries, compared to their homogeneous fractions deriving from forestry and agriculture [79–81]. In the UK (Table 5), the primary waste streams (F1–F15) are likely to produce packaging films with better strength and recyclability, albeit using less processing compared to the processed and mixed wastes from flows F16–F28. The use of processed or refined biomass might be restricted in specific cases—such as for food packaging—since regulatory requirements do not allow the use of processed material due to containing harmful chemicals [82].

3.1.3. Cellulose (C3)

Cellulose is the main structural polysaccharide of LC cell walls, that consists of a linear chain of β (1→4)-linked d-glucose units. CNF is partially degraded cellulose with diameters in nanometre scales [17,83,84]. Therefore, the higher the cellulose content in a waste material, the greater the biomass-to-CNF yield. Cellulose has a high degree of polymerisation (DP), and high DP results in better tensile strength for the CNF sheets [85–87]. As is seen from Table 5 for the UK, wood residue streams (F14–F17), possessing more cellulose than non-wood residues (F1–F13), which ought to result in better yield and film strength.

3.1.4. Hemicellulose (C4)

Hemicellulose, the second major component of the cell wall, surrounds the cellulose microfibril bundles [83]. Hemicelluloses are branched polysaccharides, containing β-(1→4)-linked backbones of glucose, mannose or xylose in an equatorial configuration [88]. The carboxyl groups in hemicellulose, by the means of electrostatic repulsion, facilitate fibre delamination, reducing fibrillation energy and increasing biomass-to-CNF yields [49]. Additionally, entrenched around cellulose microfibrils with hydrogen bonds, hemicellulose seals the fibril gap and hinders fibril aggregation upon drying, resulting in enhanced film recyclability and cost-effective transportation [19,29,49,50]. The presence of hemicellulose also enhances film properties, e.g., strength and transparency [24,50,89]. Therefore, wood residue and waste from LC streams (F14–F17 and F21–24) in Table 5 (UK scenario), due to higher hemicellulose, should provide better CNF strength and higher production yield than their derivatives, i.e., paper and cardboard in F18–F20.

3.1.5. Bulk Density (C5)

Feedstock delivery cost accounts for 30–35% of the overall costs of an LC supply chain [90]. For cost-effective supply chain, bulk density, i.e., the amount of biomass fitting inside a cubic foot of space, plays a major role [91]. In essence, the greater the bulk density of a biomass, the less space it requires for transportation, handling and storage. Higher density materials require fewer vehicles, as more weight can be placed on each vehicle, reducing the cost of transportation. As is seen from Table 5 for the UK, supply chain costs for agricultural residues derived from F1–F13 is expected to be high, owing to their relatively lower bulk density.

3.1.6. Lignin (C6)

Lignin, a heterogeneous and irregular cross-linked polymer of phenyl propane, binds to cellulose microfibrils in the biomass cell wall [83,92]. With the complex structure, lignin causes biomass recalcitrance to chemical degradation, and restricts CNF extraction [22,83]. Therefore, biomass pre-treatment is performed to remove lignin. The success of the pre-treatment relies on maximum delignification with minimum cellulose loss. Hence, lower lignin composition indicates faster biomass delignification, lesser cellulose loss and lower temperature and chemical use, thereby providing reduced processing costs and energy [20,34]. Therefore, to reduce cost and energy of delignification, paper and organic waste in F18–F20 and F25–F28 (Table 6), with lower lignin contents, are preferred for the UK.

3.1.7. Ash (C7)

Ash refers to the biomass inorganic constituents, e.g., salts of nitrogen, potassium, magnesium, phosphorus, calcium, sulphur, zinc and silicon. Ash rises as the biomass storage period increases, and hence higher ash indicates less durable biomass [93]. Increased ash reduces biomass delignification efficacy, and leads to the wear of mechanical components, e.g., centrifugal pump heads and homogenisation valves [86,94]. During large-scale CNF production, major costs and environmental impacts derive from handling, transportation and disposal of residual ash [85,95]. To illustrate, the lower ash fraction of wood residue and waste (F14–F17 and F21–24) shown in Table 6, is an indication of reduced cost and environmental impact for the wood-based packaging supply chain in the UK.

3.1.8. Particle Size (C8)

Biomass particle size affects its processability and input consumption during the CNF production process [87]. A smaller biomass particle size provides increased specific surface area (surface area of per unit mass) that reduces processing time, and chemical and energy consumption [96]. Decrease in particle size also increases biomass bulk density, reducing the cost of handling and transportation [81,97]. Therefore, size reduction is recommended before transporting the LC to the processing sites [98,99]. In Table 6 for the UK context, we consider biomass as a bulk solid except for paper and cardboard waste, and particle size data were collected from the literature. All agricultural residues (F1–F13) regarded as the finest particles (chopped in 2.42–4.22 mm) ought to consume the least processing time and inputs for CNF packaging production.

3.1.9. Cell Wall Thickness (C9)

High cell wall thickness increases biomass recalcitrance and delays mechanical disintegration, increasing energy consumption [28,49]. Studies [28,46] report that softwood, with a relatively lower cell wall thickness, requires less mechanical treatment than hardwood to produce the equivalent fibrillation level. This observation also applies to non-wood plants; for example, sunflower plants with thinner cell walls takes less fibrillation time compared to alfa, i.e., Stipa tenacissima [49]. As is seen from Table 6, waste paper and cardboard (F18–F20) are considered to have no rigid cell wall, thereby consuming less mechanical energy in CNF packaging production compared to their precursor, i.e., wood (F14–F17 and F21–24) with stiff cell walls.

3.1.10. Price (C10)

Feedstock price is an important and sensitive cost component in the biomass production [34]. High feedstock price acts as a barrier against large-scale development [55]. The price of biomass consists of the costs of labour, energy and machineries that can vary based on location, season and demand [54,66,90]. In the UK, the price of municipal and industrial wastes (F18–F28) is almost zero, making them more cost effective compared to primary forestry and agricultural residues, i.e., F1–F17 (Table 6).

3.1.11. Seasonal Variability (C11)

A major fraction of biomass supply chain costs originates from storage operation, characterized by seasonal variability of the biomass supply [100]. For example, in the UK, the year-round supply of primary forestry residue is possible with small storage operations [40]. However, supply of agricultural residues is highly prone to seasonal uncertainty as straw is collected in a narrow window [40]. The LC composition of municipal food and garden waste is also influenced by the seasonal variation, leading to the requirements for specific storage conditions [101]. Aligning with these notions, the seasonal variability of wood residues and wastes (F14–F17 and F21–24) is regarded as 'low', while high seasonal variability is considered for agricultural residues (F1–F13) and so forth (Table 6).

3.1.12. Environmental Emission (C12)

This criterion indicates whether the relocation of LCW&R use to the CNF packaging production would increase, decrease or have no impact on emissions. To understand the emission change, the EU Waste Hierarchy was used as described in Section 2.2 [70]. Thus, in Table 6, diversion of wastes from F3, F15, F17, F18, F20–21, F24–25 and F27 to bioplastic production will 'decrease' emissions, whereas, F1–2, F4, F6, F7, F9, and F10 ought to result in emission increase. The rest of the material that are left or used in soil, are considered to result in no emissions change. For enhanced understanding of the relative emissions, a consequential life cycle assessment (LCA) can be adopted [35].

3.1.13. Soil and Biodiversity Impact (C13)

Harvesting primary residues can have significant impacts on soil and biodiversity. These residues are considered as important habitats for microorganisms, fungi, insects, and birds [58,102]. Excessive extraction of forest and agricultural biomass can reduce soil productivity, moisture retention and aeration [102,103]. To comply with the sustainable harvesting guidelines, limited extraction is performed in many countries; however, these rules do not constrain secondary waste use [40,58]. The residue portions that are left or intended for land incorporation (F5, F8, F11, F13, F14, F16 in Table 6 for the UK) will have an impact on soil and biodiversity if collected to produce CNF packaging. However, the waste streams already collected for various applications do not cause these impacts.

3.2. LCW&R Performance Matrix

Tables 5 and 6 is combined and converted into a performance matrix in Figure 4, evaluating how each LCW&R stream performs across the proposed criteria for the UK context. The normalised scores are shown via green and red data bars for the beneficial and nonbeneficial criteria, respectively.

Paper and cardboard wastes for recycling (F19) provide the highest feedstock availability (C1), with no increased emissions (C12) or soil and biodiversity impact (C13), although they may result in lack of film properties due to low hemicellulose (C4) and lacking in physical composition (C2). Among the waste streams with a higher C2 level, i.e., raw and homogenous, wheat straw from livestock bedding (F1) tops in availability (C1), although it will increase emissions (C12) when moved to CNF packaging production. The yield and mechanical properties are expected to be the highest for wood residues and wastes (F14–17 and F21–24) owing to their maximum cellulose (C3) and hemicellulose (C4) compositions, yet they will consume more energy in CNF processing due to the highest lignin content (C6) and cell wall thickness (C8). Among these wood streams, extraction of the uncollected residues (F14, F16) may increase the soil and biodiversity impact (C13). The secondary waste streams treated in incineration and landfilling (F18, F21, F25, F27), come at an almost negligible feedstock price (C9) and do not increase emissions (C12) or soil and biodiversity impacts (C13); nevertheless, they may increase the processing cost, being characterised with processed and mixed materials (C2).

Table 5. Positive criteria (C1–C5) values for LCW&R streams in the UK.

LCW&R	Criteria	C1. Availability (Dry Tonnes)	C2. Physical Composition (Subjective)	C3. Cellulose (wt%)	C4. Hemicellulose (wt%)	C5. Bulk Density (kg/m³)
F1. Wheat straw → Animal bedding		3073851.20	Raw & homogenous	33-40 [40]	20-25 [40]	36.22-39.74 [81]
F2. Wheat straw → Animal feed		81534.52	Raw & homogenous	33-40 [40]	20-25 [40]	36.22-39.74 [81]
F3. Wheat straw → Heat & power		364008.70	Raw & homogenous	33-40 [40]	20-25 [40]	36.22-39.74 [81]
F4. Wheat straw → Mushroom and carrot production		278933.87	Raw & homogenous	33-40 [40]	20-25 [40]	36.22-39.74 [81]
F5. Wheat straw → Soil incorporation		314048.90	Raw & homogenous	33-40 [40]	20-25 [40]	36.22-39.74 [81]
F6. Barley straw → Animal bedding		433491.83	Raw & homogenous	31-45 [40]	27-38 [40]	33.89-38.61 [81]
F7. Barley straw → Animal feed		542612.20	Raw & homogenous	31-45 [40]	27-38 [40]	33.89-38.61 [81]
F8. Barley straw → Soil incorporation		149533.75	Raw & homogenous	31-45 [40]	27-38 [40]	33.89-38.61 [81]
F9. Oat straw → Animal bedding		61739.75	Raw & homogenous	31-48 [40]	23-38 [40]	38.61-41.69 [81]
F10. Oat straw → Animal feed		227798.37	Raw & homogenous	31-48 [40]	23-38 [40]	38.61-41.69 [81]
F11. Oat straw → Soil incorporation		6052.35	Raw & homogenous	31-48 [40]	23-38 [40]	38.61-41.69 [81]
F12. Oilseed rape straw → Heat & power		133469.86	Raw & homogenous	35-40 [40]	27-31 [40]	47.46-49.7 [81]
F13. Oilseed rape straw → Soil incorporation		106883.32	Raw & homogenous	35-40 [40]	27-31 [40]	47.46-49.7 [81]
F14. Conifer leftovers → Uncollected		1156979	Raw & homogenous	35-45 [40]	25-30 [40]	128-267 [104]
F15. Conifer leftovers → Heat & power		1156979	Raw & homogenous	35-45 [40]	25-30 [40]	128-267 [104]
F16. Broadleaf leftovers → Uncollected		22231	Raw & homogenous	40-50 [40]	25-35 [40]	128-267 [104]
F17. Broadleaf leftovers → Heat & power		22231	Raw & homogenous	40-50 [40]	25-35 [40]	128-267 [104]
F18. Paper and cardboard waste → Incineration with/out recovery		3811.08	Processed & mixed	40-50 [105,106]	0-35 [105,106]	112 [107,108]
F19. Paper and cardboard waste → Recycling & reuse		3936954.05	Processed & mixed	40-50 [105,106]	0-35 [105,106]	112 [107,108]
F20. Paper and cardboard waste → Landfilling		5062.33	Processed & mixed	40-50 [105,106]	0-35 [105,106]	112 [107,108]
F21. Wood waste → Incineration with/out recovery		2536972.89	Raw & homogeneous to processed & mixed	40-50 [40]	25-35 [40]	128-267 [104]
F22. Wood waste → Recycling & reuse		2600381.03	Raw & homogeneous to processed & mixed	40-50 [40]	25-35 [40]	128-267 [104]

Table 5. Cont.

LCW&R	Criteria	C1. Availability (Dry Tonnes)	C2. Physical Composition (Subjective)	C3. Cellulose (wt%)	C4. Hemicellulose (wt%)	C5. Bulk Density (kg/m³)
F23. Wood waste → Backfilling		88781.00	Raw & homogeneous to processed & mixed	40–50 [40]	25–35 [40]	128–267 [104]
F24. Wood waste → Landfilling		22185.97	Raw & mixed to processed & mixed	40–50 [40]	25–35 [40]	128–267 [104]
F25. Organic waste → Incineration with/out recovery		13246.16	Raw & mixed	25.7–55.4 [40,109]	7.2–43 [40,109]	200–300 [110]
F26. Organic waste → Backfilling		2058	Raw & mixed	25.7–55.4 [40,109]	7.2–43 [40,109]	200–300 [110]
F27. Organic waste → Landfilling		14452.29	Raw & mixed	25.7–55.4 [40,109]	7.2–43 [40,109]	200–300 [110]
F28. Organic waste → Composting and anaerobic digestion		682814.19	Raw & mixed	25.7–55.4 [40,109]	7.2–43 [40,109]	200–300 [110]

Table 6. Negative criteria (C6–C13) values for LCW&R streams in the UK.

LCW&R	Criteria	C6. Lignin (wt%)	C7. Ash (wt%)	C8. Cell Wall Thickness (µm)	C9. Price of Feedstock (£/tonne)	C10. Seasonal Variability (Subjective)	C11. Particle Size (mm)	C12. Environmental Emission (Subjective)	C13. Soil and Biodiversity Impact (Subjective)
F1. Wheat straw → Animal bedding		15–21 [40]	3–10 [40]	3.96 [111]	39–105 [112]	High	4.22 (chopped) [81]	Increase	No
F2. Wheat straw → Animal feed		15–21 [40]	3–10 [40]	3.96 [111]	39–105 [112]	High	4.22 (chopped) [81]	Increase	No
F3. Wheat straw → Heat & power		15–21 [40]	3–10 [40]	3.96 [111]	39–105 [112]	High	4.22 (chopped) [81]	Decrease	No
F4. Wheat straw → Mushroom and carrot production		15–21 [40]	3–10 [40]	3.96 [111]	39–105 [112]	High	4.22 (chopped) [81]	Increase	No
F5. Wheat straw → Soil incorporation		15–21 [40]	3–10 [40]	3.96 [111]	39–105 [112]	High	4.22 (chopped) [81]	Unchanged	Yes
F6. Barley straw → Animal bedding		14–19 [40]	2–7 [40]	up to 2 [113]	45–108 [112]	High	3.37 (chopped) [81]	Increase	No
F7. Barley straw → Animal feed		14–19 [40]	2–7 [40]	up to 2 [113]	45–108 [112]	High	3.37 (chopped) [81]	Increase	No
F8. Barley straw → Soil incorporation		14–19 [40]	2–7 [40]	Up to 2 [113]	45–108 [112]	High	3.37 (chopped) [81]	Unchanged	Yes

Table 6. Cont.

Criteria LCW&R	C6. Lignin (wt%)	C7. Ash (wt%)	C8. Cell Wall Thickness (μm)	C9. Price of Feedstock (£/tonne)	C10. Seasonal Variability (Subjective)	C11. Particle Size (mm)	C12. Environmental Emission (Subjective)	C13. Soil and Biodiversity Impact (Subjective)
F9. Oat straw → Animal bedding	16–19 [40]	2–7 [40]	2–3.96 [114]	50–170 [112]	High	4.15 (chopped) [81]	Increase	No
F10. Oat straw → Animal feed	16–19 [40]	2–7 [40]	2–3.96 [114]	50–170 [112]	High	4.15 (chopped) [81]	Increase	No
F11. Oat straw → Soil incorporation	16–19 [40]	2–7 [40]	2–3.96 [114]	50–170 [112]	High	4.15 (chopped) [81]	Unchanged	Yes
F12. Oilseed rape straw → Heat & power	18–23 [40]	3–8 [40]	4.91 [115]	41–80 [112]	High	2.42 (chopped) [81]	Decrease	No
F13. Oilseed rape straw → Soil incorporation	18–23 [40]	3–8 [40]	4.91 [115]	41–80 [112]	High	2.42 (chopped) [81]	Unchanged	Yes
F14. Conifer leftover → Uncollected	25–35 [40]	1–3 [40]	2–8 [116]	35–60 [55]	Low	0–63 (chipped) [104]	Unchanged	Yes
F15. Conifer leftover → Heat & power	20–25 [40]	1–3 [40]	2–8 [116]	35–60 [55]	Low	0–63 (chipped) [104]	Decrease	No
F16. Broadleaf leftovers → Uncollected	20–25 [40]	1–3 [40]	1–11 [117]	35–60 [55]	Low	0–63 (chipped) [104]	Unchanged	Yes
F17. Broadleaf leftovers → Heat & power	0–30 [105,106]	1–3 [40]	1–11 [117]	35–60 [55]	Low	0–63 (chipped) [104]	Decrease	No
F18. Paper and cardboard waste → Incineration with/out recovery	0–30 [105,106]	0–35 [118,119]	Not applicable	Negligible [40]	Low	100–300 (baled) [107]	Decrease	No
F19. Paper and cardboard waste → Recycling & reuse	0–30 [105,106]	0–35 [118,119]	Not applicable	Negligible [40]	Low	100–300 (baled) [107]	Unchanged	No
F20. Paper and cardboard waste → Landfilling	0–30 [105,106]	0–35 [118,119]	Not applicable	Negligible [40]	Low	100–300 (baled) [107]	Decrease	No
F21. Wood waste → Incineration with/out recovery	20–35 [40]	1.0–3.0 [40]	1–11 [116,117]	Negligible [40]	Low	0–63 (chipped) [104]	Decrease	No
F22. Wood waste → Recycling & reuse	20–35 [40]	1.0–3.0 [40]	1–11 [116,117]	Negligible [40]	Low	0–63 (chipped) [104]	Unchanged	No
F23. Wood waste → Backfilling	20–35 [40]	1.0–3.0 (Used same as forest residues) [40]	1–11 [116,117]	Negligible [40]	Low	0–63 (chipped) [104]	Unchanged	No

Table 6. Cont.

LCW&R Criteria	C6. Lignin (wt%)	C7. Ash (wt%)	C8. Cell Wall Thickness (μm)	C9. Price of Feedstock (£/tonne)	C10. Seasonal Variability (Subjective)	C11. Particle Size (mm)	C12. Environmental Emission (Subjective)	C13. Soil and Biodiversity Impact (Subjective)
F24. Wood waste → Landfilling	3–35 [40]	1.0–3.0 (Used same as forest residues) [40]	1–11 [116,117]	Negligible [40]	Low	0–63 (chipped) [104]	Decrease	No
F25. Organic waste → Incineration with/out recovery	3–35 [40]	2.5–20 [120,121]	0.1–11 [122]	Negligible [40]	Medium to High	10–40 (shredded) [110]	Decrease	No
F26. Organic waste → Backfilling	25–35 [40]	2.5–20 [120,121]	0.1–11 [122]	Negligible [40]	Medium to High	10–40 (shredded) [110]	Unchanged	No
F27. Organic waste → Landfilling	3–35 [40]	2.5–20 [120,121]	0.1–11 [122]	Negligible [40]	Medium to High	10–40 (shredded) [110]	Decrease	No
F28. Organic waste → Composting and anaerobic digestion	3–35 [40]	2.5–20 [120,121]	0.1–11 [122]	Negligible [40]	Medium to High	10–40 (shredded) [110]	Unchanged	No

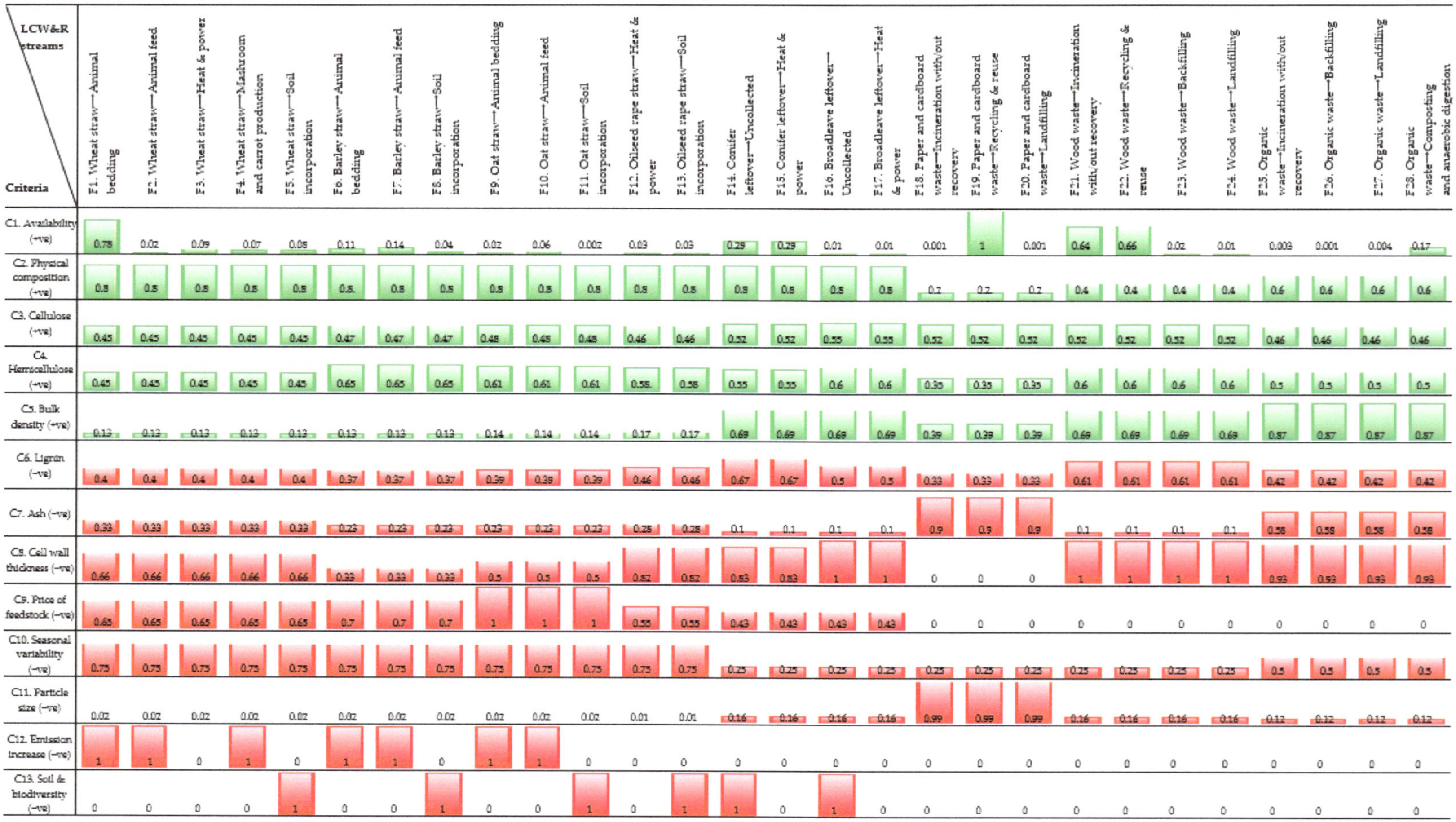

Figure 4. Performance matrix for the UK LCW&R streams. (+ve) and (−ve) are used to indicate positive and negative criteria, respectively.

4. Discussion

Sustainability assessment of LCW&R for CNF packaging production is a complex problem, requiring combinatorial consideration of technical, economic and environmental aspects [33,34,43]. These characteristics are mentioned disjointly across the literature [20,24,46,47], and a consolidated overview is absent. This study used an iterative literature review and experts' interviews, and identified 13 criteria pertaining to LC waste sustainability for CNF packaging production. The criteria list includes: availability, physical composition, cellulose content, hemicellulose content, bulk density, lignin content, ash content, cell wall thickness, price of feedstock, seasonal variability, particle size, environmental emission, and soil and biodiversity impact. These criteria data were collected for the LCW&R streams in the UK (Tables 5 and 6), and were combined into a coherent matrix to assess their performance (Figure 4). This study helps to uncover the sustainability potential of LCW&R for CNF packaging production, encompassing technical properties as well as environmental and economic criteria.

Feedstock attributes, influencing the properties and performance of the CNF packaging films, make up the technical criteria. They include physical composition, cellulose and hemicellulose contents that determine the strength, transparency, and recyclability of the packaging films. Most of the feedstock characteristics including availability, physical composition, bulk density, lignin and price, were found to have influence on the economic aspects of CNF packaging production, e.g., production yield, processing and supply chain costs, wear of machineries and raw material storage. Whereas environmental dimensions such as energy consumption, waste management, emission and soil or biodiversity impacts are correlated to certain feedstock criteria, e.g., bulk density, lignin, ash, particle size, emission, soil and biodiversity impact, and so on. Most of the sustainability assessment criteria were found to influence more than one aspect (technical, economic, environmental) of CNF packaging production (see Figure 5).

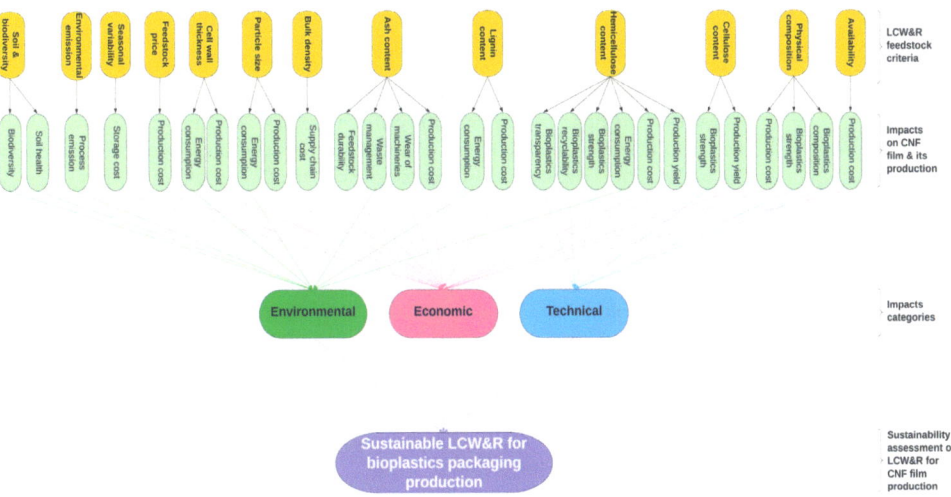

Figure 5. The thirteen feedstock criteria with impacts under technical, economic and environmental categories used for sustainability assessment of LCW&R for CNF packaging production.

Consideration of technical, economic and environmental factors under the sustainability umbrella presents a more comprehensive approach to feedstock sustainability assessment compared to the existing literature [24,35,53]. Although existing studies present a combined synopsis of feedstock criteria for other LC derivatives, e.g., biofuel and paper, a review of bioplastic feedstock criteria is a new contribution [40,85,86]. Moreover, the criteria list presented in this paper can be used to assess the existing waste material streams/flows instead of just the material itself, taking into account the environmental impact for replacing the current practices [33].

The LCW&R streams in the UK have been analysed based on the identified criteria and presented through a performance matrix (Tables 5 and 6 and Figure 4). Paper and cardboard wastes intended for recycling (F19) provides the highest feedstock availability, although they may result in lack of film mechanical properties. The highest mechanical properties can derive from wood residues and wastes (F14–17 and F21–24), but high process energy consumption will be a barrier. Moreover, the extraction of uncollected residues (F14, F16) may cause soil and biodiversity impact, although more fractions could be obtained from designated locations [58]. Diversion of secondary waste streams from incineration and landfilling (F18, F21, F25, F27) will reduce feedstock cost, environmental emissions and soil and biodiversity impacts; however, processing costs may increase due to the presence of heterogenous materials. The analysis technique used in this paper can be adopted in wide range of scenarios, requiring LC waste material diversion from existing uses to the production of CNF products including packaging.

This study explored the basic criteria for assessing sustainability of LC wastes in CNF packaging production. The results of this study will be considered in our forthcoming research on bioplastic feedstock decision analytics. Future research opportunity exists for consolidating empirical results to examine or enhance the proposed criteria. Further criteria distinctions can be considered based on: location, climatic conditions, plant species, crop cultivation, fibre location in plant, fibre age and presence of non-structural components (i.e., extractives). Moreover, the chemical pre-treatments and mechanical fibrillations used can influence the properties of the resulting CNF films, and thus can be integrated with feedstock criteria analysis to identify the overall sustainable routes for commercial CNF packaging production [42].

5. Conclusions

This study presents criteria for assessing the sustainability of LCW&R, incorporating technical, economic, and environmental aspects, for large-scale CNF packaging production. Thirteen relevant attributes were identified through an iterative literature review and expert interviews. Further, we gathered the criteria data for the UK waste streams and converted them into a performance matrix to measure the feedstock sustainability. This research will help to identify low-cost feedstocks and design biorefineries and supply chains for CNF packaging, replacing the petrochemical plastics. This study can also help in waste management decisions by identifying the waste material streams for which bioplastic packaging production and environmental emission reduction is complimentary rather than conflicting. This will support the inclusion of bioplastic processing in the national waste management plan, facilitating a circular bioeconomy.

Author Contributions: Conceptualization, S.I. and J.M.C.; methodology, S.I.; software, S.I.; validation, S.I.; formal analysis, S.I.; investigation, S.I.; data curation, S.I.; writing—original draft preparation, S.I.; writing—review and editing, S.I. and J.M.C.; visualization, S.I.; supervision, J.M.C.; project administration, S.I. and J.M.C.; funding acquisition, J.M.C. All authors have read and agreed to the published version of the manuscript.

Funding: This research was funded by the Natural Environment Research Council (NERC), grant number: NE/V010565/1. For the purpose of open access, the author has applied a Creative Commons Attribution (CC BY) licence to any Author Accepted Manuscript version arising.

Institutional Review Board Statement: Not applicable.

Informed Consent Statement: Informed consent was obtained from all interviewees involved in the study.

Data Availability Statement: The data presented in this study are available upon reasonable request from the corresponding author.

Conflicts of Interest: The authors declare no conflict of interest.

References

1. Assessment of Agricultural Plastics and their Sustainability—A Call for Action. Available online: https://www.fao.org/3/cb7856en/cb7856en.pdf (accessed on 7 January 2022).
2. Plastic Market Size, Share & Trends Report, 2022–2030. Available online: https://www.grandviewresearch.com/industry-analysis/global-plastics-market (accessed on 12 December 2022).
3. Plastics in the Bioeconomy (Issue 2). Available online: https://cdn.ricardo.com/ee/media/downloads/ed12430-bb-net-report-final-issue-2.pdf (accessed on 26 January 2022).
4. Plastic Leakage and Greenhouse Gas Emissions Are Increasing. Available online: https://www.oecd.org/environment/plastics/increased-plastic-leakage-and-greenhouse-gas-emissions.htm (accessed on 26 January 2022).
5. Barboza, L.G.A.; Cózar, A.; Gimenez, B.C.G.; Barros, T.L.; Kershaw, P.J.; Guilhermino, L. Macroplastics Pollution in the Marine Environment. In *World Seas: An Environmental Evaluation Volume III: Ecological Issues and Environmental Impacts*, 2nd ed.; Sheppard, C., Ed.; Academic Press: Cambridge, MA, USA, 2019; Volume 3, pp. 305–328. [CrossRef]
6. Peng, L.; Fu, D.; Qi, H.; Lan, C.Q.; Yu, H.; Ge, C. Micro- and nano-plastics in marine environment: Source, distribution and threats—A review. *Sci. Total Environ.* **2020**, *698*, 134254. [CrossRef] [PubMed]
7. Zhao, X.; Wang, Y.; Chen, X.; Yu, X.; Li, W.; Zhang, S.; Meng, X.; Zhao, Z.M.; Dong, T.; Anderson, A.; et al. Sustainable bioplastics derived from renewable natural resources for food packaging. *Matter* **2023**, *6*, 97–127. [CrossRef]
8. Lavrič, G.; Oberlintner, A.; Filipova, I.; Novak, U.; Likozar, B.; Vrabič-Brodnjak, U. Functional nanocellulose, alginate and chitosan nanocomposites designed as active film packaging materials. *Polymers* **2021**, *13*, 2523. [CrossRef] [PubMed]
9. Bioplastics Market Data. Available online: https://www.european-bioplastics.org/market (accessed on 26 January 2022).
10. Brizga, J.; Hubacek, K.; Feng, K. The unintended side effects of bioplastics: Carbon, land, and water footprints. *One Earth* **2020**, *3*, 45–53. [CrossRef]
11. Raj, T.; Chandrasekhar, K.; Naresh Kumar, A.; Kim, S.H. Lignocellulosic biomass as renewable feedstock for biodegradable and recyclable plastics production: A sustainable approach. *Renew. Sustain. Energy Rev.* **2022**, *158*, 112130. [CrossRef]
12. Pool, R. *The Nexus of Biofuels, Climate Change, and Human Health: Workshop Summary*; The National Academies of Sciences, Engineering, and Medicine: Washington, DC, USA, 2013.
13. Bishop, G.; Styles, D.; Lens, P.N.L. Environmental performance of bioplastic packaging on fresh food produce: A consequential life cycle assessment. *J. Clean. Prod.* **2021**, *317*, 128377. [CrossRef]
14. Garrido, F.J.O.; Piston, F.; Gomez, L.D.; Mcqueen-Mason, S.J. Biomass recalcitrance in barley, wheat and triticale straw: Correlation of biomass quality with classic agronomical traits. *PLoS ONE* **2018**, *13*, e0205880. [CrossRef]
15. Abe, M.M.; Martins, J.R.; Sanvezzo, P.B.; Macedo, J.V.; Branciforti, M.C.; Halley, P.; Botaro, V.R.; Brienzo, M. Advantages and disadvantages of bioplastics production from starch and lignocellulosic components. *Polymers* **2021**, *13*, 2484. [CrossRef]
16. Davis, G.; Song, J.H. Biodegradable packaging based on raw materials from crops and their impact on waste management. *Ind. Crops Prod.* **2006**, *23*, 147–161. [CrossRef]
17. Rojas, J.; Bedoya, M.; Ciro, Y. Current trends in the production of cellulose nanoparticles and nanocomposites for biomedical applications. In *Cellulose—Fundamental Aspects and Current Trends*; Poletto, M., Ed.; IntechOpen: London, UK, 2015; pp. 193–228. [CrossRef]
18. Petroudy, S.D. Physical and mechanical properties of natural fibers. In *Advanced High Strength Natural Fibre Composites in Construction*; Fan, M., Fu, F., Eds.; Woodhead Publishing: Cambridge, UK, 2017; pp. 59–83. [CrossRef]
19. Arola, S.; Malho, J.M.; Laaksonen, P.; Lille, M.; Linder, M.B. The role of hemicellulose in nano fibrillated cellulose networks. *Soft Matter* **2013**, *9*, 1319–1326. [CrossRef]
20. Lavoine, N.; Desloges, I.; Dufresne, A.; Bras, J. Microfibrillated cellulose—Its barrier properties and applications in cellulosic materials: A review. *Carbohydr. Polym.* **2012**, *90*, 735–764. [CrossRef] [PubMed]
21. Khalil, H.P.S.; Davoudpour, Y.; Saurabh, C.K.; Hossain, M.S.; Adnan, A.S.; Dungani, R.; Paridah, M.T.; Sarker, M.Z.I.; Fazita, M.R.N.; Syakir, M.I.; et al. A review on nanocellulosic fibres as new material for sustainable packaging: Process and applications. *Renew. Sustain. Energy Rev.* **2016**, *64*, 823–836. [CrossRef]
22. Rajinipriya, M.; Nagalakshmaiah, M.; Robert, M.; Elkoun, S. Importance of agricultural and industrial waste in the field of nanocellulose and recent industrial developments of wood based nanocellulose: A review. *ACS Publ.* **2018**, *6*, 2807–2828. [CrossRef]
23. Shanmugam, K.; Doosthosseini, H.; Varanasi, S.; Garnier, G.; Batchelor, W. Nanocellulose films as air and water vapour barriers: A recyclable and biodegradable alternative to polyolefin packaging. *Sustain. Mater. Technol.* **2019**, *22*, e00115. [CrossRef]
24. Kontturi, K.S.; Lee, K.Y.; Jones, M.P.; Sampson, W.W.; Bismarck, A.; Kontturi, E. Influence of biological origin on the tensile properties of cellulose nanopapers. *Cellulose* **2021**, *28*, 6619–6628. [CrossRef]
25. Azeredo, H.M.C.; Rosa, M.F.; Mattoso, L.H.C. Nanocellulose in bio-based food packaging applications. *Ind. Crops Prod.* **2017**, *97*, 664–671. [CrossRef]

26. Kalia, S.; Boufi, S.; Celli, A.; Kango, S. Nanofibrillated cellulose: Surface modification and potential applications. *Colloid Polym. Sci.* **2014**, *292*, 5–31. [CrossRef]
27. Gómez, H.C.; Serpa, A.; Velásquez-Cock, J.; Gañán, P.; Castro, C.; Vélez, L.; Zuluaga, R. Vegetable nanocellulose in food science: A review. *Food Hydrocoll.* **2016**, *57*, 178–186. [CrossRef]
28. Nechyporchuk, O.; Belgacem, M.N.; Bras, J. Production of cellulose nanofibrils: A review of recent advances. *Ind. Crops Prod.* **2016**, *93*, 2–25. [CrossRef]
29. Ang, S.; Ghosh, D.; Haritos, V.; Batchelor, W. Recycling cellulose nanofibers from wood pulps provides drainage improvements for high strength sheets in papermaking. *J. Clean. Prod.* **2021**, *312*, 127731. [CrossRef]
30. Vikman, M.; Vartiainen, J.; Tsitko, I.; Korhonen, P. Biodegradability and Compostability of Nanofibrillar Cellulose-Based Products. *J. Polym. Environ.* **2015**, *23*, 206–215. [CrossRef]
31. Statista. Flexible Packaging Global Production Volume 2017–2022. Available online: https://www.statista.com/statistics/719097/production-volume-of-the-global-flexible-packaging-industry/ (accessed on 14 February 2023).
32. Stark, N.M. Opportunities for Cellulose Nanomaterials in Packaging Films: A Review and Future Trends. *J. Renew. Mater.* **2016**, *4*, 313. [CrossRef]
33. Bishop, G.; Styles, D.; Lens, P.N.L. Land-use change and valorisation of feedstock side-streams determine the climate mitigation potential of bioplastics. *Resour. Conserv. Recycl.* **2022**, *180*, 106185. [CrossRef]
34. Bussemaker, M.J.; Day, K.; Drage, G.; Cecelja, F. Supply chain optimisation for an ultrasound-organosolv lignocellulosic biorefinery: Impact of technology choices. *Waste Biomass Valorization* **2017**, *8*, 2247–2261. [CrossRef] [PubMed]
35. Tonini, D.; Hamelin, L.; Astrup, T.F. Environmental implications of the use of agro-industrial residues for biorefineries: Application of a deterministic model for indirect land-use changes. *GCB Bioenergy* **2016**, *8*, 690–706. [CrossRef]
36. Badgujar, K.C.; Bhanage, B.M. Dedicated and waste feedstocks for biorefinery: An approach to develop a sustainable society. In *Waste Biorefinery: Potential and Perspectives*; Bhaskar, T., Pandey, A., Mohan, S.V., Lee, D.J., Khanal, S.K., Eds.; Elsevier: Amsterdam, The Netherlands, 2018; pp. 3–38. [CrossRef]
37. Piemonte, V.; Gironi, F. Land-use change emissions: How green are the bioplastics? *Environ. Prog. Sustain. Energy* **2011**, *30*, 685–691. [CrossRef]
38. Jonoobi, M.; Mathew, A.P.; Oksman, K. Natural resources and residues for production of bionanomaterials. In *Handbook of Green Materials: 1 Bionanomaterials: Separation Processes, Characterization and Properties*; Oksman, K., Mathew, A.P., Bismarck, A., Rojas, O., Sain, M., Eds.; World Scientific: Singapore, 2014; pp. 19–33. [CrossRef]
39. United Kingdom Roadmap for Lignocellulosic Biomass and Relevant Policies for a Bio-Based Economy in 2030. Available online: https://www.s2biom.eu/images/Publications/WP8_Country_Outlook/Final_Roadmaps_March/S2Biom-UNITED-KINGDOM-biomass-potential-and-policies.pdf (accessed on 4 January 2022).
40. Lignocellulosic Feedstock in the UK. Available online: https://www.nnfcc.co.uk/files/mydocs/LBNet%20Lignocellulosic%20feedstockin%20the%20UK_Nov%202014.pdf (accessed on 4 January 2022).
41. Availability of Cellulosic Residues and Wastes in the EU—International Council on Clean Transportation. Available online: https://www.theicct.org/publications/availability-cellulosic-residues-and-wastes-eu (accessed on 16 July 2022).
42. Balea, A.; Fuente, E.; Tarrés, Q.; Pèlach, M.À.; Mutjé, P.; Delgado-Aguilar, M.; Blanco, A.; Negro, C. Influence of pretreatment and mechanical nanofibrillation energy on properties of nanofibers from Aspen cellulose. *Cellulose* **2021**, *28*, 9187–9206. [CrossRef]
43. Bosworth, S.C. Perennial grass biomass production and utilization. In *Bioenergy: Biomass to Biofuels and Waste to Energy*; Dahiya, A., Ed.; Academic Press: Cambridge, MA, USA, 2020; pp. 89–105. [CrossRef]
44. Malucelli, L.C.; Lacerda, L.G.; Dziedzic, M.; da Silva Carvalho Filho, M.A. Preparation, properties and future perspectives of nanocrystals from agro-industrial residues: A review of recent research. *Rev. Environ. Sci. Biotechnol.* **2017**, *16*, 131–145. [CrossRef]
45. Bian, H.; Yang, Y.; Tu, P.; Bian, H.; Yang, Y.; Tu, P.; Chen, J.Y. Value-added utilization of wheat straw: From cellulose and cellulose nanofiber to all-cellulose nanocomposite film. *Membranes* **2022**, *12*, 475. [CrossRef]
46. Stelte, W.; Sanadi, A.R. Preparation and characterization of cellulose nanofibers from two commercial hardwood and softwood pulps. *Ind. Eng. Chem. Res.* **2009**, *48*, 11211–11219. [CrossRef]
47. Woiciechowski, A.L.; José, C.; Neto, D.; Porto De Souza Vandenberghe, L.; De Carvalho Neto, P.; Novak Sydney, A.C.; Letti, A.J.; Karp, S.G.; Alberto, L.; Torres, Z.; et al. Lignocellulosic biomass: Acid and alkaline pretreatments and their effects on biomass recalcitrance-Conventional processing and recent advances. *Bioresour. Technol.* **2020**, *304*, 122848. [CrossRef]
48. Blanco, A.; Monte, M.C.; Campano, C.; Balea, A.; Merayo, N.; Negro, C. Nanocellulose for Industrial Use: Cellulose Nanofibers (CNF), Cellulose Nanocrystals (CNC), and Bacterial Cellulose (BC). In *Handbook of Nanomaterials for Industrial Applications*; Hussain, C.M., Ed.; Elsevier: Amsterdam, The Netherlands, 2018; pp. 74–126. [CrossRef]
49. Chaker, A.; Alila, S.; Mutjé, P.; Vilar, M.R.; Boufi, S. Key role of the hemicellulose content and the cell morphology on the nanofibrillation effectiveness of cellulose pulps. *Cellulose* **2013**, *20*, 2863–2875. [CrossRef]
50. Iwamoto, S.; Abe, K.; Yano, H. The effect of hemicelluloses on wood pulp nano fibrillation and nanofiber network characteristics. *Biomacromolecules* **2008**, *9*, 1022–1026. [CrossRef]
51. Khalil, H.P.S.; Adnan, A.S.; Yahya, E.B.; Olaiya, N.G.; Safrida, S.; Hossain, M.S.; Balakrishnan, V.; Gopakumar, D.A.; Abdullah, C.K.; Oyekanmi, A.A.; et al. A review on plant cellulose nanofibre-based aerogels for biomedical applications. *Polymers* **2020**, *12*, 1759. [CrossRef] [PubMed]

52. Pre-Treatments to Enhance the Enzymatic Saccharification of Lignocellulose: Technological and Economic Aspects. Available online: https://www.bbnet-nibb.co.uk/resource/pre-treatments-to-enhance-the-enzymatic-saccharification-of-lignocellulose-technological-and-economic-aspects/ (accessed on 14 May 2022).
53. van Dyken, S.; Bakken, B.H.; Skjelbred, H.I. Linear mixed-integer models for biomass supply chains with transport, storage and processing. *Energy* **2010**, *35*, 1338–1350. [CrossRef]
54. Agrocycle Factsheet: Straw Production and Value Chains. Available online: https://www.nnfcc.co.uk/files/mydocs/Straw%20factsheet.pdf (accessed on 4 June 2022).
55. Use of Sustainably Sourced Residue and Waste Streams for Advanced Biofuel Production in the European Union: Rural Economic Impacts and Potential for Job Creation. Available online: https://www.nnfcc.co.uk/files/mydocs/14_2_18%20%20ECF%20Advanced%20Biofuels_NNFCC%20published%20v2.pdf (accessed on 23 March 2022).
56. Understanding Waste Streams: Treatment of Specific Waste. Available online: https://www.europarl.europa.eu/EPRS/EPRS-Briefing-564398-Understanding-waste-streams-FINAL.pdf (accessed on 2 February 2022).
57. Straw Prices Soar, Piling Pressure on Northern European Livestock Farmers. Available online: https://www.euractiv.com/section/agriculture-food/news/straw-prices-soar-piling-pressure-on-northern-europe-livestock-farmers/ (accessed on 15 April 2022).
58. Titus, B.D.; Brown, K.; Helmisaari, H.S.; Vanguelova, E.; Stupak, I.; Evans, A.; Clarke, N.; Guidi, C.; Bruckman, V.J.; Varnagiryte-Kabasinskiene, I.; et al. Sustainable forest biomass: A review of current residue harvesting guidelines. *Energy Sustain. Soc.* **2021**, *11*, 10. [CrossRef]
59. Guidance on Classification of Waste According to EWC-Stat Categories—Supplement to the Manual for the Implementation of the Regulation (EC) No 2150/2002 on Waste Statistics. Available online: https://ec.europa.eu/eurostat/documents/342366/351806/Guidance-on-EWCStat-categories-2010.pdf/0e7cd3fc-c05c-47a7-818f-1c2421e55604 (accessed on 25 March 2022).
60. UK Statistics on Waste—gov.uk. Office for National Statistics. Available online: https://www.gov.uk/government/statistics/uk-waste-data (accessed on 1 September 2022).
61. Agriculture in the United Kingdom Data Sets. Available online: https://www.gov.uk/government/statistical-data-sets/agriculture-in-the-united-kingdom (accessed on 3 September 2022).
62. The Main Components of Yield in Wheat. Available online: https://ahdb.org.uk/knowledge-library/the-main-components-of-yield-in-wheat (accessed on 17 December 2022).
63. The Main Components of Yield in Barley. Available online: https://ahdb.org.uk/knowledge-library/the-main-components-of-yield-in-barley (accessed on 17 December 2022).
64. Oat Growth Guide: An Output from Optimising Growth to Maximise Yield and Quality. Available online: https://www.hutton.ac.uk/sites/default/files/files/publications/Oat-Growth-Guide.pdf (accessed on 17 December 2022).
65. Plant Biomass: Miscanthus, Short Rotation Coppice and Straw. Available online: https://www.gov.uk/government/statistics/area-of-crops-grown-for-bioenergy-in-england-and-the-uk-2008-2020/section-2-plant-biomass-miscanthus-short-rotation-coppice-and-straw (accessed on 7 August 2022).
66. Straw and Forage Study. Available online: https://www.gov.scot/binaries/content/documents/govscot/publications/factsheet/2018/04/straw-and-forage-study-sruc-research-report/documents/straw-forage-study-sruc-report-2017-2018-pdf/straw-forage-study-sruc-report-2017-2018-pdf/govscot%3Adocument/Straw%2Band%2Bforage%2Bstudy%2B-%2BSRUC%2Breport%2B2017-2018.pdf (accessed on 5 August 2022).
67. 25-Year Forecast of Softwood Timber Availability. Available online: https://www.forestresearch.gov.uk/publications/25-year-forecast-of-softwood-timber-availability/ (accessed on 30 January 2022).
68. Forestry Statistics 2018—Forest Research. Available online: https://www.forestresearch.gov.uk/tools-and-resources/statistics/forestry-statistics/forestry-statistics-2021/ (accessed on 30 January 2022).
69. Compost Moisture Content. Available online: http://www.carryoncomposting.com/416920216 (accessed on 20 November 2022).
70. Arena, U.; di Gregorio, F. A waste management planning based on substance flow analysis. *Resour. Conserv. Recycl.* **2014**, *85*, 54–66. [CrossRef]
71. Owen, E.; Jayasuriyat, C.N. Use of crop residues as animal feeds in developing countries. *Res. Dev. Agric.* **1989**, *6*, 129–138.
72. Islam, S.; Ponnambalam, S.G.; Lam, H.L. A novel framework for analysing the green value of food supply chain based on life cycle assessment. *Clean Technol. Environ. Policy* **2017**, *19*, 93–103. [CrossRef]
73. Olinto, A.C.; Islam, S. Optimal aggregate sustainability assessment of total and selected factors of industrial processes. *Clean Technol. Environ. Policy* **2017**, *19*, 1791–1797. [CrossRef]
74. Hughes, S.R.; Qureshi, N. Biomass for biorefining: Resources, allocation, utilization, and policies. In *Biorefineries: Integrated Biochemical Processes for Liquid Biofuels*; Qureshi, N., Hodge, D.B., Vertès, A.A., Eds.; Elsevier: Amsterdam, The Netherlands, 2014; pp. 37–58. [CrossRef]
75. Bhatia, S.K.; Otari, S.V.; Jeon, J.M.; Gurav, R.; Choi, Y.K.; Bhatia, R.K.; Pugazhendhi, A.; Kumar, V.; Rajesh Banu, J.; Yoon, J.J.; et al. Biowaste-to-bioplastic (polyhydroxyalkanoates): Conversion technologies, strategies, challenges, and perspective. *Bioresour. Technol.* **2021**, *326*, 124733. [CrossRef] [PubMed]
76. Is Resource Availability Slowing You Down? Available online: https://www.google.com/search?q=Is+Resource+Availability+Slowing+you+Down%3F&rlz=1C1GCEU_enGB842GB842&oq=Is+Resource+Availability+Slowing+you+Down%3F&aqs=chrome..69i57j69i60.1250j0j4&sourceid=chrome&ie=UTF-8 (accessed on 5 February 2022).

77. Abdel-Shafy, H.I.; Mansour, M.S.M. Solid waste issue: Sources, composition, disposal, recycling, and valorization. *Egypt. J. Pet.* **2018**, *27*, 1275–1290. [CrossRef]
78. The Effect of Age and Recycling on Paper Quality. Available online: https://scholarworks.wmich.edu/cgi/viewcontent.cgi?article=5957&context=masters_theses (accessed on 5 February 2022).
79. Awoyale, A.A.; Lokhat, D.; Okete, P. Investigation of the effects of pretreatment on the elemental composition of ash derived from selected Nigerian lignocellulosic biomass. *Sci. Rep.* **2021**, *11*, 21313. [CrossRef]
80. Di Pretoro, A.; Montastruc, L.; Manenti, F.; Joulia, X. Flexibility assessment of a biorefinery distillation train: Optimal design under uncertain conditions. *Comput. Chem. Eng.* **2020**, *138*, 106831. [CrossRef]
81. Tumuluru, J.S.; Tabil, L.G.; Song, Y.; Iroba, K.L.; Meda, V. Grinding energy and physical properties of chopped and hammer-milled barley, wheat, oat, and canola straws. *Biomass Bioenergy* **2014**, *60*, 58–67. [CrossRef]
82. Food Contact Materials Authorisation Guidance. Available online: https://www.food.gov.uk/business-guidance/regulated-products/food-contact-materials-guidance (accessed on 15 December 2022).
83. Abdel-Hamid, A.M.; Solbiati, J.O.; Cann, I.K.O. Insights into lignin degradation and its potential industrial applications. *Adv. Appl. Microbiol.* **2013**, *82*, 1–28. [CrossRef] [PubMed]
84. Klemm, D.; Heublein, B.; Fink, H.P.; Bohn, A.; Klemm, D.; Fink, H.-P. Cellulose: Fascinating biopolymer and sustainable raw material. *Angew. Chem. Int. Ed.* **2005**, *44*, 3358–3393. [CrossRef]
85. Abara Mangasha, L. Review on Effect of Some Selected Wood Properties on Pulp and Paper Properties. *J. For. Environ.* **2019**, *1*, 16–22. [CrossRef]
86. Anupam, K.; Lal, P.S.; Bist, V.; Sharma, A.K.; Swaroop, V. Raw material selection for pulping and papermaking using TOPSIS multiple criteria decision-making design. *Environ. Prog. Sustain. Energy* **2014**, *33*, 1034–1041. [CrossRef]
87. Zhang, Q.; Zhang, P.; Pei, Z.J.; Wang, D. Relationships between cellulosic biomass particle size and enzymatic hydrolysis sugar yield: Analysis of inconsistent reports in the literature. *Renew. Energy* **2013**, *60*, 127–136. [CrossRef]
88. Scheller, H.V.; Ulvskov, P. Hemicelluloses. *Annu. Rev. Plant Biol.* **2010**, *61*, 263–289. [CrossRef] [PubMed]
89. Tenhunen, T.M.; Peresin, M.S.; Penttilä, P.A.; Pere, J.; Serimaa, R.; Tammelin, T. Significance of xylan on the stability and water interactions of cellulosic nanofibrils. *React. Funct. Polym.* **2014**, *85*, 157–166. [CrossRef]
90. Kumar, A.; Sokhansanj, S.; Flynn, P.C. Development of a multicriteria assessment model for ranking biomass feedstock collection and transportation systems. *Appl. Biochem. Biotechnol.* **2006**, *129*, 71–87. [CrossRef] [PubMed]
91. Bulk Density Impacts on the Supply Chain. Available online: https://generainc.com/bulk-density-impacts-on-the-supply-chain/ (accessed on 15 December 2022).
92. Sannigrahi, P.; Pu, Y.; Ragauskas, A. Cellulosic biorefineries-unleashing lignin opportunities. *Curr. Opin. Environ. Sustain.* **2010**, *2*, 383–393. [CrossRef]
93. Ogden, C.A.; Ileleji, K.E.; Johnson, K.D. Fuel property changes of switchgrass during one-year of outdoor storage. *Biomass Bioenergy* **2019**, *120*, 359–366. [CrossRef]
94. Pennells, J.; Godwin, I.D.; Amiralian, N.; Martin, D.J. Trends in the production of cellulose nanofibers from non-wood sources. *Cellulose* **2020**, *27*, 575–593. [CrossRef]
95. James, A.K.; Thring, R.W.; Helle, S.; Ghuman, H.S. Ash Management Review—Applications of Biomass Bottom Ash. *Energies* **2012**, *5*, 3856–3873. [CrossRef]
96. Gharpuray, M.M.; Lee, Y.H.; Fan, L.T. Structural modification of lignocellulosics by pretreatments to enhance enzymatic hydrolysis. *Biotechnol. Bioeng.* **1983**, *25*, 157–172. [CrossRef]
97. Cheng, Z.; Leal, J.H.; Hartford, C.E.; Carson, J.W.; Donohoe, B.S.; Craig, D.A.; Xia, Y.; Daniel, R.C.; Ajayi, O.O.; Semelsberger, T.A. Flow behavior characterization of biomass Feedstocks. *Powder Technol.* **2021**, *387*, 156–180. [CrossRef]
98. Richard Hess, J.; Wright, C.T.; Kenney, K.L. Cellulosic biomass feedstocks and logistics for ethanol production. *Biofuels Bioprod. Biorefin.* **2007**, *1*, 181–190. [CrossRef]
99. Uniform-Format Solid Feedstock Supply System: A Commodity-Scale Design to Produce an Infrastructure-Compatible Bulk Solid from Lignocellulosic Biomass-Executive Summary. Available online: https://inldigitallibrary.inl.gov/sites/sti/sti/4408280.pdf (accessed on 3 April 2022).
100. Rentizelas, A.A.; Tolis, A.J.; Tatsiopoulos, I.P. Logistics issues of biomass: The storage problem and the multi-biomass supply chain. *Renew. Sustain. Energy Rev.* **2009**, *13*, 887–894. [CrossRef]
101. Denafas, G.; Ruzgas, T.; Martuzevičius, D.; Shmarin, S.; Hoffmann, M.; Mykhaylenko, V.; Ogorodnik, S.; Romanov, M.; Neguliaeva, E.; Chusov, A.; et al. Seasonal variation of municipal solid waste generation and composition in four East European cities. *Resour. Conserv. Recycl.* **2014**, *89*, 22–30. [CrossRef]
102. Residue Management Consideration for This Fall. Available online: https://crops.extension.iastate.edu/blog/mahdi-al-kaisi/residue-management-consideration-fall (accessed on 20 October 2022).
103. Stump Harvesting: Interim Guidance on Site Selection and Good Practice (Issue April). Available online: https://cdn.forestresearch.gov.uk/2022/02/fc_stump_harvesting_guidance_april09.pdf (accessed on 22 October 2022).
104. Gruduls, K.; Bardule, A.; Zalitis, T.; Lazdiņš, A. Characteristics of wood chips from loging residues and quality influencing factors. *Res. Rural. Dev.* **2013**, *2*, 49–54.
105. Byadgi, S.A.; Kalburgi, P.B. Production of bioethanol from waste newspaper. *Procedia Environ. Sci.* **2016**, *35*, 555–562. [CrossRef]

106. Xu, H.; Huang, L.; Xu, M.; Qi, M.; Yi, T.; Mo, Q.; Zhao, H.; Huang, C.; Wang, S.; Liu, Y. Preparation and Properties of Cellulose-Based Films Regenerated from Waste Corrugated Cardboards Using [Amim]Cl/CaCl$_2$. *ACS Omega* **2020**, *5*, 23743–23754. [CrossRef]
107. Tanguay-Rioux, F.; Héroux, M.; Legros, R. Physical properties of recyclable materials and implications for resource recovery. *Waste Manag.* **2021**, *136*, 195–203. [CrossRef]
108. Material Bulk Densities. Available online: https://wrap.org.uk/resources/report/material-bulk-densities (accessed on 27 July 2022).
109. White, J.K. The application of LDAT to the HPM2 challenge. *Proc. Inst. Civ. Eng. Waste Resour. Manag.* **2008**, *161*, 137–146. [CrossRef]
110. Kristanto, G.A.; Zikrina, M.N. Analysis of the effect of waste's particle size variations on biodrying method. *AIP Conf. Proc.* **2017**, *1903*, 040009. [CrossRef]
111. Singh, S.; Dutt, D.; Tyagi, C.H. Complete characterization of wheat straw. *BioResources* **2011**, *6*, 154–177. [CrossRef]
112. Hay and Straw Prices. Available online: https://ahdb.org.uk/dairy/hay-and-straw-prices (accessed on 27 July 2022).
113. Laborel-Préneron, A.; Magniont, C.; Aubert, J.E. Characterization of barley straw, hemp shiv and corn cob as resources for bioaggregate based building materials. *Waste Biomass Valorization* **2018**, *9*, 1095–1112. [CrossRef]
114. Kärkönen, A.; Korpinen, R.; Järvenpää, E.; Aalto, A.; Saranpää, P. Properties of oat and barley hulls and suitability for food packaging materials. *J. Nat. Fibers* **2022**, *19*, 13326–13336. [CrossRef]
115. Mazhari Mousavi, S.M.; Hosseini, S.Z.; Resalati, H.; Mahdavi, S.; Rasooly Garmaroody, E. Papermaking potential of rapeseed straw, a new agricultural-based fiber source. *J. Clean. Prod.* **2013**, *52*, 420–424. [CrossRef]
116. The Cell Wall Ultrastructure of Wood Fibres-Effects of the Chemical Pulp Fibre Line. Available online: https://www.diva-portal.org/smash/get/diva2:7109/FULLTEXT01.pdf (accessed on 22 October 2022).
117. Effects of Cell Wall Structure on Tensile Properties of Hardwood. Available online: https://www.diva-portal.org/smash/get/diva2:409533/FULLTEXT02.pdf (accessed on 22 October 2022).
118. Ma, Y.; Hummel, M.; Määttänen, M.; Särkilahti, A.; Harlin, A.; Sixta, H. Upcycling of wastepaper and cardboard to textiles. *Green Chem.* **2016**, *18*, 858–866. [CrossRef]
119. Properties of Paper. Available online: https://www.paperonweb.com/paperpro.htm (accessed on 7 October 2022).
120. Ash Content of Grasses for Biofuel. Available online: http://www.carborobot.com/Download/Papers/Bioenergy_Info_Sheet_5.pdf (accessed on 7 October 2022).
121. Sadef, Y.; Javed, T.; Javed, R.; Mahmood, A.; Alwahibi, M.S.; Elshikh, M.S.; AbdelGawwa, M.R.; Alhaji, J.H.; Rasheed, R.A. Nutritional status, antioxidant activity and total phenolic content of different fruits and vegetables' peels. *PLoS ONE* **2022**, *17*, e265566. [CrossRef] [PubMed]
122. Plant Cell Wall. Available online: https://www.botanicaldoctor.co.uk/learn-about-plants/cell-wall (accessed on 7 October 2022).

Disclaimer/Publisher's Note: The statements, opinions and data contained in all publications are solely those of the individual author(s) and contributor(s) and not of MDPI and/or the editor(s). MDPI and/or the editor(s) disclaim responsibility for any injury to people or property resulting from any ideas, methods, instructions or products referred to in the content.

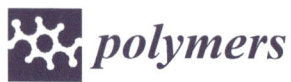 

Article

Preparation and Characterization of Novel Green Seaweed Films from *Ulva rigida*

Uruchaya Sonchaeng [1], Phanwipa Wongphan [1], Wanida Pan-utai [2], Yupadee Paopun [3], Wiratchanee Kansandee [3], Prajongwate Satmalee [4], Montakan Tamtin [5], Prapat Kosawatpat [6] and Nathdanai Harnkarnsujarit [1,7,*]

[1] Department of Packaging and Materials Technology, Faculty of Agro-Industry, Kasetsart University, Bangkok 10900, Thailand
[2] Department of Applied Microbiology, Institute of Food Research and Product Development, Kasetsart University, Bangkok 10900, Thailand
[3] Scientific Equipment and Research Division, Kasetsart University Research and Development Institute, Kasetsart University, Bangkok 10900, Thailand; rdiwnk@ku.ac.th (W.K.)
[4] Department of Food Chemistry and Physics, Institute of Food Research and Product Development, Kasetsart University, Bangkok 10900, Thailand; ifrpws@ku.ac.th
[5] Kung Krabaen Bay Royal Development Study Center, Department of Fisheries, Ministry of Agriculture and Cooperatives, Chantha Buri 22120, Thailand
[6] Phetchaburi Coastal Aquaculture Research and Development Center, Coastal Aquaculture Research and Development Division, Department of Fisheries, Ministry of Agriculture and Cooperatives, Phetchaburi 76100, Thailand; prapat1120@gmail.com
[7] Center for Advanced Studies for Agriculture and Food, Kasetsart University, Bangkok 10900, Thailand
* Correspondence: nathdanai.h@ku.ac.th; Tel.: +662-562-5045

Citation: Sonchaeng, U.; Wongphan, P.; Pan-utai, W.; Paopun, Y.; Kansandee, W.; Satmalee, P.; Tamtin, M.; Kosawatpat, P.; Harnkarnsujarit, N. Preparation and Characterization of Novel Green Seaweed Films from *Ulva rigida*. *Polymers* 2023, 15, 3342. https://doi.org/10.3390/polym15163342

Academic Editors: Beata Kaczmarek and Marcin Wekwejt

Received: 4 July 2023
Revised: 29 July 2023
Accepted: 4 August 2023
Published: 8 August 2023

Copyright: © 2023 by the authors. Licensee MDPI, Basel, Switzerland. This article is an open access article distributed under the terms and conditions of the Creative Commons Attribution (CC BY) license (https://creativecommons.org/licenses/by/4.0/).

Abstract: *Ulva rigida* green seaweed is an abundant biomass consisting of polysaccharides and protein mixtures and a potential bioresource for bioplastic food packaging. This research prepared and characterized novel biodegradable films from *Ulva rigida* extracts. The water-soluble fraction of *Ulva rigida* was extracted and prepared into bioplastic films. ^{1}H nuclear magnetic resonance indicated the presence of rhamnose, glucuronic and sulfate polysaccharides, while major amino acid components determined via high-performance liquid chromatography (HPLC) were aspartic acid, glutamic acid, alanine and glycine. Seaweed extracts were formulated with glycerol and triethyl citrate (20% and 30%) and prepared into films. *Ulva rigida* films showed non-homogeneous microstructures, as determined via scanning electron microscopy, due to immiscible crystalline component mixtures. X-ray diffraction also indicated modified crystalline morphology due to different plasticizers, while infrared spectra suggested interaction between plasticizers and *Ulva rigida* polymers via hydrogen bonding. The addition of glycerol decreased the glass transition temperature of the films from −36 °C for control films to −62 °C for films with 30% glycerol, indicating better plasticization. Water vapor and oxygen permeability were retained at up to 20% plasticizer content, and further addition of plasticizers increased the water permeability up to 6.5 g·mm/m^2·day·KPa, while oxygen permeability decreased below 20 mL·mm/m^2·day·atm when blending plasticizers at 30%. Adding glycerol efficiently improved tensile stress and strain by up to 4- and 3-fold, respectively. Glycerol-plasticized *Ulva rigida* extract films were produced as novel bio-based materials that supported sustainable food packaging.

Keywords: algae; biomaterial; biopolymer; bioplastic; food packaging; edible film

1. Introduction

Green seaweeds, widely distributed in coastal areas, are rich sources of polysaccharides and amino acids. *Chlorophyceae* seaweeds from the *Ulva* genus are world renowned because of the bloom phenomenon 'green tides' that negatively impacts tourism areas and marine life [1,2]. Currently, *Ulva* biomass is harvested to produce compost, fertilizers and biofuels.

The biomass has good nutritional qualities and contains proteins, dietary fiber and complex polysaccharide components. Previous investigations indicated that *Ulva* seaweeds contain cellulose and hemicellulose as major components, with the potential to form networks and utilization as sustainable packaging [1–7]. However, these cellulosic materials are rarely soluble in water, suggesting the formation of film networks from a non-water-soluble fraction. Conversely, the present study demonstrated the film formation from the water-soluble fraction of *Ulva rigida*. The development of *Ulva* materials into biodegradable films and packaging would support sustainable development goals while increasing the value of this abundant biomaterial.

Biopolymers are attractive alternative materials to replace conventional petroleum-based packaging films which cause severe environmental problems. The accumulation of non-biodegradable plastic waste leads to pollution in landfills. Developments using biopolymer films including chitosan, alginate and cellulose derivatives have recently attracted attention [6,8–11]. However, biopolymers have limited mechanical and barrier properties. Plasticizers can be incorporated into film components to improve mechanical properties and flexibility [12,13].

New sources of biomaterials are now increasingly investigated to seek alternative renewable resources for packaging materials. Novel polymeric packaging from seaweeds has been investigated and developed, e.g., *Laminaria japonica* and *Sargassum natans* [8], *Alaria esculenta* and *Saccharina latissimi* [10] and *Furcellaria lumbricalis* and *Gigartina skottsbergii* seaweeds [9]. Seaweeds contain mixtures of organic (e.g., carbohydrates, proteins and fibers) and inorganic substances (e.g., sodium, potassium and calcium) with polysaccharide composition of cellulose, alginate (anionic polysaccharides) and several sulfated polysaccharides as the main cell wall components. Sulfated polysaccharides are commonly characterized by the presence of sulfate (SO_4^{2-}) functional groups, named 'ulvans' in green seaweeds. These component mixtures are soluble and insoluble in different solvents [3]. Insoluble substances form phase separation due to their immiscibility, which causes non-homogeneous film materials and results in poor mechanical strength and high permeability of volatile compounds [12,14]. Water and solvent extraction of seaweed components leads to similar properties of the extracts that are suitable for conversion into polymeric films to achieve more homogeneous structures. *Ulva rigida* contains high amounts of cell wall polysaccharides including cellulose and water-soluble polysaccharides containing mainly ulvan sulfate groups. The major component is a disaccharide formed by β-D-glucuronic acid (1,4)-L-rhamnose-3-sulfate [3]. Mixed hydrocolloid components make *Ulva rigida* a potential resource for film and packaging materials, but detailed studies are lacking. The formation of *Ulva* biopolymer films requires the addition of plasticizers to improve deformability and stretchability.

This research prepared and characterized biopolymer films from *Ulva rigida* crude extracts. Films were formulated with different plasticizers (glycerol and triethyl citrate). The morphology, chemical structures, mechanical and barrier properties of the films were determined. Findings support the utilization of *Ulva rigida* as a novel renewable resource for biodegradable packaging to enhance the value of sustainable materials.

2. Materials and Methods

2.1. Preparation and Extraction of Ulva rigida

Ulva rigida (donated from the Phetchaburi Coastal Aquaculture Research and Development Center, Department of Fisheries, Phetchaburi, Thailand) was collected after 21 days of cultivation in seawater at salinity 30–32 ppm (Figure 1A). Fresh *Ulva rigida* was harvested and washed before drying in a hot air oven at 60 °C for 3–6 h. Dried biomass was milled, giving particle size of 0.5 mm. Water-soluble substances were extracted according to Hamouda et al. [7] with minor modifications. The oven-dried biomass was added with distilled water at a biomass/solvent ratio of 1:20 ($\%w/v$) under controlled temperature of 90 °C and extraction time of 120 min. The supernatant was separated by high-speed refrigerated centrifugation (Model 6000, Kubota, Tokyo, Japan) at $10,000\times g$ for 10 min at

25 °C. An amount of 3 mL of ethanol was added to the supernatant before storing at 4 °C for 24 h. The precipitate was recovered and dried overnight at 50 °C.

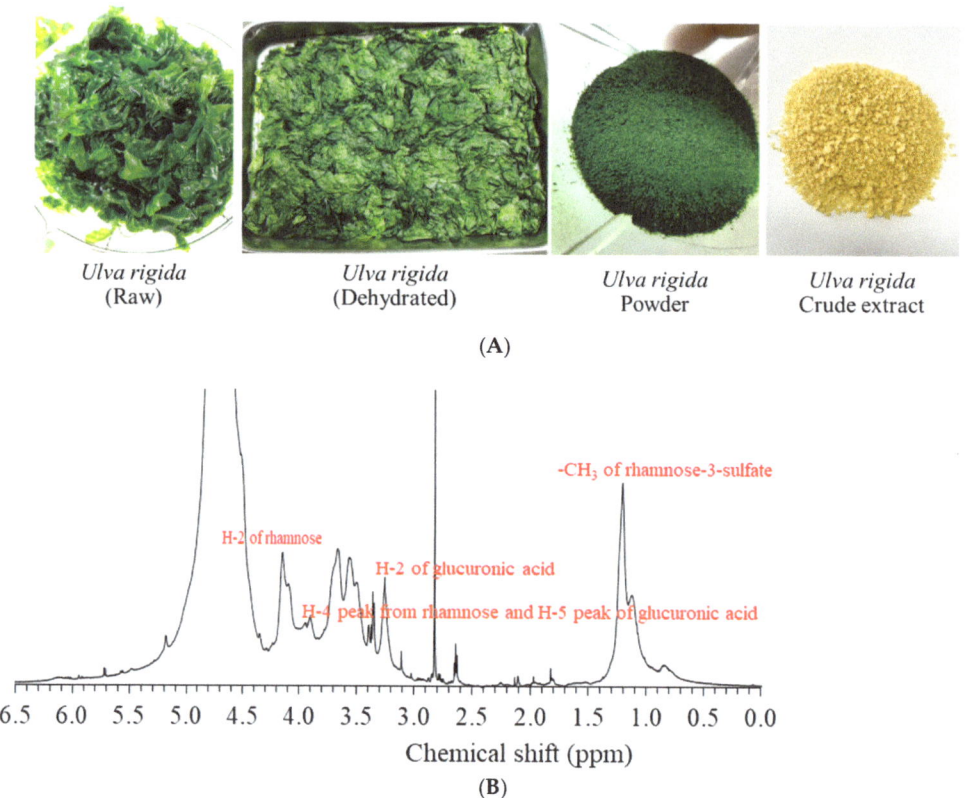

Figure 1. (**A**) Appearance of fresh, dehydrated, powdered and extract of *Ulva rigida*; (**B**) The ^{1}H NMR spectrum of *Ulva rigida* extract in D_2O.

2.2. Analysis of the Biochemical Composition of Ulva rigida Extracts

Oven-dried *Ulva rigida* biomass was analyzed to determine the biochemical composition following AOAC methods [15]. Moisture content was determined via oven-drying at 105 °C to constant weight. Ash content was determined via ignition of the dried samples in an electric furnace at 550 °C. Protein content was determined via the Kjeldahl method using a nitrogen conversion factor of 6.25. Total lipid content was determined using a modified Bligh and Dyer method [16]. Briefly, the samples were suspended in distilled water, methanol and chloroform at a ratio of 0.8:2.0:1.0 and mixed well. The mixture was ultrasonically homogenized for 15 min and then separated via centrifuging at $6000 \times g$ rpm for 15 min. The lipid phase was collected, and the cell debris was extracted until the cells had no color. The lipid extract was then filtered to remove contaminated cell debris and dried to constant weight at 80 °C.

Amino acids in the *Ulva rigida* biomass were extracted and quantified according to the method of Al-Dhabi and Valan Arasu [17] with slight modifications. Briefly, 100 mg of biomass was mixed with 5 mL of 6 N HCl. The slurry sample mixture was incubated at 110 °C for 24 h and then cooled. The sample was then diluted to pH 2.2 with 10 M NaOH and filtered through a 0.45 PTFE syringe filter. The filtrate was determined via high-performance liquid chromatography (HPLC) with a fluorescent detector (HP 1260, Agilent Technologies, Waldbronn, Germany) using superficially porous particles (SPP) technology

4.6 × 100 mm AdvanceBio AAA (2.7 μm) column (Agilent). The oven temperature of the column was set at 30 °C. The sample was derivatized with o-phthalaldehyde (OPA) and 3-mercaptopropionic acid and injected into the column. The gradient mobile phase consisted of a mixture of solvent A (40 mM NaH_2PO_4) and solvent B (methanol, acetonitrile and water). Flow rate was set at 0.7 mL min^{-1}, and peaks were measured as Ex/Em at 340/450 nm. Amino acid standard solution (AA-S-18, Sigma-Aldrich, Singapore) was used as the external standard to calculate the amino acid composition.

2.3. ^1H Nuclear Magnetic Resonance (NMR) of Ulva rigida Extracts

The ^1H NMR spectra were carried out on Ascend™ 600/Avance III HD (Bruker, Billerica, MA, USA). Sample powder (16 mg) was dissolved in D_2O. The solution was operated at 600 MHz running TopSpin 3.6.4 software (Bruker).

2.4. Preparation of Ulva rigida Films

Ulva films were prepared by dissolving 1.5% (w/w) of crude polysaccharides into distilled water. The formulations of the suspensions were prepared using different plasticizers, namely glycerol (Asian Scientific Co., Ltd., Samutprakarn, Thailand), triethyl citrate (pure Ph. Eur., NF, PanReac AppliChem ITW Reagents, Darmstadt, Germany), or a combination of glycerol and triethyl citrate in a 1:1 ratio. Concentrations of plasticizer at 0%, 20% and 30% (w/w of crude polysaccharide weight) were investigated. The suspensions were prepared into films via solution casting with continuous stirring at room temperature (25 ± 3 °C) for 3 h using a magnetic stirrer (IKA Magnetic Stirrers C-MAG HS 7, IKA® Works (Thailand) Co., Ltd., Bangkok, Thailand). Air bubbles were removed via an ultrasonic bath (Sonorex Digitec DT 255 H-RC, Bandelin Electronic GmbH & Co. KG, Berlin, Germany) for 30 min. Blend suspensions (50 ± 5 g) were poured onto polystyrene Petri dishes (diameter 140 mm) and dried at 50 °C for 15 h in a hot air oven. The dried films were removed from the plate and stored in a temperature–humidity-controlled chamber (Climate Chamber Binder KBF 720, Binder GmbH, Tuttlingen, Germany) at 50% relative humidity before analyses for at least 48 h. Thickness of films was determined using a micrometer (model ID-C112BS, Mitutoyo, Kanagawa, Japan).

2.5. Morphology of Ulva rigida Films

2.5.1. Microstructure

The surface and cross-section morphology of the films was observed by a FEI Quanta 450 Scanning Electron Microscope (SEM) (Thermo Fisher Scientific, Waltham, MA, USA) at 15 kV and magnification of 500× and 3000×, respectively. Film samples were immersed and cracked in liquid nitrogen, and specimens were covered with gold using a sputter coater (Quorum Technology Polaron Range SC7620, East Sussex, UK) to facilitate electrical conductivity.

2.5.2. X-ray Diffraction Analysis

Crystallinity was determined using X-ray diffraction (Diffractometer D8, Bruker AXS, Karlsruhe, Germany) at a scanning rate of 0.8/s and a 0.02° step size. The scanned region ranged from 4° to 40° using voltage and current of 40 kV voltage and 40 mA, respectively.

2.6. Fourier Transform Infrared Spectroscopy (FTIR) of Ulva rigida Films

FTIR spectra of the film samples were recorded using a Model 400 Fourier transform infrared spectrometer (Perkin Elmer, Beaconsfield, UK) with attenuated total reflectance (ATR) mode. Absorbance spectra were obtained at 500–4000 cm^{-1} wavelength with anvil geometry of 45° at 4 cm^{-1} resolution and 64 scanning times. The spectra were standardized with the spectrum of air. Test was carried out in eight replications.

2.7. Thermal Stability and Properties of Ulva rigida Films

2.7.1. Thermogravimetric Analysis (TGA)

Thermal stability of the films was determined via thermogravimetric analysis (TGA 2 STARe System, Mettler Toledo, Greifensee, Switzerland). Film pieces (10–15 mg) were placed in aluminum pans, sealed and scanned over the range 25–900 °C under a nitrogen atmosphere at flow rate of 20 mL/min with heating rate of 10 °C/min. First derivative graphs were derived from the weight loss values of the samples.

2.7.2. Differential Scanning Calorimetry (DSC)

Thermal properties of the films were determined using a differential scanning calorimeter (DSC 1, STARe System, Mettler Toledo, Greifensee, Switzerland). Films (1–2 mg) were placed in aluminum pans under nitrogen flux with a flow rate of 50 mL/min. The films were heated from 25 to 80 °C with a heating rate of 10 °C/min to remove moisture in the sample, followed by cooling from 80 to −80 °C at a cooling rate of 10 °C/min. Finally, the films were heated to 300 °C at a rate of 10 °C/min.

2.8. Surface and Barrier Properties of Ulva rigida Films

2.8.1. Water Contact Angle

Hydrophobicity of the film surface was determined via contact angle measurement (Dataphysics OCA 15EC, Dataphysics Instruments GmbH, Filderstadt, Germany). A 3 µL droplet of distilled water was dropped on the film surface using a microsyringe connected to a computer system. Images were immediately taken, and contact angle values were averaged from nine samples using the SCA 20 software version 2 (Dataphysics). Data were averaged from at least 5 samples.

2.8.2. Water Vapor Permeability (WVP)

Water vapor transmission rate (WVTR) was tested using the standard cup method following ASTM E96-80. The samples were kept at 25 ± 2 °C and 50 ± 2% RH in a humidity chamber (Binder KBF 720, Binder GmbH, Tuttlingen, Germany). Triplicate film samples were cut into a circle (7 cm diameter), placed on a metal cup containing silica gel and sealed with an O-ring using paraffin wax. The cups were weighed periodically until constant weight. Water vapor permeability was calculated from triplicate samples using Equation (1).

$$WVP = (WVTR \times thickness)/\Delta P \qquad (1)$$

where WVTR is derived from the slope of the linear regression line obtained from plotting the weight gain against time, and ΔP is the vapor pressure difference.

2.8.3. Oxygen Permeability (OP)

Oxygen permeability was calculated from the oxygen transmission rate (OTR) determined using an oxygen permeation analyzer (Model 8500, Illinois Instruments, Johnsburg, IL, USA) according to ASTM D3985-81. The OP was calculated from duplicate samples using Equation (2).

$$OP = (OTR \times thickness)/\Delta P \qquad (2)$$

where ΔP is the oxygen partial pressure across the film.

2.9. Mechanical Properties of Ulva rigida Films

Mechanical properties were determined using an Instron Universal Testing Machine (Model 5965, Instron, Norwood, MA, USA) according to ASTM D882-88. The samples were cut into rectangular pieces (150 mm × 25 mm) and tested in duplicate at a speed of 500 mm/min. Distance between the gap was 5 cm. Results were plotted between tensile stress (MPa) and tensile strain (%).

2.10. Statistical Analysis

Statistical analyses were conducted to determine significant differences among film sample data via analysis of variance (ANOVA) using IBM SPSS Statistics version 22.0 (IBM Corp., Armonk, NY, USA) and Duncan's multiple range test, with significance set at $p < 0.05$.

3. Results and Discussion

3.1. Characterization of Ulva rigida Seaweed Extracts

3.1.1. Proximate Analysis and Amino Acid Composition

Proximate analysis results showed that *Ulva rigida* seaweed extracts consisted of 22.42 ± 0.72% carbohydrate, 19.01 ± 0.92% protein, 5.67 ± 0.44% lipid, 5.53 ± 0.20% crude fiber and 40.21 ± 0.90% ash. The amino acid profile of *Ulva rigida* extract is shown in Table 1. Major amino acid components consisted of aspartic acid, glutamic acid, alanine and glycine. Seaweed is a source of polysaccharides with protein (10 to 27%) and minor contents of lipid (0.2–3%) [2,10]. Shuuluka et al. [5] indicated major amino acid components of *Ulva rigida* as aspartic acid, alanine and glutamic acid (13.0 ± 1.1, 12.3 ± 0.7 and 9.4 ± 1.0 g/100 g protein, respectively). Similarly, Brain-Isasi et al. [2] reported levels of glutamic acid, aspartic acid and alanine as 16.31 ± 0.82, 14.56 ± 0.73 and 9.89 ± 0.51 g/100 g protein, respectively, as the major amino acid components. Differences in amino acid components depend on several factors including season and environment [2,5]. These amino acids are small molecular weight substances that plasticize the film matrix and improve deformability and flexibility.

Table 1. Amino acid composition of *Ulva rigida* seaweed extract as determined via HPLC.

Amino Acid Composition (mg/100 g)	
Aspartic acid	2055.87 ± 47.90
Glutamic acid	1433.68 ± 28.79
Serine	823.04 ± 19.28
Histidine	252.36 ± 5.08
Glycine	1034.79 ± 18.34
Threonine	531.57 ± 15.39
Arginine	769.26 ± 16.84
Alanine	1097.61 ± 25.23
Tyrosine	517.68 ± 6.31
Cystine	745.40 ± 16.80
Valine	345.38 ± 5.88
Methionine	290.81 ± 2.44
Phenylalanine	629.89 ± 16.22
Isoleucine	214.56 ± 2.47
Leucine	749.68 ± 19.07
Lysine	561.83 ± 13.86
Tryptophan	181.37 ± 3.51
Proline	670.66 ± 34.40

3.1.2. Nuclear Magnetic Resonance (NMR)

The ^{1}H NMR spectrum of *Ulva rigida* extract in D_2O is shown in Figure 1B. A strong ^{1}H resonance at 1.22 ppm, corresponding to the methyl groups of rhamnose-3-sulfate [4]. Kidgell et al. [1] also reported similar proton resonance peaks in ulvans at 3.3 ppm and 3.7 ppm attributed to H-2 and H-3/H-4 of glucuronic acid, respectively, while peaks at 3.8–3.9 ppm were due to combination of the H-4 peak from rhamnose and H-5 peak of glucuronic acid. The major ^{1}H NMR peak at 4.2 ppm was from H-2 of rhamnose.

3.2. Film Microstructure

Surface microstructures of the films are shown in Figure 2A. All films showed numerous finely dispersed particles spreading on the surface. Triethyl citrate at both concentrations showed greatly increased amounts of fine particles. Adding glycerol reduced the number of fine particles. Some larger clumps embedded beneath the film surface with size ranging from 15 to 40 µm. The control film showed higher numbers of large clumps. Figure 2B shows cross-section images of the matrices as fine particles less than 10 µm. Triethyl citrate-plasticized films contained the highest numbers of fine particles. The non-homogeneity of the film matrices reflected the immiscibility of film components. No cracks were found in all films including the control (Figure 2A,B) suggesting strong bonding of the polymeric networks [8]. The presence of protein in *Ulva rigida* enhanced the bonding and strength of the polymer networks via several interactions including electrostatic interactions, hydrophobic interactions and hydrogen bonds [10,18]. Some of these components formed aggregates due to favorable thermodynamics, while the recrystallization of polymer components caused fine crystallites. Accordingly, the phase separation of aggregates and crystallites occurred, causing non-homogeneous matrices.

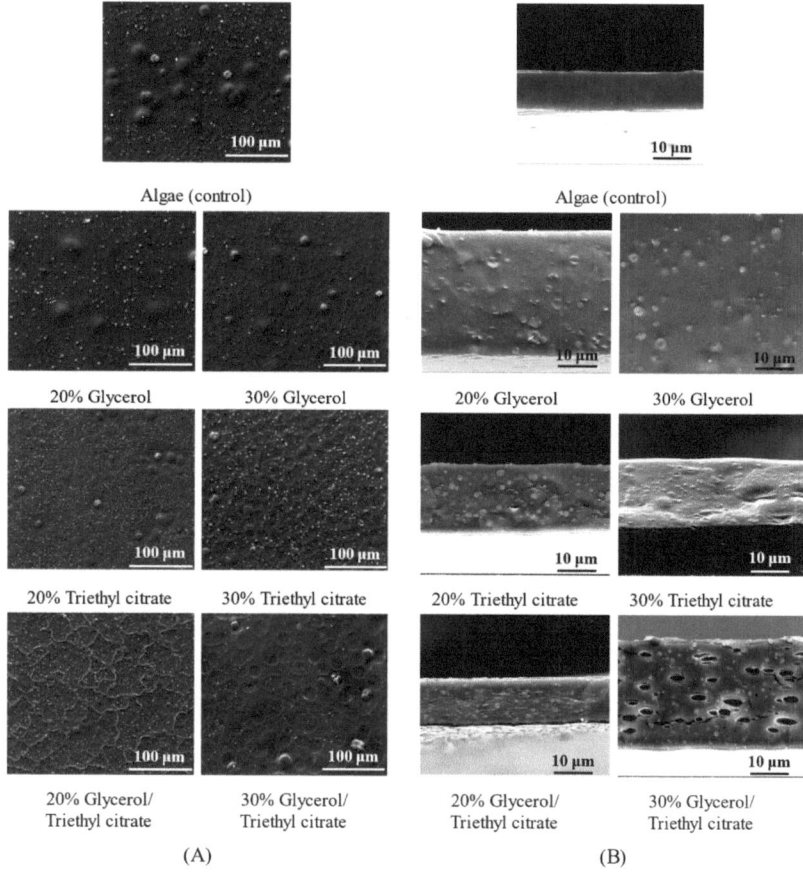

Figure 2. Microstructures as (**A**) surface and (**B**) cross-section of *Ulva rigida* films containing glycerol, triethyl citrate and their blends (glycerol/triethyl citrate) at 20% and 30%.

3.3. Fourier Transform Infrared Spectroscopy

Infrared (IR) absorption spectra of the films are shown in Figure 3. The absorption peaks and corresponding functional groups are shown in Supplementary Table S1. The control film had maximum absorption peaks at 1026 cm^{-1} and 1055 cm^{-1}, ascribed to C-O stretching vibrations in C-O-C that were sensitive to the amounts of disorder amorphous and order crystalline structures of the polymers, respectively [19,20]. The addition of both plasticizers (both glycerol and triethyl citrate) increased the intensity of the absorption peak at 1026 cm^{-1} and merged the peak at 1055 cm^{-1}, reflecting the amorphous structures of the plasticizers. The absorption peaks at 980 and 848 cm^{-1} were assigned to asymmetrical and symmetrical stretching vibration of the C-O-S bond due to the presence of sulfate polysaccharides, respectively [21].

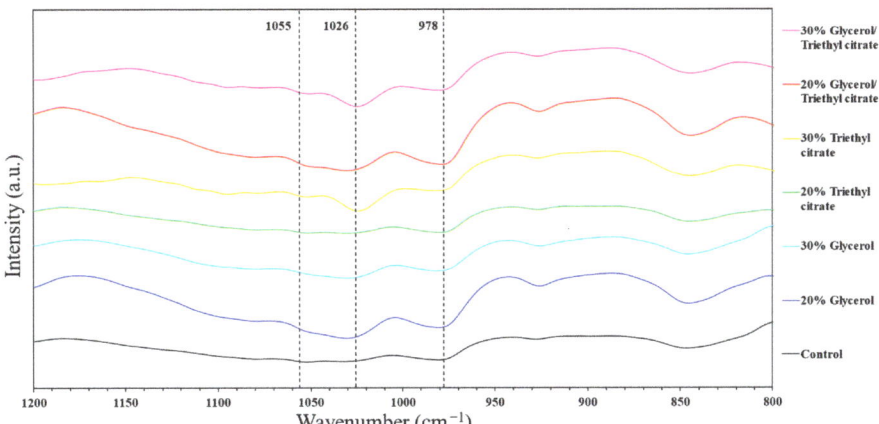

Figure 3. FTIR spectra of *Ulva rigida* films containing glycerol, triethyl citrate and their blends (glycerol/triethyl citrate) at 20% and 30%.

Glycerol had similar IR absorption spectra to the control film in the fingerprint region (500–1500 cm^{-1}) due to similar vibration of C-H bending (1500–1200 cm^{-1}) in CH_2 and C-O stretching (980–1250 cm^{-1}). Higher concentrations of triethyl citrate increased absorption at 1735 cm^{-1} attributed to C = O stretching vibration of ester bonds in the triethyl citrate structure [12]. The band at 1425 cm^{-1} was associated with vibration of C-OH deformation in O–C–O symmetric stretching vibration of the carboxylate group in seaweed polymers [10]. The IR wide absorption band at 1629 cm^{-1} was due to carboxyl groups in *Ulva* spp. which overlapped with C = C stretching vibration of the methacrylic group [4]. The IR spectra showed peaks around 1200–1250 cm^{-1} and 840–845 cm^{-1}, ascribed to S = O stretching and C–O–S stretching characteristics of polysaccharides in *Ulva* spp. [6]. The wide absorption peak between 3000 and 3700 cm^{-1} was ascribed to O-H stretching vibration, and attributed to intra- and inter-molecular hydrogen bonding. Small absorption peaks between 2850 and 3050 cm^{-1} were attributed to C-H stretching vibration [10,20,22] and were stronger in films containing plasticizers. Adding glycerol intensified the spectrum at 3240 cm^{-1}, suggesting hydrogen bonding between the polymer and glycerol, while adding triethyl citrate gave a wider peak with stronger absorption at 3495 cm^{-1}, due to hydrogen bonding with polymer components. The results indicated interactions between plasticizers and polymers, which contributed to plasticization effects of the films.

3.4. Thermogravimetric Analysis

Figure 4A shows weight loss, reflecting thermal degradation and volatilization of film components. The onset temperature, weight loss in each stage of degradation and char residue is shown in Supplementary Table S2. The first sharp weight loss started at 60 °C and was due to evaporation of absorbed water, corresponding with mass loss of 20–28% depending on plasticizer types and concentrations. Glycerol-plasticized films had the highest mass loss, reflecting the highest levels of absorbed water attributed to high hydrophilicity from large numbers of hydroxyl groups. The second weight loss of the control film started at 215 °C, corresponding with an approximately 10% weight reduction followed by a sharp weight loss due to volatilization of polymer components [3,6,8,12]. Triethyl citrate-plasticized films had a similar weight reduction to the control, while films with glycerol showed a higher weight reduction, corresponding with the amounts of plasticizers and due to volatilization of glycerol [12,23]. The final residue left after the thermal decomposition process decreased with the addition of plasticizer, indicating the breaking of polymer–polymer interactions with the addition of glycerol and triethyl citrate.

The first derivative weight loss of the films is shown in Figure 4B. The control film showed a peak with a shoulder, suggesting evaporation of multi-components as free and bound water below 100 °C [12]. Films containing plasticizers showed sharper peaks at 100 °C due to water evaporation, suggesting higher levels of water absorption. Mixtures of plasticizers showed two extra volatilization peaks between 135 and 200 °C that did not occur in glycerol- and triethyl citrate-plasticized films. Mixtures of the plasticizers stabilized the film components, giving higher degradation temperatures. The sharpest degradation started at 200 °C in all films. The control and triethyl citrate-plasticized films had a major degradation peak at 222 °C, while glycerol-plasticized films showed a sharp peak at 235 °C. Shifting of the degradation peak at around 230 °C in the blend plasticizers reflected the major role of glycerol on the degradation temperature in protein and polysaccharide components [6,10,23]. A small shoulder was found at 265 °C for the control and triethyl citrate films, and at 285 °C for the films with glycerol and the mixtures. Degradation at around 300 °C was due to polymer components. These findings reflected the interactions between glycerol and the polymers, which had a major effect on thermal stability.

3.5. Differential Scanning Calorimetry (DSC)

The DSC thermograms showed an endothermic shift in heat flow, suggesting the glass transition temperature (T_g) of the films (Figure 5A). The control film had onset T_g at −36 °C. *Ulva rigida* extract consisted of small molecular weight solids, particularly amino acids (Table 1). These small molecules plasticized the polysaccharide film matrices resulting in low T_g values. The addition of both plasticizers (glycerol and triethyl citrate) and their mixtures reduced the T_g of the films. Glycerol decreased T_g values more than triethyl citrate, corresponding with T_g values of −53 °C and −62 °C for films containing 20% and 30% glycerol, respectively. Increasing plasticizer contents gave a lower T_g due to higher plasticization effects, with a larger magnitude of endothermic shift found in glycerol-plasticized films. These results reflected stronger plasticization effects of glycerol in seaweed extract films. Glycerol has an extremely low T_g value of −86 °C [24], while the molecular weight of triethyl citrate is higher than glycerol, corresponding with a lower mole number at the same weight. A smaller plasticizer size, namely glycerol, also enhanced the dispersion and interactions with the polymer in the matrices [13,25]. Glycerol molecules consist of three hydroxyl groups (-OH) per mole, which readily form hydrogen bonding with polysaccharides. Mixtures of glycerol and triethyl citrate gave intermediate T_g between their blends. The plasticizers decreased the T_g by disrupting the intermolecular chain movement of the polymers, thereby improving the segmental mobility of the polymers. A hydrophilic plasticizer such as glycerol facilitates more interactions between itself and the polymer, resulting in an increase in the free volume of matrix, thereby reducing the T_g. Hence, the results clearly indicated interaction between glycerol and triethyl citrate with the polymer matrices causing T_g reduction.

Figure 5B shows endothermic peaks, reflecting phase transitions of film components starting above 70 °C due to water evaporation. The sharp endothermic peaks between 150 and 220 °C reflected the melting of the crystallites, as discussed in the X-ray diffractions results. The control film showed a sharp peak at 160 °C followed by smaller peaks at 183 °C and 212 °C. Glycerol-plasticized films showed a single but wider peak than the control at higher temperature. Triethyl citrate showed a wide peak between 75 and 175 °C, followed by a sharp peak around 200 °C and a smaller peak at higher temperatures (similarly to the control). The results suggested that triethyl citrate enhanced the non-homogeneous crystallization of film components, corresponding with wide and multiple melting temperatures.

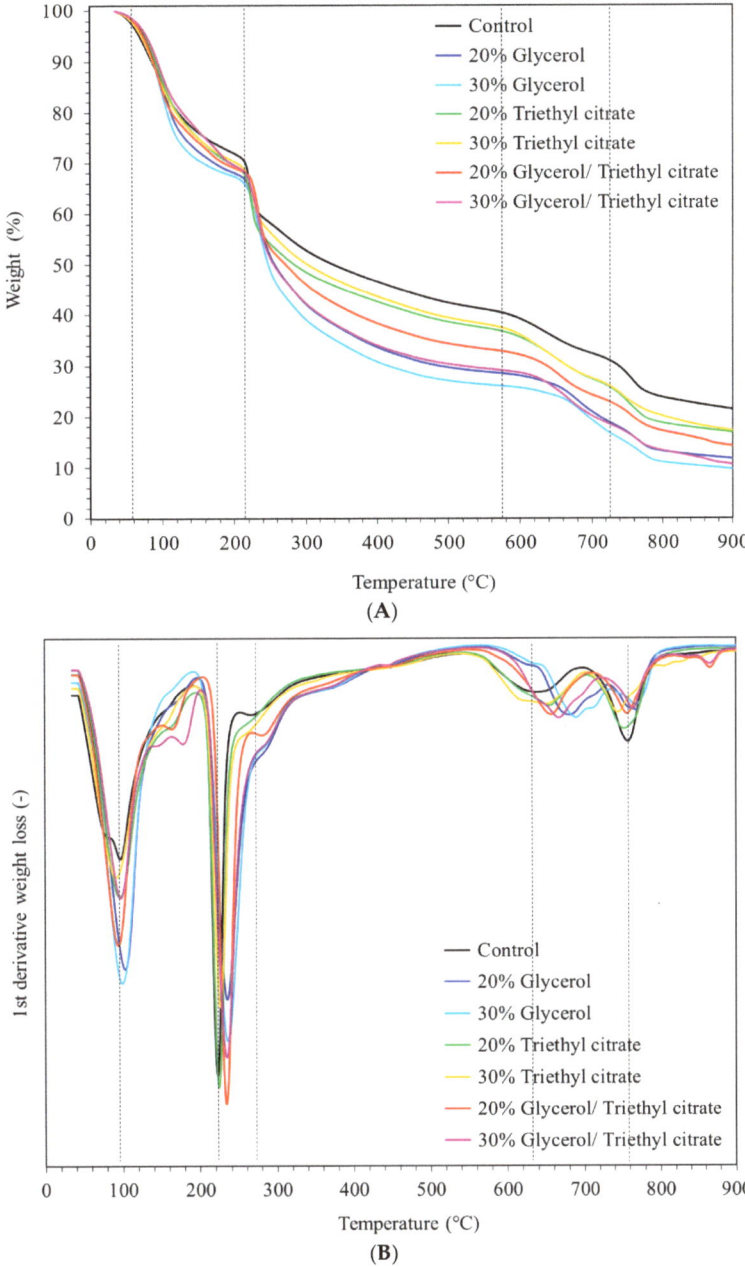

Figure 4. Thermal degradation as (**A**) weight percent and (**B**) first derivative weight loss of *Ulva rigida* films containing glycerol, triethyl citrate and their blends (glycerol/triethyl citrate) at 20% and 30%.

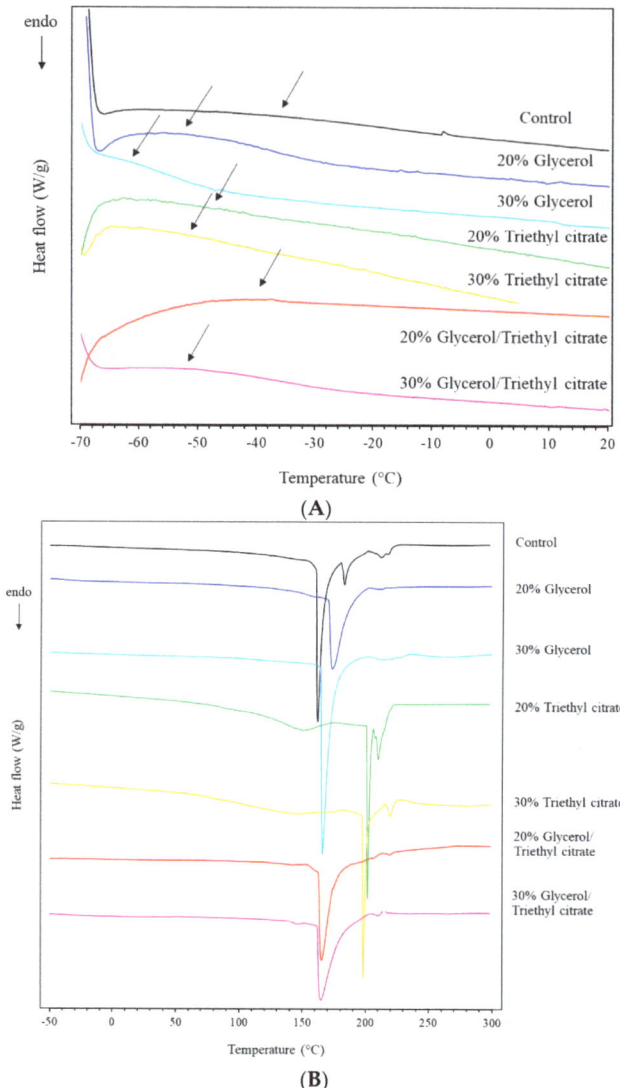

Figure 5. DSC thermograms (**A**) enlarged at low temperature range and (**B**) between −50 °C and 300 °C in the second scan of *Ulva rigida* films containing glycerol, triethyl citrate and their blends (glycerol/triethyl citrate) at 20% and 30%. The arrows indicate endothermic shift, suggesting the glass transition temperature (T_g) of the films.

3.6. X-ray Diffraction

X-ray diffractograms of films with different plasticizers are shown in Figure 6. The films had diffraction angles at 2θ = 11.7, 14.6, 20.8, 22.5, 25.5, 29.2, 29.6, 31.1 and 31.7°. Glycerol clearly caused a broad peak between diffraction angles in the range of 16–25°, reflecting amorphous components. Conversely, triethyl citrate gave a lower intensity of the diffraction curve, suggesting reduced amorphous components. The DSC analysis (Figure 5B) also indicated that triethyl citrate-plasticized films showed large and wide melting peaks, reflecting the higher levels of crystalline components. The sharp peak at 14.6° (corresponding to the (110) crystalline plane) was much more intense, suggesting a preferential orientation of the (110) crystal plane parallel to the film surface [9]. The intensity of the peaks at 2θ = 11.7, 20.8, 29.2, 31.3 and 33.7° decreased when adding 20% plasticizers (glycerol and triethyl citrate and their mixtures), while a further increase in plasticizers to 30% increased peak intensity of these aforementioned peaks. Adding plasticizers increased peak intensity at 2θ = 14.6, 25.5, 29.6 and 31.7°. The reduction in crystallinity was attributed to amorphous complexes formed through intermolecular interactions between film-forming substrates and plasticizers [11]. The results suggested that glycerol and triethyl citrate impacted the morphology of the crystalline structures in seaweed films.

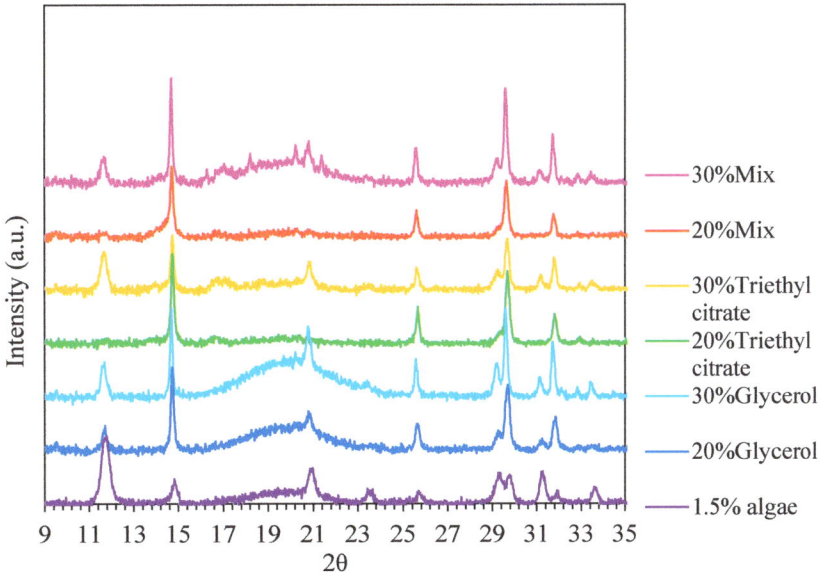

Figure 6. X-ray diffractograms of *Ulva rigida* films containing glycerol, triethyl citrate and their blends (glycerol/triethyl citrate) at 20% and 30%.

3.7. Surface Hydrophobicity and Barrier Properties

Surface hydrophobicity was determined according to water contact angle (CA), as shown in Figure 7A. CA of the control film was 96°, which is considered a hydrophobic surface (CA > 90°) [11]. Adding plasticizers clearly decreased CA, indicating the decreasing hydrophobicity of the film surface. Plasticization with glycerol greatly enhanced the hydrophilicity of the matrices, as reflected by higher degree of water absorption; however, CA was higher than CA of films containing triethyl citrate. Lower CA values of triethyl citrate-plasticized films reflected higher surface energy. The wettability of films depends on surface energy and surface roughness. A lower surface roughness reduced hydrophobic surface wettability [14,26,27]. The results suggested that adding triethyl citrate reduced the surface roughness of the films, corresponding with lower CA values. Plasticizer mixtures gave intermediate CA values between glycerol and triethyl citrate. Accordingly, the wettability of seaweed films mainly depended on surface roughness.

Water vapor permeability (WVP) and oxygen permeability (OP) are important parameters in food packaging that affect the quality and stability of packaged products. The WVP changed insignificantly when adding 20% plasticizers, while further increasing the plasticizer content increased WVP (Figure 7B). The hydroxyl groups and oxygen atoms in ester structures of citrate readily absorbed water, increasing diffusion of water vapor through the matrices [28]. However, plasticizers also decreased T_g, which increased the molecular mobility of the matrices and crystallization. The formation of ordered and tightly packed crystalline structures inhibited the diffusion of volatile molecules including water vapor and gas [25,29]. Accordingly, the WVP values of the films insignificantly increased with the addition of 20% plasticizers. However, increasing the plasticizer content to 30% enhanced mobility and diffusion, which increased WVP. Similarly, adding 20% plasticizers led to insignificant changes in OP (Figure 7C). Increasing plasticizers also increased diffusion rates through the polymers due to increasing molecular mobility. However, blending plasticizers at 30% reduced OP due to high crystallinity which inhibited oxygen diffusion.

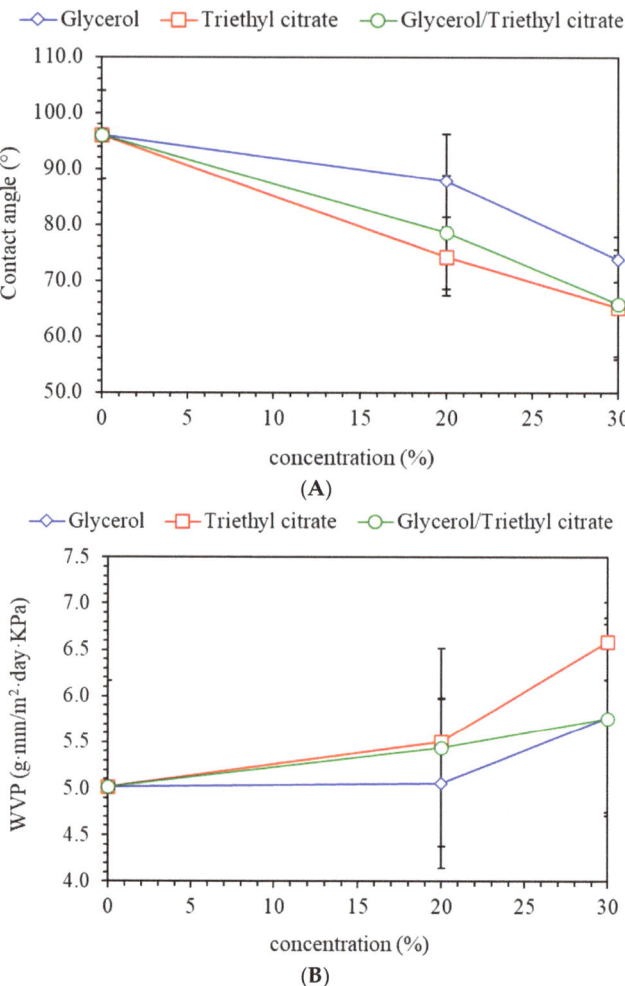

Figure 7. *Cont.*

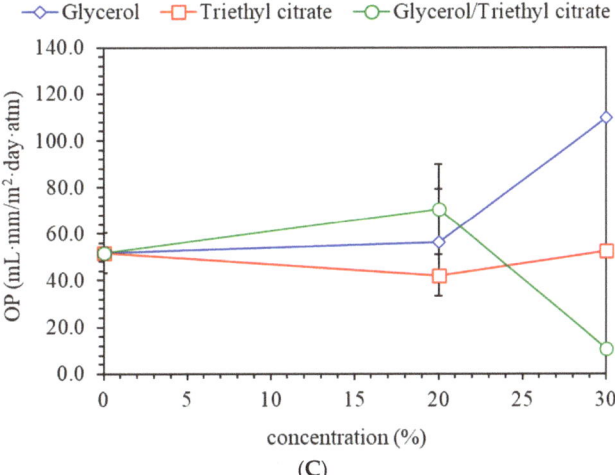

Figure 7. (**A**) Contact angle, (**B**) water vapor permeability (WVP) and (**C**) oxygen permeability (OP) of *Ulva rigida* films containing glycerol, triethyl citrate and their blends (glycerol/triethyl citrate) at 20% and 30%.

3.8. Tensile Properties

Mechanical properties were determined as tensile stress and tensile strain curves (Figure 8). The control film was weak and least flexible due to low plasticization effects. Polysaccharides consist of numerous hydroxyl groups that form strong inter- and intra-molecular hydrogen bonding. Consequently, the networks were rigid and required plasticization to improve deformability [13,25]. The extract consisted of small-molecular-weight amino acids, particularly aspartic, glutamic, alanine and glycine, which plasticized the matrices (Table 1). However, the results clearly demonstrated that plasticization efficiency was not sufficient to form flexible networks, and higher plasticizer contents were required.

Figure 8 shows that adding glycerol and triethyl citrate significantly improved tensile stress. Glycerol clearly increased tensile strain, reflecting the enhanced elongation of the films by up to 3-fold, while tensile stress was enhanced by up to 4-fold. Adding triethyl citrate at 20% greatly enhanced tensile stress but reduced strain values, indicating lower deformability and higher rigidity. Triethyl citrate induced crystallization of film components. The formation of immiscible rigid particles in the matrices caused non-homogeneous polymer networks, as also shown by SEM (Figure 2). These rigid particles acted as reinforcement, improving tensile stress by up to 8.5-fold (films with 20% triethyl citrate). However, the non-homogeneity decreased the area for distribution of applied external stress, leading to a lower extension ability [30]. Increasing triethyl citrate to 30% reduced strength because the large rigid particles increased non-homogeneity and decreased adhesion between the polymer networks [8,14]. Blending of plasticizers showed no synergistic improvement in mechanical strength. Glycerol (20–30%) greatly improved tensile properties and produced flexible seaweed films.

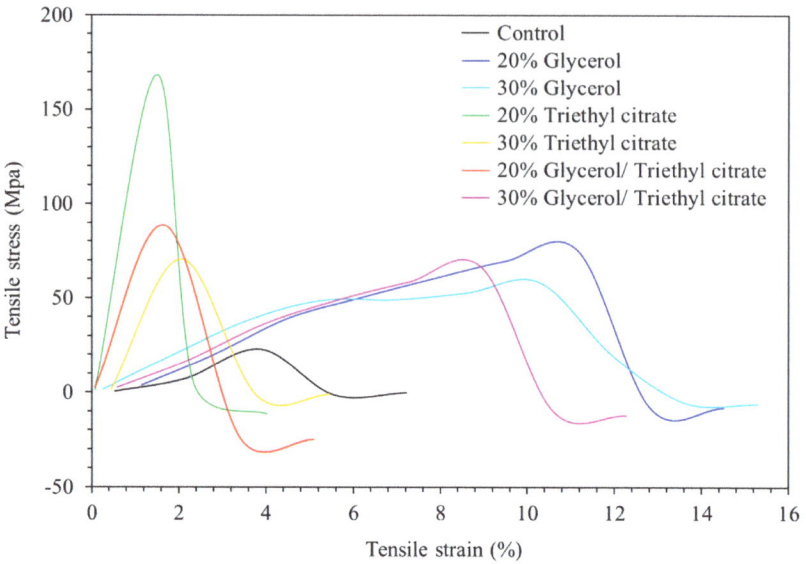

Figure 8. Tensile stress–strain curves of *Ulva rigida* films containing glycerol, triethyl citrate and their blends (glycerol/triethyl citrate) at 20% and 30%.

4. Conclusions

The water-soluble fraction of *Ulva rigida* consisted of polysaccharides and amino acid mixtures. Major amino acids were aspartic acid, glutamic acid, alanine and glycine, while polysaccharide structures consisted of rhamnose and sulfate derivatives. Native *Ulva rigida* extracts can form film networks; however, mechanical properties were poor, with very limited extensivity. Adding plasticizers such as glycerol and triethyl citrate improved the mechanical properties. These plasticizers interacted with *Ulva rigida* components via hydrogen bonding and modified the crystal morphology. Glycerol showed greater plasticization effects, resulting in a higher reduction in glass transition temperature and higher tensile strain, indicated by improved film elongation by up to 3-fold. Triethyl citrate greatly improved tensile stress but limited elongation. Film barrier properties were unaltered at up to 20% plasticizers, while increasing plasticizers to 30% generally increased water vapor and oxygen permeability due to increasing molecular mobility. The plasticization of *Ulva rigida* polymers produced efficient bioplastic films as a novel bioresource for sustainable food packaging.

Supplementary Materials: The following supporting information can be downloaded at: https://www.mdpi.com/article/10.3390/polym15163342/s1, Table S1. FTIR peaks and corresponding functional group; Table S2. TGA onset temperature, weight loss in each stage of degradation and char residue.

Author Contributions: Conceptualization, U.S., P.W. and N.H.; methodology, P.W. and N.H.; validation, P.W. and N.H.; formal analysis, P.W. and N.H.; investigation, P.W., W.P.-u. and N.H.; writing—original draft preparation, U.S., P.W. and N.H.; writing—review and editing, W.P.-u., Y.P., W.K., P.S., M.T., P.K. and N.H.; supervision, N.H.; funding acquisition, N.H. All authors have read and agreed to the published version of the manuscript.

Funding: This research was funded by Kasetsart University Research and Development Institute (KURDI) FF(KU) 17.65.

Institutional Review Board Statement: Not applicable.

Data Availability Statement: The data that support the findings of this study are available on request from the corresponding author.

Conflicts of Interest: The authors declare no conflict of interest.

References

1. Kidgell, J.T.; Glasson, C.R.; Magnusson, M.; Vamvounis, G.; Sims, I.M.; Carnachan, S.M.; Taki, A.C. The molecular weight of ulvan affects the in vitro inflammatory response of a murine macrophage. *Int. J. Biol. Macromol.* **2020**, *150*, 839–848. [CrossRef] [PubMed]
2. Brain-Isasi, S.; Carú, C.; Lienqueo, M.E. Valorization of the green seaweed *Ulva rigida* for production of fungal biomass protein using a hypercellulolytic terrestrial fungus. *Algal Res.* **2021**, *59*, 102457. [CrossRef]
3. Leiro, J.M.; Castro, R.; Arranz, J.A.; Lamas, J. Immunomodulating activities of acidic sulphated polysaccharides obtained from the seaweed *Ulva rigida* C. Agardh. *Int. Immunopharmacol.* **2007**, *7*, 879–888. [CrossRef]
4. Morelli, A.; Chiellini, F. Ulvan as a new type of biomaterial from renewable resources: Functionalization and hydrogel preparation. *Macromol. Chem. Phys.* **2010**, *211*, 821–832. [CrossRef]
5. Shuuluka, D.; Bolton, J.J.; Anderson, R.J. Protein content, amino acid composition and nitrogen-to-protein conversion factors of *Ulva rigida* and *Ulva capensis* from natural populations and *Ulva lactuca* from an aquaculture system, in South Africa. *J. Appl. Phycol.* **2013**, *25*, 677–685. [CrossRef]
6. Wahlström, N.; Edlund, U.; Pavia, H.; Toth, G.; Jaworski, A.; Pell, A.J.; Richter-Dahlfors, A. Cellulose from the green macroalgae *Ulva lactuca*: Isolation, characterization, optotracing, and production of cellulose nanofibrils. *Cellulose* **2020**, *27*, 3707–3725. [CrossRef]
7. Hamouda, R.A.; Hussein, M.H.; El-Naggar, N.E.; Karim-Eldeen, M.A.; Alamer, K.H.; Saleh, M.A.; El-Azeem, R.M.A. Promoting Effect of Soluble Polysaccharides Extracted from *Ulva* spp. on *Zea mays* L. Growth. *Molecules* **2022**, *27*, 1394. [CrossRef]
8. Doh, H.; Dunno, K.D.; Whiteside, W.S. Preparation of novel seaweed nanocomposite film from brown seaweeds *Laminaria japonica* and *Sargassum natans*. *Food Hydrocoll.* **2020**, *105*, 105744. [CrossRef]
9. Šimkovic, I.; Gucmann, F.; Mendichi, R.; Schieroni, A.G.; Piovani, D.; Dobročka, E.; Hricovíni, M. Extraction and characterization of polysaccharide films prepared from *Furcellaria lumbricalis* and *Gigartina skottsbergii* seaweeds. *Cellulose* **2021**, *28*, 9567–9588. [CrossRef]
10. Cebrián-Lloret, V.; Metz, M.; Martínez-Abad, A.; Knutsen, S.H.; Ballance, S.; López-Rubio, A.; Martínez-Sanz, M. Valorization of alginate-extracted seaweed biomass for the development of cellulose-based packaging films. *Algal Res.* **2022**, *61*, 102576. [CrossRef]
11. Yang, Y.; Yu, X.; Zhu, Y.; Zeng, Y.; Fang, C.; Liu, Y.; Jiang, W. Preparation and application of a colorimetric film based on sodium alginate/sodium carboxymethyl cellulose incorporated with rose anthocyanins. *Food Chem.* **2022**, *393*, 133342. [CrossRef]
12. Teixeira, S.C.; Silva, R.R.A.; de Oliveira, T.V.; Stringheta, P.C.; Pinto, M.R.M.R.; Soares, N.D.F.F. Glycerol and triethyl citrate plasticizer effects on molecular, thermal, mechanical, and barrier properties of cellulose acetate films. *Food Biosci.* **2021**, *42*, 101202. [CrossRef]
13. Sothornvit, R.; Krochta, D.J. Plasticizer effect on oxygen permeability of β-lactoglobulin films. *J. Agric. Food Chem.* **2000**, *48*, 6298–6302. [CrossRef]
14. Phothisarattana, D.; Wongphan, P.; Promhuad, K.; Promsorn, J.; Harnkarnsujarit, N. Blown film extrusion of PBAT/TPS/ZnO nanocomposites for shelf-life extension of meat packaging. *Colloids Surf. B Biointerfaces* **2022**, *214*, 112472. [CrossRef] [PubMed]
15. AOAC (Association of Official Analytical Chemistry). *Official Methods of Analysis of the Association of Analytical Chemists International*; Association of Official Analytical Chemistry: Rockville, MD, USA, 2005.
16. Breil, C.; Vian, M.A.; Zemb, T.; Kunz, W.; Chemat, F. "Bligh and Dyer" and Folch Methods for Solid–Liquid–Liquid Extraction of Lipids from Microorganisms. Comprehension of Solvatation Mechanisms and towards Substitution with Alternative Solvents. *Int. J. Mol. Sci.* **2017**, *18*, 708. [CrossRef]
17. Al-Dhabi, N.A.; Valan Arasu, M. Quantification of Phytochemicals from Commercial Spirulina Products and Their Antioxidant Activities. *Evid. Based Complement. Alternat. Med.* **2016**, *2016*, 7631864. [CrossRef]
18. Yuan, D.; Meng, H.; Huang, Q.; Li, C.; Fu, X. Preparation and characterization of chitosan-based edible active films incorporated with Sargassum pallidum polysaccharides by ultrasound treatment. *Int. J. Biol. Macromol.* **2021**, *183*, 473–480. [CrossRef]
19. Wongphan, P.; Panrong, T.; Harnkarnsujarit, N. Effect of different modified starches on physical, morphological, thermomechanical, barrier and biodegradation properties of cassava starch and polybutylene adipate terephthalate blend film. *Food Packag. Shelf Life* **2022**, *32*, 100844. [CrossRef]
20. Higaki, Y.; Takahara, A. Structure and properties of polysaccharide/imogolite hybrids. *Polym. J.* **2022**, *54*, 473–479. [CrossRef]
21. Santos, P.R.M.; Johny, A.; Silva, C.Q.; Azenha, M.A.; Vázquez, J.A.; Valcarcel, J.; Silva, A.F. Improved Metal Cation Optosensing Membranes through the Incorporation of Sulphated Polysaccharides. *Molecules* **2022**, *27*, 5026. [CrossRef] [PubMed]
22. Promsorn, J.; Harnkarnsujarit, N. Pyrogallol loaded thermoplastic cassava starch based films as bio-based oxygen scavengers. *Ind. Crops Prod.* **2022**, *186*, 115226. [CrossRef]
23. Phothisarattana, D.; Harnkarnsujarit, N. Migration, aggregations and thermal degradation behaviors of TiO_2 and ZnO incorporated PBAT/TPS nanocomposite blown films. *Food Packag. Shelf Life* **2022**, *33*, 100901. [CrossRef]

24. Bachler, J.; Handle, P.H.; Giovambattista, N.; Loerting, T. Glass polymorphism and liquid–liquid phase transition in aqueous solutions: Experiments and computer simulations. *Phys. Chem. Chem. Phys.* **2019**, *21*, 23238–23268. [CrossRef] [PubMed]
25. Roos, Y.H.; Drusch, S. *Phase Transitions in Foods*; Academic Press: Cambridge, MA, USA, 2015.
26. Jin, H.; Tian, L.; Bing, W.; Zhao, J.; Ren, L. Bioinspired marine antifouling coatings: Status, prospects, and future. *Prog. Mater. Sci.* **2022**, *124*, 100889. [CrossRef]
27. Zhang, W.; Wang, D.; Sun, Z.; Song, J.; Deng, X. Robust superhydrophobicity: Mechanisms and strategies. *Chem. Soc. Rev.* **2021**, *50*, 4031–4061. [CrossRef]
28. Coma, V.; Sebti, I.; Pardon, P.; Pichavant, F.H.; Deschamps, A. Film properties from crosslinking of cellulosic derivatives with a polyfunctional carboxylic acid. *Carbohydr. Polym.* **2003**, *51*, 265–271. [CrossRef]
29. Benvenuti, M.; Mangani, S. Crystallization of soluble proteins in vapor diffusion for x-ray crystallography. *Nat. Protoc.* **2007**, *2*, 1633–1651. [CrossRef]
30. Promsorn, J.; Harnkarnsujarit, N. Oxygen absorbing food packaging made by extrusion compounding of thermoplastic cassava starch with gallic acid. *Food Control.* **2022**, *142*, 109273. [CrossRef]

Disclaimer/Publisher's Note: The statements, opinions and data contained in all publications are solely those of the individual author(s) and contributor(s) and not of MDPI and/or the editor(s). MDPI and/or the editor(s) disclaim responsibility for any injury to people or property resulting from any ideas, methods, instructions or products referred to in the content.

Article

Performance Evaluation of Hot Mix Asphalt (HMA) Containing Polyethylene Terephthalate (PET) Using Wet and Dry Mixing Techniques

Nisma Agha [1,*], Arshad Hussain [1,*], Agha Shah Ali [1] and Yanjun Qiu [2]

1. School of Civil and Environmental Engineering (SCEE), National University of Sciences and Technology (NUST), Islamabad 44000, Pakistan
2. School of Civil Engineering, Southwest Jiaotong University, Chengdu 610031, China
* Correspondence: nismaagha.tn19@student.nust.edu.pk (N.A.); drarshad@nit.nust.edu.pk (A.H.); Tel.: +92-333-5766087 (N.A.); +92-341-9756251 (A.H.)

Citation: Agha, N.; Hussain, A.; Ali, A.S.; Qiu, Y. Performance Evaluation of Hot Mix Asphalt (HMA) Containing Polyethylene Terephthalate (PET) Using Wet and Dry Mixing Techniques. *Polymers* **2023**, *15*, 1211. https://doi.org/10.3390/polym15051211

Academic Editors: Beata Kaczmarek and Marcin Wekwejt

Received: 30 September 2022
Revised: 23 January 2023
Accepted: 26 February 2023
Published: 27 February 2023

Copyright: © 2023 by the authors. Licensee MDPI, Basel, Switzerland. This article is an open access article distributed under the terms and conditions of the Creative Commons Attribution (CC BY) license (https://creativecommons.org/licenses/by/4.0/).

Abstract: This study evaluates the performance of Polyethylene Terephthalate (PET)-modified hot mix asphalt. Aggregate, bitumen of grade 60/70 and crushed plastic bottle waste were utilized in this study. Polymer Modified Bitumen (PMB) was prepared using a high shear laboratory type mixer rotating at a speed of 1100 rpm with varying PET content of 2%, 4%, 6%, 8% and 10%, respectively. Overall, the results of preliminary tests suggested that bitumen hardened with the addition of PET. Following optimum bitumen content determination, various modified and controlled HMA samples were prepared as per wet and dry mixing techniques. This research presents an innovative technique to compare the performance of HMA prepared via dry and wet mixing techniques. Performance evaluation tests, which include the Moisture Susceptibility Test (ALDOT-361-88), Indirect Tensile Fatigue Test (ITFT-EN12697-24) and Marshall Stability and Flow Tests (AASHTO T245-90), were conducted on controlled and modified HMA samples. The dry mixing technique yielded better results in terms of resistance against fatigue cracking, stability and flow; however, the wet mixing technique yielded better results in terms of resistance against moisture damage. The addition of PET at more than 4% resulted in a decreased trend for fatigue, stability and flow due to the stiffer nature of PET. However, for the moisture susceptibility test optimum PET content was noted to be 6%. Polyethylene Terephthalate-modified HMA is found to be the economical solution for high volume road construction and maintenance, besides having other significant advantages such as increased sustainability and waste reduction.

Keywords: eco-friendly; recycle; plastic bottles; polyethylene terephthalate (PET); hot mix asphalt (HMA); polymer-modified bitumen (PMB); fatigue life; moisture susceptibility; stability; flow

1. Introduction

Plastic is a material that is consumed at an increasing rate every year. Its use is so common because of its good electrical and mechanical insulating properties, good chemical resistance, low density and easy processing along with its main advantage of less initial cost but the main problem lies in the disposal of this commonly used material. The improper disposal of plastic thus results in plastic pollution. Due to the non-biodegradable nature of plastic, it poses a serious threat to land and various waterbodies, because of which a considerable percentage of marine and land creatures are exposed to life-threatening situations [1]. Of all the various plastic types, plastic bottles exist untreated in large quantities at garbage dump sites, which threatens environmental safety. Plastic bottles, due to their chief chemical component Polyethylene Terephthalate (PET), are often regarded as PET bottles. The non-biodegradable nature of plastic bottles makes them lethal to animals when ingested. This is because the intermolecular bonding structure of PET is of the kind that does not allow it to decompose, digest or corrode. If improperly disposed of, these plastic materials pollute water bodies and become a cause of water-borne diseases. Consequently, water bodies are hindered by plastics in terms of their flow, which creates problems of pollution and suspension. These adverse environmental impacts of plastics necessitate their recycling or their use for some other purpose that can be beneficial as well as environmentally friendly.

Besides environmental concerns, economic development and sustainability are of crucial importance to countries' growth and overall national security. An important factor in improving a country's economic status is that of the road infrastructure system and more importantly the length of existing paved roads, often used as an index to assess the extent of a country's development. The presence of a proper road transport network not only minimizes the transportation cost, both in terms of time and financial aspects, but also aids in the interlinking of several regions within the country and in a better understanding of neighboring countries at a global level [2]. Transportation links (roads) are the prerequisite of a transport system. They carry traffic and continuously face load repetitions and, moreover, the damaging effect of climate results in various road defects.

One of the major deteriorations caused by the environment is the exposure of pavements to water, hence causing moisture damage. Moisture damage is defined as the loss in strength and durability of asphalt mixtures caused by the presence of water. The damage gains momentum as more moisture permeates and gradually causes the mastic to weaken, making it even more susceptible to damage during cyclic loading [3]. Poor drainage conditions or excessive rain can be the reasons for this uncontrolled exposure. Thus, pavements become more prone to cracking and other pavement distresses such as stripping, rutting, bleeding, corrugation and shoving, cracking, raveling and other localized failures [3].

Fatigue cracking in the bound layer is a major indication of structural failure in a pavement. The magnitude, frequency and duration of load application were identified to have major effect on pavement performance in terms of surface cracking [4]. Repeated traffic loads on pavements cause fatigue failure and because of that a series of interconnected cracks are formed on the road surface termed as fatigue cracking. For thin pavements, the initiation of cracking is from the bottom of the pavement as the tensile stress is greatest at that portion of the pavement. This cracking then propagates to the surface, which is commonly known as bottom-up fatigue cracking. For thick pavements, the cracking initiates from the top of the surface and from the surface areas of high localized tensile stresses, which exist because of the pavement–tire interaction. Moreover, the aging of the asphalt binder also plays an important role in top-down fatigue cracking in the case of thick Hot Mix Asphalt (HMA) pavements [5], see Figure 1.

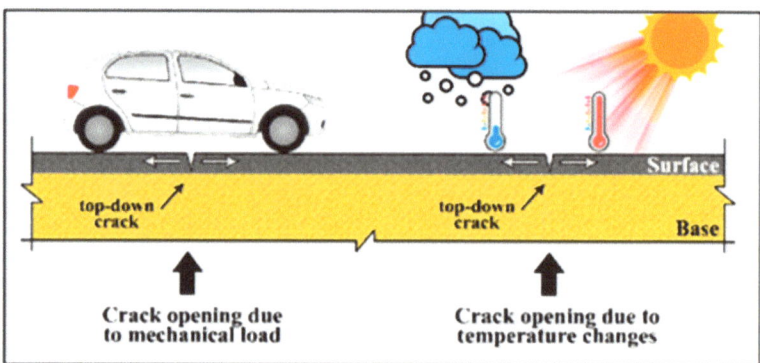

Figure 1. Pavement cracking.

Fatigue cracking reduces the overall life of pavements and results in major ride discomfort. These surface distresses also result in an increase in maintenance costs. Therefore, remedial measures that can address the issue of surface distresses because of increased traffic, vehicular loading and environmental deterioration can save the economy in terms of low initial project and rehabilitation costs. The utilization of plastics in road construction can serve the purpose to a substantial extent. The use of plastics in pavements has exhibited some very desirable results as far as pavement response to load repetitions and resistance to moisture damage is concerned. With the use of plastics in roads, the durability of the pavement structure can be increased, and waste plastic material can also be utilized that reduces the environmental threat. It is, therefore, the most suitable option to utilize waste plastic material in a way that contributes significantly in the field of road construction and plays a significant part in reducing environmental pollution.

Besides the eco-friendly disposal of plastic waste by incorporating it in road construction to mitigate the early deterioration of roads, the use of recycled plastic waste is also a cost-effective approach [6]. Waste polymers can be incorporated in HMA pavements through three techniques, wet mixing, dry mixing and aggregate replacement. In the wet mixing technique, PET is thoroughly mixed with hot liquid asphalt binder to form a mixture prior its mixing with aggregate. In the dry mixing process, to uniformly disperse the polymers, substantial mixing and shearing are required. For this method, polymers as solid granular particles or in the form of chips are mixed with the aggregates followed by the addition of bitumen [7]. As per Mishra and Gupta, the dry process is simple, economical and environmentally friendly, while the wet process requires more effort in terms of investment and machinery, and hence is not commonly used [8]. However, in this research for both mentioned techniques PET is used as a substitute material for a portion of the bitumen in Job Mix Formula (JMF) calculations [9,10].

In all the mixes not containing any plastic content, fatigue life was found to be minimum. In contrast, as per Casey et al., the addition of a plastic particle ingredient drastically increases the fatigue life of asphalt mixes. It is illustrated by lower stress levels ranging from 250 kPa to 350 kPa that the fatigue life of mixes with 1% plastic is increased twofold. Since PET particles do not melt due to their higher melting point, this results in the PET particles existing as partially rigid materials in the mixture, thus improving the flexibility of the mixture [11]. With an improved mixture, the crack creation and propagation in asphalt mixtures is postponed, which eventually leads to an increased higher fatigue life. Baghaee et al., in his research conducted in 2013, evaluated the indirect tensile fatigue strength of PET-modified HMA mixtures according to EN 12697, and compared them to results of controlled mix. It was concluded that, due to the modified mixture's improved flexibility, crack initiation and propagation was delayed, which eventually indicated the higher fatigue life of flexible pavements [12].

As per Yilmaz et al., water damages the pavement in such a way that it decreases the strength and quality of asphalt concrete. Asphalt pavement is susceptible to moisture damage by the loss of the bond between binder and fine and coarse aggregate. This loss in bond results in the propagation of cracks along the pavement length. Moisture damage further aggravates as water penetrates through cracks and debilitates the mastic, making it more vulnerable to moisture amid the cyclic loading of vehicles [3]. Ferreira et al. conducted research in Brazil in 2022 and concluded PET produced positive effects on the moisture resistance of HMA; this led to the greater tensile strength of conditioned samples resulting in a higher Tensile Strength Ratio (TSR). Ferreira et al. further suggested that an improvement in TSR is indicative of increased asphalt binder–aggregate adhesion under moist conditions [13], see Figure 2.

For regions associated with extensive rainfall pavement deterioration, Ferreira et al. also proposed that resistance to moisture damage can be enhanced by the replacement of natural sand with recycled micro polyethylene terephthalate, thus demonstrating the feasibility of the practical application of blending PET in concrete asphalt paving [13].

Silva et al. suggested improved resistance against moisture damage when micronized PET was incorporated as a binder modifier in varying contents of 0, 4, 5 and 6% (by bitumen weight) [14]. Ghabchi et al. concluded that the micro nature of PET improves the viscosity and binder–aggregate adhesion; this is reflected as an increase in resistance to rutting, cracking and damage induced by moisture [15]. Studies by Likitlersuang et al. also suggested that resistance against moisture damage is improved due to an increase in adhesion between the binder and aggregate in asphalt under moisture exposition [16].

Figure 2. Moisture-induced pavement distresses.

Successful past studies by Bamigboye et al. and Tayyab et al. on HMA modification by crumb tires, glass waste, biofuel, slag, fiber, lime, fly ash and sasobit also suggest PET as a potential modifier [17,18].

Burak Sengoz and Giray Isikyakar in 2008 showed that, based on the type of polymer and its content, the morphology and properties of the modified bitumen and the mechanical properties of HMA modified by polymer change. Samples with low polymer content exhibited the continuous dispersion of polymer in bitumen; however, the continuous dispersion of polymer has not been observed in samples containing high polymer content. Improvement in the conventional properties, by polymer modification, includes penetration, susceptibility to temperature, softening point, etc [19]. Ali et al. stated when grinded plastic is used as a modifier in bitumen in a replacement ratio of increments of 0.5% up to 2%, the results of index properties such as the flash point, softening point, fire point and penetration varied as compared to those of virgin bitumen. Moreover, regardless of the replacement ratio, modified bitumen yielded satisfactory performance [20]. The penetration values of Polymer Modified Bitumen (PMB) decreased as the PET content increased in the test conducted by Sojobi et al., which is an indication of increased stiffness and softening point; also, the tests indicated that because of the more stable asphalt, ductility values also increased with increasing PET content [21].

The engineering properties of Stone Mastic Asphalt (SMA) mixture incorporating waste PET were studied by Ahmad et al. in 2017. The study evaluated the mechanical properties of the asphalt mixture blended with PET in proportions of 0%, 2%, 4%, 6%, 8% and 10%. Based on the research by Ahmad et al., bituminous mixtures incorporating PET showed improved properties such as increases in stiffness, stability and viscosity. In other words, pavement can better withstand fatigue damage, thermal cracking, rutting and stripping. Asphalt modification by polyethylene yielded better resistance against fatigue and deformation [22]. In a previous study by Al-Hadidy et al. in 2009, the effect of polyethylene on the low temperature performance and moisture sensitivity of SMA mixtures was studied. It was concluded that such modified mixtures produced satisfactory performance in areas with extreme variance of temperature and heavy rain zones [23]. Studies conducted by Nishanthini et al. in 2020 suggested that resistance to water damage and fatigue cracking decreased when HMA was modified by 5–10% PET (by weight of binder) and less than 18% (by weight of fine aggregate). Marshall stability also improved upon such HMA modification. [24]

Jegatheesa et al. in 2018 stated that, for both Polymer Coated Aggregate (PCA) and Polymer Modified Bitumen (PMB), Marshall stability values showed an increase with increasing PET content. However, Air Voids (AV), which showed an increase in PCA with the increasing PET content, were seen to decrease in the case of PMB [25]. In a related study, Baghaee et al. found increased Marshall stability values for the addition of PET up to 0.6% (aggregate replacement). Furthermore, results obtained by Baghaee et al. showed that by the addition of plastic into the mixture the internal friction was reduced, eventually resulting in higher flow [12]. A study by Ahmadinia et al. in 2011 showed that stiffer mixtures were obtained with the addition of PET greater than 6 percent, which resulted in increased Marshall stability and decreased flow values, although increasing PET content beyond 6% resulted in a decline in the stability of the asphaltic mix, which consequently resulted in increased flow [26].

Another study by Kalantar et al. in 2012 also suggests that with the addition of polymer the flow of the mixture increases [7]. Ahmadinia et al. in 2012 suggested that adding waste PET between 2–10% by weight of bitumen content in SMA mixtures results in the improved adhesion of PET granules between the asphalt HMA and PET-modified HMA prepared by wet and dry mixing techniques based on various performance characteristics, which include moisture susceptibility, indirect tensile fatigue strength, stability and flow. The use of plastic waste encourages reduced plastic waste and promotes reduced initial costs; therefore, cost analysis was also carried out. Table 1 summarizes all the past studies conducted specifically for the PET modification of bitumen using different modification strategies.

Table 1. Past studies incorporating PET in asphalt mixes.

S. No.	Authors	Year/Country	PET Replacement Technique	Particle Nature	PET Modification Percentage	Conclusion
1.	Baghaee et al.	2013-Malaysia	Aggregate replacement	PET chips	0, 0.2, 0.4, 0.6, 0.8 and 1% (by aggregate weight)	Improved fatigue life of pavement upon addition of PET
2.	Ferreira et al.	2022-Brazil	Sand replacement	Crushed PET	2, 4, 8% (by sand weight) 8% (by sand volume)	Improved ITS and TSR of HMA improved when modified with PET; however, Resilient Modulus (RM) decreased
3.	Silva et al.	2018-Brazil	Binder additive	Micronized PET	0, 4, 5 and 6% (by bitumen weight)	Addition of PET improved results for Resilient Modulus, Indirect tensile strength, Lottman, Fatigue and Flow Number
4.	Ghabchi et al.	2021-United States	Binder additive	Micronized PET	0, 5, 10, 15 and 20% (by bitumen weight)	higher resistance to moisture-induced damage upon addition of PET
5.	Ali et al.	2014-Pakistan	Binder additive	Ground PET	0, 0.5, 1.0, 1.5, 2.0	Inclusion of PET improved test results for flash point, fire point, softening point and penetration test
6.	Sojobi et al.	2015-Nigeria	Binder modification Aggregate replacement	Molten plastic waste	0,5, 10, 20 (Polymer modified mix) 10, 20, 30 (polymer-coated)	Plastic content for PMB is likely to decrease penetration while increases softening point, ductility and viscosity. Stability and AV increased

Table 1. Cont.

S. No.	Authors	Year/Country	PET Replacement Technique	Particle Nature	PET Modification Percentage	Conclusion
7.	Ahmad et al.	2017-Malaysia	Binder additive		0, 2, 4, 6, 8 and 10% (by bitumen weight)	Improved stiffness, viscosity and rutting on addition of PET
8.	Jegatheesa et al.	2018-Sri Lanka	Binder additive	PET fibers with a nominal diameter of 0.5 mm and a length of 4.0 to 6.0 mm	5, 10, 15, 20, 25, 30, 35 and 40% (by bitumen weight)	Inclusion of PET improves Marshall stability and bulk properties
9.	Baghaee et al.	2013- Malaysia	Aggregate replacement	Crushed PET	0, 0.2, 0.4, 0.6, 0.8 and 1% (by aggregate weight)	On addition of PET, stability, flow and fracture resistance increased. Stiffness and specific gravity
10.	Ahmadinia et al.	2011-Malaysia	Binder additive	Sieve 1.18 mm passing; #40 retained	0, 2, 4, 6, 8, 10% (by bitumen weight)	Increased stiffness, Air voids and stability (up to 6% PET). Decreased bulk specific gravity
11.	Ahmadinia et al. in 2012	2012-Malaysia	Binder additive	Sieve 1.18 mm passing; #40 retained	0, 2, 4, 6, 8, 10% (by bitumen weight)	Addition of PET increased stiffness, resistance against rutting and provided lower binder drain-down

2. Materials and Methods

Under the effect of vehicular loading and climatic changes, asphaltic pavements are prone to deteriorations. With increases in population and economic growth, demand for high quality roads has also increased. This demand is a big challenge to a country's economy. This research aims at addressing the challenge faced by the construction industry and countries' economies by suggesting a means to utilize plastic waste in road construction. The main objective of this research is to suggest the best mixing technique of PET in HMA mixtures by studying and comparing trends in the properties of virgin HMA mixtures with those of PET-modified HMA mixtures. The ITF strength and resistance to moisture damage are two key properties that are covered in this research besides assessing flow and stability. Figure 3 represents research methodology adopted for testing and analysis for this research.

A testing matrix of the research was made for six different types of mixes that included one control mix and the remaining five were modified for varying PET content as shown in Table 2. Table 2 includes a series of various numbers of samples; the first mention of samples indicates the number of samples prepared for Type-x, the second mention indicates the number of samples prepared for Type-y and the third mention indicates the number of samples prepared for Type-z.

2.1. Binder

The base bitumen with a 60/70 penetration grade was selected in the current research due to adequate performance in cold to moderate climate conditions and was procured from Attock Refinery Limited (ARL), Rawalpindi, Pakistan. Conventional tests conforming to ASTM standards were performed to characterize the properties of base bitumen. The results obtained have been tabulated in Table 3.

2.2. Aggregate

Limestone aggregate procured from the Taxila (District Rawalpindi) quarry site was used in the current research. In order to find out the index properties of the aggregate, tests were conducted in conformity with the relevant test standards and are presented along with the results in Table 4.

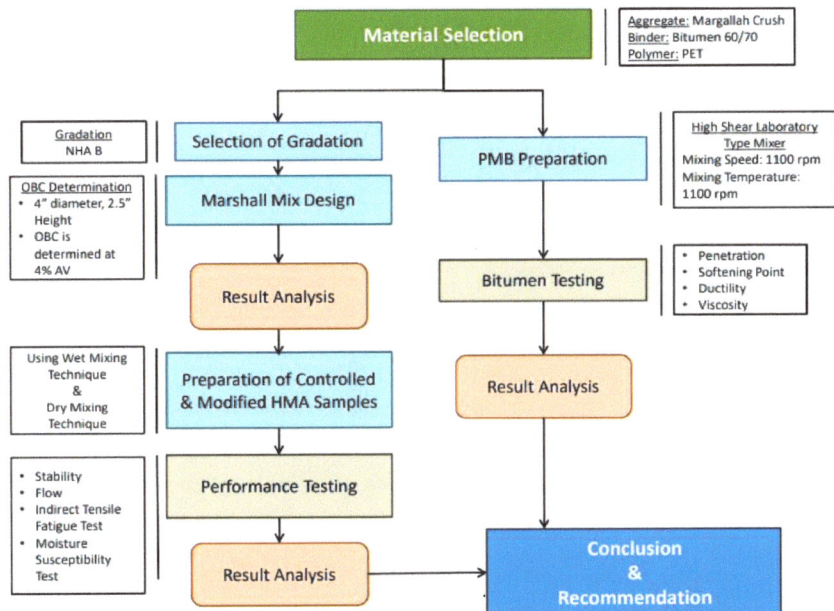

Figure 3. Research methodology flowchart.

Table 2. Performance testing matrix.

S. No.	PET (% Bitumen)	Number of Samples Required		
		ITFT	Moisture Susceptibility	Stability and Flow
1	0	3+0+0	6+0+0	3+0+0
2	2	0+3+3	0+6+6	0+3+3
3	4	0+3+3	0+6+6	0+3+3
4	6	0+3+3	0+6+6	0+3+3
5	8	0+3+3	0+6+6	0+3+3
6	10	0+3+3	0+6+6	0+3+3
	TOTAL	3+15+15	6+30+30	3+15+15

Type-x = controlled HMA samples. Type-y = modified HMA samples prepared using wet mixing technique. Type-z = modified samples prepared using dry mixing technique.

Table 3. Properties of base bitumen.

S. No.	Test Description	Specification	Results	Limits
1	Penetration test @ 25 °C	ASTM D 5-06	67	60–70
2	Flash Point (°C)	ASTM D 92	273 °C	~280 °C
3	Fire Point (°C)	ASTM D 92	375 °C	~320 °C
4	Softening Point (°C)	ASTM D 36–95	44.7 °C	35–45 °C
5	Ductility Test (cm)	ASTM 113-99	118 cm	>100 cm
6	Viscosity Test (Pa-sec)	ASTM D 88–94	2.98	≤3
7	Specific Gravity	ASTM D 70	1.02	0.97–1.02

Table 4. Properties of aggregate.

S. No.	Test Description	Specification		Results	Limits
1	Elongation Index (EI)	ASTM D 4791		11.20%	≤15%
2	Flakiness Index (FI)	ASTM D 4791		1.1%	≤15%
3	Aggregate Absorption	Fine:	ASTM C 128	1.6%	≤3%
		Coarse:	ASTM C 127	0.7%	≤3%
4	Impact Value	BS 812		15.23%	≤30%
5	Los Angles Abrasion	ASTM C 131		23.13%	≤45%
6	Specific Gravity	Fine:	ASTM C 128	2.12	-
		Coarse:	ASTM C 127	2.71	-

The gradation of the aggregate was chosen to conform with Pakistan National Highway Authority (NHA) standard Gradation-B for Asphaltic Concrete for Wearing Course (ACWC) as presented in Figure 4.

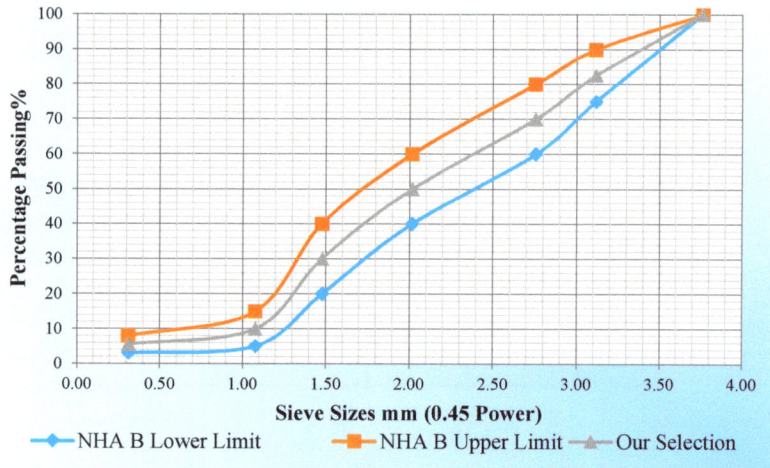

Figure 4. The NHA class-B gradation curve.

In addition to the gradation of aggregate, Table 5 shows blend ratios of 1200 g modified and controlled HMA samples. All quantities are measured in grams.

Table 5. The HMA blend ratios.

	Controlled HMA Mix					
Aggregate	Bitumen Content					
	3.0%	3.5%	4.0%	4.5%	5.0%	5.5%
1/2	204	203	202	201	200	198
3/8	146	145	144	143	143	142
#4	233	232	230	229	228	227
#8	233	232	230	229	228	227
#16	233	232	230	229	228	227
#200	52	52	52	52	51	51
pan	64	64	63	63	63	62
Bitumen	36	42	48	54	60	66

Table 5. *Cont.*

Aggregate	Modified HMA Mix					
	PET *w/w* of OBC					
	0%	2%	4%	6%	8%	10%
1/2	201	201	201	201	201	201
3/8	144	144	144	144	144	144
#4	230	230	230	230	230	230
#8	230	230	230	230	230	230
#16	230	230	230	230	230	230
#200	52	52	52	52	52	52
pan	63	63	63	63	63	63
Bitumen	51.6	50.6	49.5	48.5	47.5	46.4
PET		1.03	2.06	3.10	4.13	5.16

2.3. Polyethylene Terephthalate (PET)–Modifier

Locally, waste plastic bottles, which were handpicked from streets, roads and public areas and sold by weight to local mechanical recyclers, were procured in the form of crushed pellets of size < 2.36 mm (see Figure 5). Typical properties of PET are tabulated in Table 6.

Figure 5. Crushed PET pellets.

Table 6. Typical properties of PET.

S. No.	Property	Specification
1	Chemical Formula	$(C_{10}H_8O_4)$ n
2	Melting Point	260 °C
3	Typical Injection Mold Temperature	74–91 °C
4	Heat Deflection Temperature	70 °C at 0.46 MPa
5	Tensile Strength	152 MPa
6	Flexural Strength	221 MPa
7	Specific Gravity	1.56
8	Shrink Rate	0.1–0.3%

3. Sample Preparation

3.1. Preparation of PMB

Through the means of high shear laboratory type mixer revolving at around a speed of 1100 rpm, PET-modified bitumen samples were prepared. The base bitumen was brought to a fluid condition by heating it up to 155–160 °C. Later, PET was added gradually to the heated bitumen and mixing was continued for 2 h. Later, PMB samples of 2%, 4%, 6%, 8% and 10% PET content were stored in small containers and covered with aluminum foil for later testing.

3.2. Marshal Mix Design-OBC Determination

The ASTM D-6927 standard for Marshall Mix Design was used to prepare controlled HMA samples. To determine the Optimum Bitumen Content (OBC), National Asphalt Pavement Association procedure was adopted. As per the procedure, HMA samples were prepared with varying bitumen content of 3.0%, 3.5%, 4.0%, 4.5%, 5.0% and 5.5%. A total of 1200 g of cylindrical HMA sample of 4″ diameter and 2.5″ height was prepared at a mixing temperature of 160 °C and compacted at 135 °C by giving 75 blows on each side of sample. Volumetric properties, which include Flow (mm), Stability (kN), Voids in Mineral Aggregate (VMA) (%), Voids Filled with Asphalt (VFA) (%) and Air Voids (AV) (%) of Marshall samples, were calculated and are plotted on graphs as shown in Figure 6. The HMA samples with higher bitumen content tend to have higher flow, VMA, VFA and lower stability and AV.

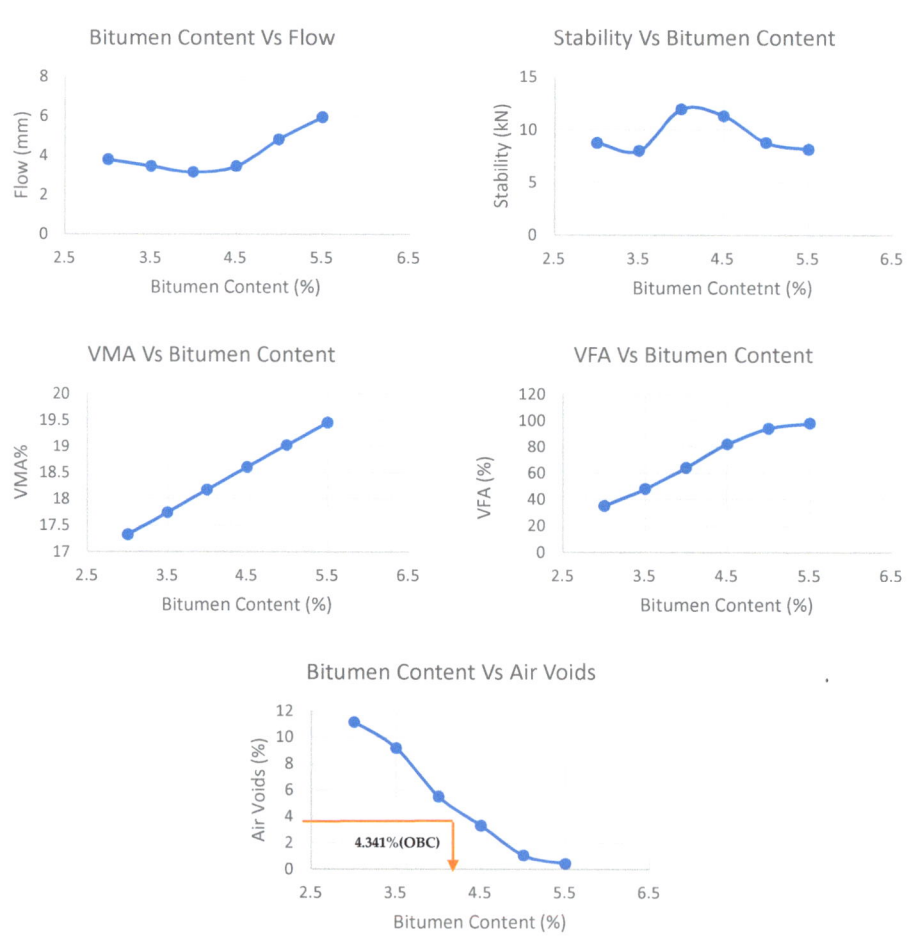

Figure 6. Volumetric properties of HMA samples.

As per the National Asphalt Pavement Association, OBC was determined against 4% AV, which turned out to be 4.341%. All the other volumetric properties were then noted against 4% AV. Table 7 summarizes the volumetric properties of HMA at OBC.

Table 7. Volumetric properties of HMA at OBC.

S. No.	Property	Value
1	Stability	11.7 kN
2	Flow	3.4 mm
3	VFA	70%
4	VMA	18.4%

3.2.1. Controlled HMA Sample Preparation

As per ASTM D-6927, 1200 g of HMA samples were prepared by first heating aggregate and virgin bitumen up to 110 °C in an oven prior to mixing at 160 °C; the mix was later compacted to cylindrical samples at 135 °C by giving 75 blows on each side. Controlled HMA samples (Type-x) were prepared to assess properties at a later stage keeping in view Table 2.

3.2.2. Modified HMA Sample Preparation–Wet Mixing Technique

A total of 1200 g of HMA samples were prepared by first heating aggregate and PMB (in% replacement of OBC) up to 110 °C in an oven prior to mixing at 160 °C; the mix was later compacted to cylindrical samples at 135 °C by giving 75 blows on each side. Modified HMA samples using the wet mixing technique (Type-y) were prepared keeping in view Table 2.

3.2.3. Modified HMA Sample Preparation–Dry Mixing Technique

Keeping in view the high melting point of PET, the dry mixing technique was also considered. Aggregate was initially heated to a high temperature of 260 °C; later, crushed pellets of PET of a size < 2.36 mm were added and mixed thoroughly such that the aggregate was well coated with polymer; the mix was then cooled to 160 °C. Bitumen was separately heated up to 110 °C in an oven and then added to the heated aggregate-polymer mix at 160 °C. After the proper mixing of bitumen with aggregate-polymer, the mix was compacted to cylindrical samples at 135 °C by giving 75 blows on each side. Modified HMA samples using the dry mixing technique (Type-z) were prepared keeping in view Table 2.

4. Laboratory Testing–Results and Discussion

4.1. PMB Testing

Various preliminary tests were conducted on the prepared modified bitumen samples as per the mentioned standards. The results of the virgin bitumen were also plotted for each index property to obtain a comparison between the controlled and modified samples of bitumen. Figure 7 represents trends in respective properties against each PET (%) content.

It is observed that an increase in PET content in PMB results in a gradual decline in ductility and penetration; however, softening point and viscosity increase. The increase in softening point suggests that bitumen resistance against the deformation of modified asphalt is improved. Preliminary testing conducted on PMB concluded that the properties of virgin bitumen altered noticeably when modified via PET.

4.2. Indirect Tensile Fatigue Test

The Indirect Tensile Fatigue Test (ITFT) was adopted to evaluate fatigue life in terms of cyclic loading in the Universal Testing Machine (UTM) following standard EN12697-24. The test was conducted in a stress-controlled condition such that the load was applied in a vertical direction resulting in horizontal tensile stress. The sample failed by splitting in the vertical plane due to increased strain within the sample [27]. Type-x, Type-y and Type-z of the HMA samples were tested for fatigue. As per EN12697-24, the thickness of the sample was 51 mm ± 1 mm and the diameter was 100 ± 3 mm (for a maximum aggregate size of 25 mm). Samples were conditioned for 4 h at 25 °C in a temperature-controlled chamber, prior to testing under a load of 3500 N, with a loading time of 0.1 s and rest time of 0.4 s. Figure 8 shows the device diagram and load symmetry of the ITFT. Testing was carried out at 25 °C. The test finished once the sample fractured, and the machine stopped itself. Figure 9 shows a fractured sample on completion of the loading cycles. For the sake of conclusion, the number of cycles leading to failure was noted.

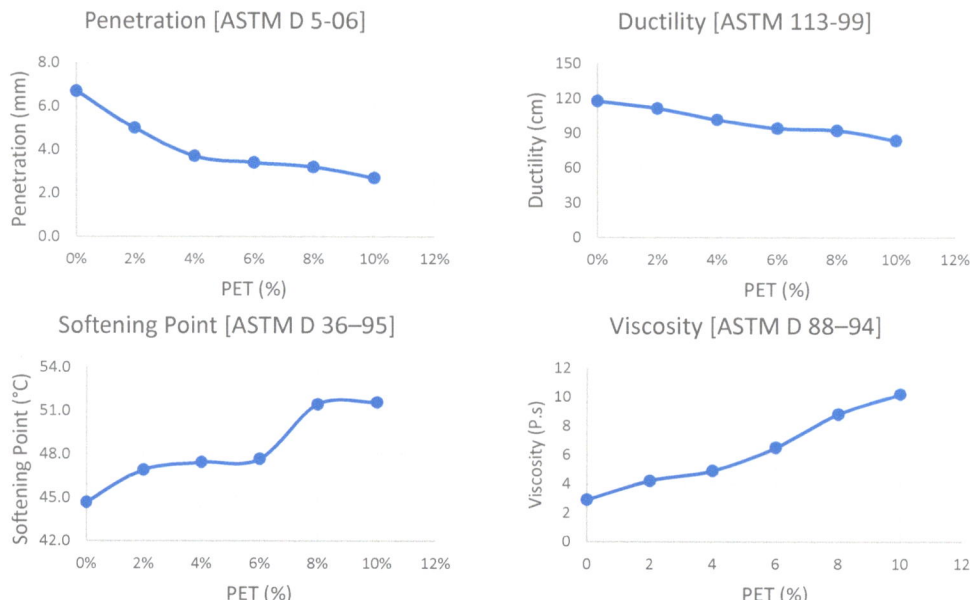

Figure 7. Preliminary Testing on PMB.

According to Baghaee et al., with the addition of plastic the fatigue life of asphalt mixes increases appreciably [12]. To get a traditional fatigue plot, loading cycles are plotted against varying PET content as shown Figure 10. This also represents a comparison of ITF loading cycles against each type of HMA sample (see Table 2) for varying PET content (%). The hypothesis is validated by the obtained results; fatigue life in terms of loading cycles increases dramatically with the increase in PET content from 0% to 2%; the increase is gradual as the PET content is further increased to 4% and 6% in the cases of wet mixing and dry mixing, respectively. However, fatigue life decreases as the PET content is further increased to 6% and 8% in Type-y and Type-z samples, respectively. It is to be noted that the dry mixing technique yielded better results in terms of fatigue life as compared to the wet mixing technique for each PET content.

As per regression analysis, the wet mixing technique produced a stronger relationship between average loading cycles, which signified fatigue life and variation in PET content. The R^2 value for the wet mixing technique was 0.829 while for the dry mixing technique it was 0.613.

4.3. Moisture Susceptibility Test

In the past, various tests have been conducted to assess pavement susceptibility to moisture damage. As per Yilmaz et al., no test to date has accomplished any world-wide standardized acceptance to assess the extent of moisture-induced impairment. As a matter of fact, any test that can compare the test results of damp and dry HMA samples or establish a relation between the two can be utilized to assess the impact of moisture on HMA [3]. The Alabama Department of Transportation (ALDOT) provides a procedure to assess the resistance of HMA samples to moisture-induced damage as per standard ALDOT-361-88. This strategy assesses the change in diametral tensile strength caused by the saturation impact and conditioning of HMA samples on exposure to water [28].

A sample diameter of 4″ (100 mm) and thickness of 2.5″ (63 mm) were used for moisture susceptibility testing in this research. For the controlled mix type (Type-x) two sets of samples were tested, conditioned and unconditioned samples. Similar types of samples were prepared to assess the moisture susceptibility of PET-modified HMA samples (Type-y and Type-z) with varying PET content of 2%, 4%, 6%, 8% and 10%, respectively (see Figure 11a). The samples were conditioned by placing them in a water bath at 60 °C for 24 h. (see Figure 11b). The samples were later placed at 25 °C for 4 h in a temperature-controlled chamber (see Figure 11c) and finally tested in UTM where the load was applied at 50 mm/min (see Figure 11d) Finally, maximum load leading to failure was noted in kN. Figure 11e shows fractured sample.

Figure 8. The ITFT load assembly.

Figure 9. Fatigue-fractured sample.

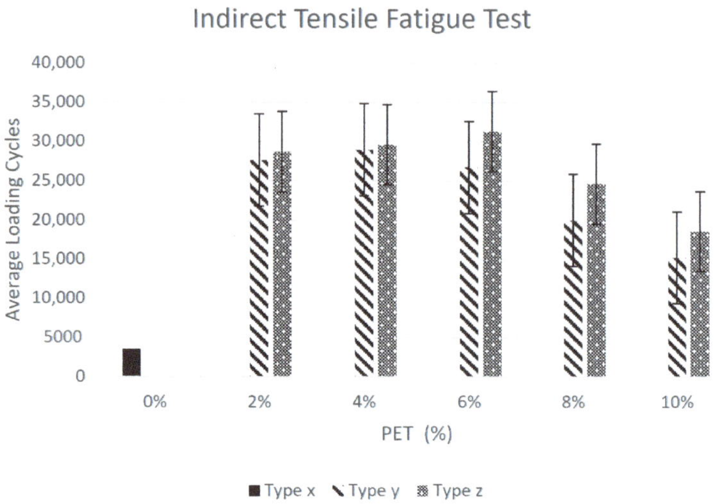

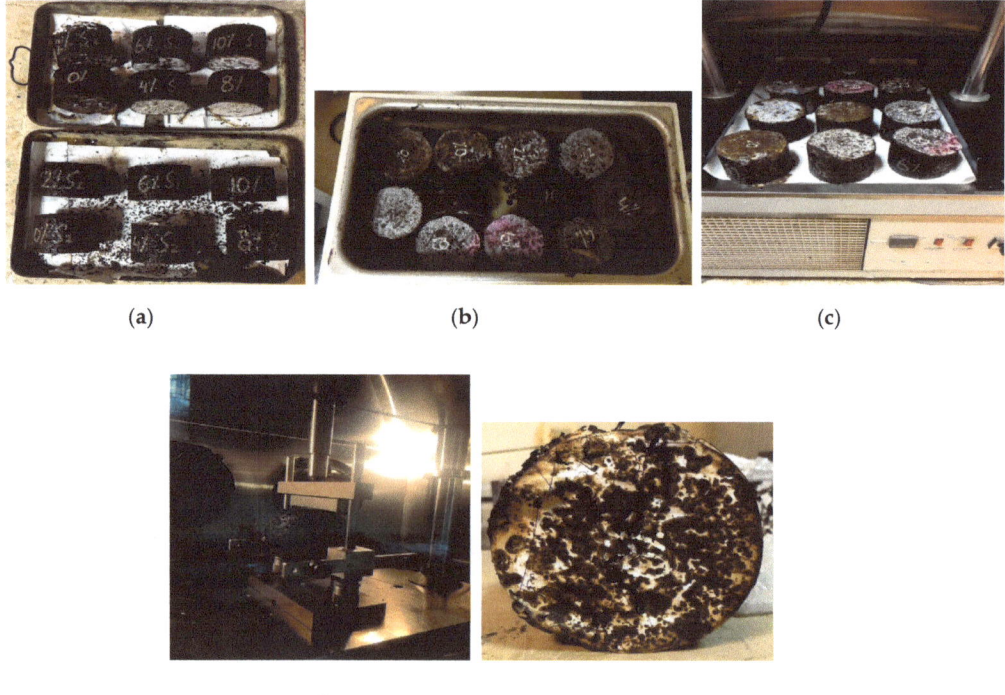

Figure 10. Indirect tensile fatigue test.

Figure 11. Moisture susceptibility testing: (**a**) Unconditioned samples 's-1' and conditioned samples 's-2'; (**b**) Conditioning of samples in water bath for 24 h at 60 °C; (**c**) Unconditioned and conditioned samples kept at 25 °C for 4 h in a temperature-controlled chamber prior to testing; (**d**) Samples tested in UTM where the load was applied at 50 mm/min; (**e**) Fractured sample.

As per standard ALDOT-361-88, indirect tensile strength is measured for both conditioned and unconditioned specimens using Equation (1).

$$S_t = \frac{2000*P}{\Pi*D*T} \tag{1}$$

Here:

S_t = tensile strength (kPa);
D = sample diameter (mm);
T = sample thickness (mm);
P = maximum load (N).

The indirect tensile strength of conditioned and unconditioned samples was compared and presented in the form of TSR using Equation (2).

$$TSR = \frac{S_2}{S_1} \tag{2}$$

Here:

S_2 = average tensile strength of conditioned sample;
S_1 = average tensile strength of un-conditioned sample.

A minimum TSR value greater than 80%, adopted by many roadway agencies' specifications [29], was observed for all three sample types. As per the research findings of Ferreira et al., Silva et al. and Ghabchi et al., the resistance against the moisture damage of PET-modified asphalt mixtures was noticeably higher as compared to controlled HMA mixtures [13–15]. Figure 12 shows a graphical representation of average tensile strength ratios for all three types of HMA samples containing varying PET content (%). The results as shown in Figure 12 also validate the hypothesis, which states that pavement tends to be less susceptible to moisture damage considering improved TSR (%), which increases considerably with the increase in PET content. The improvement in TSR (%) is gradual as PET content is increased from 0% to 2%. Furthermore, the increase in TSR (%) was noted to be gradual as the PET content was further increased to 4% in both mixing techniques. However, pavement resistance to moisture damage showed a declining trend as PET content was further increased beyond 4%. It is to be noted that the wet mixing technique yielded better results in terms of moisture susceptibility as compared to the dry mixing technique for each PET content.

As per regression analysis, the wet mixing technique produced a stronger relationship between the tensile strength ratio and variation in PET content. The R^2 value for the wet mixing technique was 0.734 while for the dry mixing technique it was 0.201.

4.4. Stability and Flow Test

As per AASHTO T245-90, the standard flow test and Marshall stability test were performed on a cylindrical sample. The specimen was placed in a water bath with a temperature of 60 °C for a period of 30 min, followed by samples being damp-dried and placed in the Marshall apparatus. Marshall stability is the maximum load applied for a given strain rate of 2 in. per minute, which brings about failure. During the stability test, the Marshall flow that occurs at a failure point is determined by a gauge that observes the vertical deformations, in mm, occurring in a specimen. The following Figure 13 shows a graphical representation of Stability (kN) and Flow (mm) with varying content of PET.

As per regression analysis, satisfactory results were obtained to establish the validity of both the wet and dry mixing techniques. The R^2 value for the wet mixing technique in establishing the relationship between stability and variation in PET content was 0.922 while for the dry mixing technique it was 0.816; however, the R^2 value for the dry mixing technique in establishing the relationship between flow and variation in PET content was 0.884 while for the wet mixing technique it was 0.865.

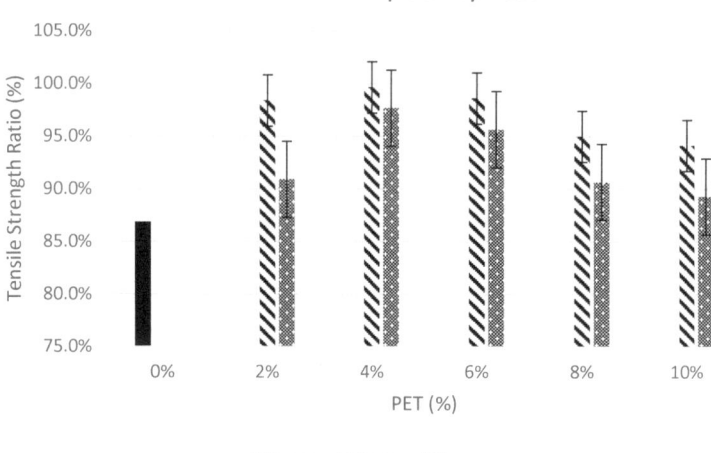

Figure 12. Moisture susceptibility test.

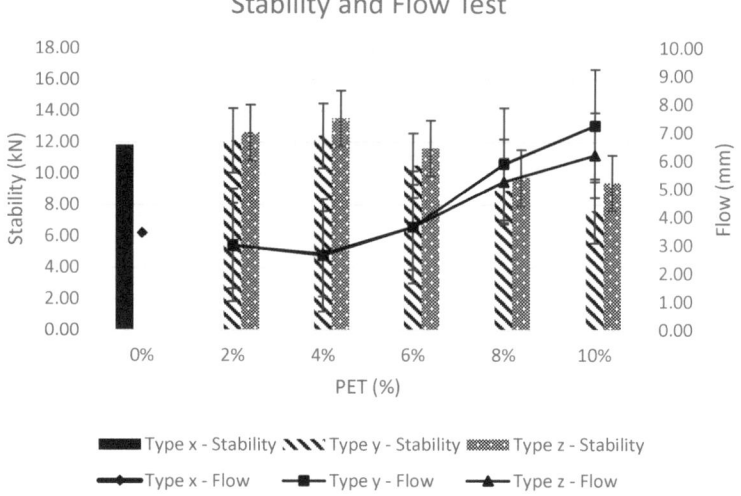

Figure 13. Stability and flow test.

As per the research findings of Baghaee et al. and Ahmadinia et al., the stability of the asphalt mix increases with the addition of PET. The stability of PET-modified HMA is better as compared to controlled HMA samples due to improvement in adhesion among aggregate-binder-plastic pellets, [12,30]; however, in contrast to the findings of Ahmadinia et al. conducted in 2012, stability only increases to a maximum value corresponding to PET content of 4% [30] for both types of HMA mixes, i.e., Type-y and Type-z, and decreases gradually for higher PET contents. It is to be noted that Type-z HMA samples prepared by the dry mixing technique gave higher stability values at each PET concentration. At PET content $\geq 6\%$, the stability value falls to a value even less than Type-x stability.

High flow value is generally an indicator for a mix that is susceptible to permanent deformation under traffic; correspondingly, mixes with low flow values are indicative of the presence of higher air voids than normal values and pavement is likely to face premature cracking owing to the brittleness of the mix during the pavement's service life. The research findings of Kalantar et al. conducted in 2012 and Baghaee et al. conducted in 2013 suggested that flow increases as PET content is increased, [7,12]; however, test results of Type-y and Type-z indicated a decrease in flow values up to the PET content of 4%, which later increased when PET content was increased beyond 4%. A similar trend was noted by Ali et al. in his research conducted in 2013 [20]; however, in the current research, the point of inflection for PET was noted to be 6%.

5. Environmental and Economic Sustainability

The usage of plastic in general and plastic bottles especially is on a surge throughout the world. Despite reducing-reusing-recycling efforts, plastic bottle waste is increasing day by day. As per the WWF study, total waste in Pakistan is estimated to be around 250 million tons of which a majority portion consists of food scraps, plastic bags and PET bottles. Plastic bottles produce 15.5% of total waste in Pakistan [31]. Pollution caused by PET is serious and growth in its consumption is alarming considering the long degradation period of around 300 years [32]. Furthermore, ever rising petroleum rates on a biannual basis [33], the depletion of natural resources [34] and the ban of limestone quarrying in the Margallah Hills [35], adverse environmental pollution, flooding, the lowering of water aquifers, high construction costs and lesser funds, the damage caused to pavements on exposure to moisture [3] and excessive and frequent vehicular loading [12] are a few of the many challenges faced by the road construction industry. Keeping his research findings in view, Baghaee et al. concluded that utilizing waste plastic as pavement-quality enhancer is rewarding, improving pavement life and serviceability while also preventing plastic waste from polluting the environment [12]. Ferreira et al. also concluded that the utilization of alternative pavement construction materials should be encouraged; this reduces the excessive utilization of raw materials and concurrently contributes to sustainable practices in engineering works [13].

To assess cost reduction by using PET in HMA mixtures, cost comparison was carried out for virgin HMA and PET-modified HMA mix. A standard road width of 3.6 m for high-speed, high-volume highways [36] was assumed with an Asphaltic Concrete for Wearing Course (ACWC) thickness of 50 mm [37]. It was assumed for the lower layers of the two types of HMA mixtures, inclusive of Asphalt Concrete Base Course (ACBC), Granular Base Course (GBC), Granular Sub-Base Course (GSBC) that the preparation and compaction costs of the subgrade were completely similar and hence their cost was completely ignored. From the Marshall Mix design, the density of asphalt was found to be 2327 kg/m3 for the HMA mix. To estimate cost, the NHA's section on the Composite Schedule of Rates (CSR)–2014 of the District of Rawalpindi was used [38]. Figure 14 indicates slight decreases of 1.30% and 1.95% in the cost for PET-modified HMA Type-y containing 4% and 6% PET, respectively, when compared with the job mix formula of virgin HMA for a one-kilometer road section. However, due to increased fuel consumption owing to high heating temperatures for the Type-z-modified HMA mix, the cost is 6.57% and 5.67% higher with 4% and 6% PET content. Kristjánsdóttir et al. in 2007 suggested the use of WMA technology, which can be incorporated successfully to lower the temperature at which asphalt mixtures are produced and paved [39]. It is to be noted that the cost analysis is only carried out for material costs involved at initial stages of road construction. When life-cycle cost assessment involving monetary and non-monetary costs is carried out, savings (%) will increase noticeably.

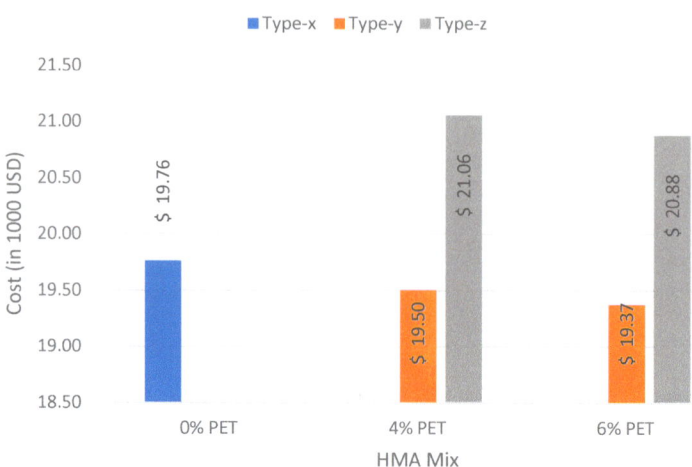

Figure 14. Cost comparison.

Table 8 illustrates various factors that have been considered for the cost analysis of the modified and unmodified mix.

Table 8. Cost analysis data.

Basic Parameters	
Segment length	1000 m
Segment width	3.65 m
Segment thickness	0.05 m
Material volume	180 m³
Material density	2327 kg/m³
Material mass	418,860 kg~418.86 ton
USD 1	PKR 226.52 (December 2022)
Fuel price (Diesel)	PKR 235.30/liter (December 2022)

Material	Type-x	Type-y		Type-z	
		4% PET	6% PET	4% PET	6% PET
Required binder (ton)	18.2	17.5	17.1	17.5	17.5
Cost of binder (PKR)	2,000,325	1,920,312	1,880,305	1,920,312	1,920,312
Required aggregate (ton)	400.7	400.7	400.7	400.7	400.7
Cost of aggregate (PKR)	1,541,081	1,541,081	1,541,081	1,541,081	1,541,081
Fuel required (liter)	3972.3	3972.3	3972.3	5561.2	5561.2
Cost of Fuel (PKR)	934,672	934,672	934,672	1,308,541	1,308,541
PET required	-	0.73	1.09	0.73	1.09
PET cost	-	21,814	32,721	21,814	32,721

6. Conclusions

This research evaluates the fatigue, moisture susceptibility, stability and flow properties of HMA modified by the addition of different percentages of PET to improve poor pavement serviceability due to high traffic volumes and weather conditions and to also assuage the growing pollution caused by plastic bottle waste. Therefore, the main objective of this research was developing environmentally friendly pavements by encouraging the reuse of waste material in the industry.

1. The PET modification of bitumen resulted in a higher softening point, penetration and viscosity of bitumen, which suggests that PET-modified pavement is less likely to

be deformed. However, the ductility of PET-modified bitumen decreases gradually with increasing PET%.
2. With the addition and increase in PET content up to optimum level, the results for resistance against fatigue loading and moisture damage, as well as the results for stability and flow testing, improved. However, the performance testing results worsened once PET content increased further; this may be due to the decreased aggregate-binder bonding and increased stiffness of the modified HMA mix.
3. The dry mixing technique yielded better results for ITFT, stability and flow; however, the wet mixing technique produced better results for moisture susceptibility testing.
4. The optimum PET content that produced the best results for ITFT was 6% (dry mixing); however, for moisture susceptibility testing (wet mixing), flow and stability, the optimum PET turned out to be 4%.
5. Cost comparison suggests a 1.40% and 2.10% decrease in cost for the 4% and 6% PET-modified HMA mix prepared by the wet mixing technique, respectively. For the dry mixing technique cost increases by 5.55% and 4.60% for 4% and 6% PET content, respectively. The PET modification of pavement can be used to cut down life cycle costs. This will not only help to reduce road construction costs in the long run but can also help to mitigate environmental problems such as solid waste disposal and the depletion of natural resources.

This research can further be tested for dual wheel track laboratory equipment to assess the performance of PET-modified pavements under the rutting action of vehicles. The use of WMA technology might be helpful to reduce additional costs associated with the dry mixing of PET into HMA. Therefore, further testing is required to demonstrate the effectiveness of chemicals related to WMA technology and the interaction of such chemicals with the PET-HMA matrix. The current study can also be tested further by incorporating RAP in PET-modified HMA by following the homogenization treatment method for in-plant hot-mix recycled asphalt mixtures as conducted by Jie et al. in 2022 [40]. Furthermore, the current study could also be developed further by studying the change in properties of PET-modified HMA by partially replacing aggregate with slag as per the study conducted by Ahmad Goli in 2022 [41].

Author Contributions: Conceptualization, N.A. and A.H.; Methodology, N.A. and A.H.; Validation, A.H. and Q.Y.; Formal Analysis, N.A.; Investigation, N.A. and A.S.A.; Data curation, N.A.; Writing—original draft, N.A. and A.S.A.; Writing—review and editing, N.A. and A.H.; Visualization, N.A.; Supervision, A.H. and Q.Y.; Project administration, A.H. and Q.Y.; Funding acquisition, N.A. All authors have read and agreed to the published version of the manuscript.

Funding: This research received no funding.

Institutional Review Board Statement: Not applicable.

Informed Consent Statement: Not applicable.

Data Availability Statement: The data presented in this study is available on request from the corresponding author.

Acknowledgments: The authors owe immense gratitude to Attock Refinery Ltd., local plastic shredders and the aggregate crushing plant of Taxila for providing materials for our research work whenever requested.

Conflicts of Interest: The authors declare that they have no known competing financial interest or personal relationships that could have appeared to influence the work reported in this paper.

References

1. Kehinde, O.; Ramonu, O.J.; Babaremu, K.O.; Justin, L.D. Plastic wastes: Environmental hazard and instrument for wealth creation in Nigeria. *Heliyon* **2020**, *6*, e05131. [CrossRef]
2. Aldagheiri, M.I. The Role of the Transport Road Network in the Economic Development of Saudi Arabia. *WIT Trans. Built Environ.* **2009**, *107*, 275–285.
3. Yilmaz, A.; Karahancer, S. Water Effect on Deteriorations of Asphalt Pavements. *J. Sci. Technol.* **2012**, *2*, 1–6.

4. Monismith, C.L.S.Y. *Asphalt Mixture Fatigue Testing*; Wuhan University of Technology: Wuhan, China, 2013.
5. Pavement Interactive Pavement Interactive. Available online: Avementinteractive.org/reference-desk/pavement-management/pavement-distresses/fatigue-cracking/#:~:text=In%20thin%20pavements%2C%20cracking%20initiates,or%20\T1\textquotedblleftclassical\T1\textquotedblright%20fatigue%20cracking (accessed on 16 June 2022).
6. Gade, A.; Tapse, A.; Bonde, S. A Cost-Effective Approach Towards Road Construction—Kondave a Case Study. In *Avement Materials and Associated Geotechnical Aspects of Civil Infrastructures*; Springer: Cham, Switzerland, 2019.
7. Kalantar, Z.N.; Karim, M.R.; Mahrez, A. A review of using waste and virgin polymer in pavement. *Constr. Build. Mater.* **2012**, *33*, 55–62. [CrossRef]
8. Mishra, B.; Gupta, M.K. Use of plastic waste in bituminous mixes by wet and dry methods. In Proceedings of the Institution of Civil Engineers, Uttar Pradesh, India, 9 June 2020.
9. Rahman, W.M.N.W.A.; Wahab, A.F.A. Green Pavement using Recycled Polyethylene Terephthalate (PET) as Partial Fine Aggregate Replacement in Modified Asphalt. *Procedia Eng.* **2013**, *53*, 124–128. [CrossRef]
10. Earnest, M.D. Performance Characteristics of Polyethylene Terephthalate (PET) Modified Asphalt. Spring 2015. Available online: https://digitalcommons.georgiasouthern.edu/etd/1260/ (accessed on 14 June 2022).
11. Casey, D.; Mcnally, C.; Gibney, A.; Gilchrist, M. Development of a recycled polymer modified binder for use in stone mastic asphalt. *Resour. Conserv. Recycl.* **2008**, *52*, 1167–1174. [CrossRef]
12. Baghaee, T.M.; Karim, M.; Mehrtash, S. Utilization of waste plastic bottles in asphalt mixture. *J. Eng. Sci. Technol.* **2013**, *8*, 264–271.
13. Ferreira, J.W.d.S.; Marroquin, J.F.R.; Felix, J.F.; Farias, M.M.; Casagrande, M.D.T. The feasibility of recycled micro polyethylene terephthalate (PET) replacing natural sand in hot-mix asphalt. *Constr. Build. Mater.* **2022**, *330*, 127276. [CrossRef]
14. Silva, J.A.A.; Rodrigues, J.K.G.; de Carvalho, M.W.; Lucena, L.C.F.L.; Cavalcante, E.H. Mechanical performance of asphalt mixtures using polymer-micronized PET-modified binder. *Road Mater. Pavement Des.* **2018**, *19*, 1001–1009. [CrossRef]
15. Ghabchi, R.; Dharmarathna, C.P.; Mihandoust, M. Feasibility of using micronized recycled Polyethylene Terephthalate (PET) as an asphalt binder additive: A laboratory study. *Constr. Build. Mater.* **2021**, *292*, 123377. [CrossRef]
16. Likitlersuang, S.; Chompoorat, T. Laboratory investigation of the performances of cement and fly ash modified asphalt concrete mixtures. *Int. J. Pavement Res. Technol.* **2016**, *9*, 337–344. [CrossRef]
17. Bamigboye, G.O.; Bassey, D.E.; Olukanni, D.O.; Ngene, B.U.; Adegoke, D.; Odetoyan, A.O.; Kareem, M.A.; Enabulele, D.O.; Nworgu, A.T. Waste materials in highway applications: An overview on generation and utilization implications on sustainability. *J. Clean. Prod.* **2021**, *283*, 124581. [CrossRef]
18. Tayyab, S.; Hussain, A.; Fazal, H.; Khattak, A. Performance Evaluation of Fatigue and Fracture Resistance of WMA Containing High Percentages of RAP. *Civ. Eng. J.* **2021**, *7*, 1529–1545. [CrossRef]
19. Sengoz, B.; Isikyakar, G. Evaluation of the properties and microstructure of SBS and EVA polymer modified bitumen. *Constr. Build. Mater.* **2008**, *22*, 1897–1905. [CrossRef]
20. Ali, T.; Iqbal, N.; Ali, M. Sustainability Assessment of Bitumen with Polyethylene as Polymer. *IOSR J. Mech. Civ. Eng.* **2013**, *10*, 1–6. [CrossRef]
21. Sojobi, A.O.; Nwobodo, S.E.; Aladegboye, O.J. Recycling of polyethylene terephthalate (PET) plastic bottle wastes in bituminous asphaltic concrete. *Cogent Eng.* **2016**, *3*, 1133480. [CrossRef]
22. Ahmad, A.F.; Razali, A.R.; Razelan, I.M. Utilization of polyethylene terephthalate (PET) in asphalt pavement: A review. *IOP Conf. Ser. Mater. Sci. Eng.* **2017**, *203*, 012004. [CrossRef]
23. Al-Hadidy, A.I.; Tan, Y. Effect of polyethylene on life of flexible pavements. *Constr. Build. Mater.* **2009**, *23*, 1456–1464. [CrossRef]
24. Nishanthini, J.; Terrance, R.M.; Wasala, B.M. Mechanical properties of modified hot mix asphalt containing polyethylene terephthalate fibers as binder additive and carbonized wood particles as fine aggregate replacement. *Asian Transp. Stud.* **2020**, *6*, 100029.
25. Jegatheesa, N.; Rengarasu, T.; Bandara, W. Effect of Polyethylene Terephthalate (PET) Fibres as Binder Additive in Hot Mix Asphalt Concrete. In *Transaction of Annual Sessions of the Institution of Engineers Sri Lanka*; Institution of Engineers: Colombo, Sri Lanka, 2018.
26. Ahmadinia, E.; Zargar, M.; Karim, M.R.; Abdelaziz, M.; Shafigh, P. Using waste plastic bottles as additive for stone mastic asphalt. *Mater. Des.* **2011**, *32*, 4844–4849. [CrossRef]
27. Hussain, A.; Din, S.; Khan, M.A.; Haq, F.; Asim, R. Moisture Damage and Fatigue Evaluation of Hot Mix Asphalt (HMA) Containing Reclaimed Asphalt Pavement (RAP) and Polythene Bags (LDPE). *Asian J. Nat. Appl. Sci.* **2017**, *6*, 63.
28. Tran, N.; Taylor, A.; Timm, D.; Robbins, M.; Powell, B.; Dongre, R. Comprehensive Laboratory Performance Evaluation. *National Center for Asphalt Technology* September 2010. Available online: https://eng.auburn.edu/research/centers/ncat/files/reports/2010/rep10-05.pdf (accessed on 14 June 2022).
29. Alabama Department of Transportation [ALDOT-361-88] Resistance of Compacted Hot-Mix Asphalt to Moisture Induced Damage. In *Aldot Procedures-Testing Manual*; Alabama Department of Transportation: Montgomery, AL, USA, 2008; pp. 1–5.
30. Ahmadinia, E.; Zargar, M.; Karim, M.R.; Abdelaziz, M.; Ahmadinia, E. Performance evaluation of utilization of waste Polyethylene Terephthalate (PET) in stone mastic asphalt. *Constr. Build. Mater.* **2012**, *36*, 984–989. [CrossRef]
31. Mukheed, M.; Alisha, K. Plastic pollution in Pakistan: Environmental and health Implications. *J. Pollut. Eff. Contr.* **2020**, *4*, 251–258.

32. WWF Tackling Plastic Pollution in Pakistan. 2021. Available online: https://www.wwfpak.org/issues/plastic_pollution/ (accessed on 30 May 2022).
33. Market Rate System Bi-Annual Period. Government of Punjab, 2019–2022. Available online: https://finance.punjab.gov.pk/mr-2019 (accessed on 1 June 2022).
34. UNEP. *Rising Demand for Sand Calls for Resource Governance*; UNEP: Geneva, Switzerland, 2019.
35. Khan, S. *Supreme Court Bans Stone Crushing at Margalla Hills*; International-The News: Islamabad, Pakistan, 2020.
36. AASHTO. *A Policy on Geometric Design of Highways and Streets*, 6th ed.; American Association of State Highway and Transportation Official: Washington, DC, USA, 2011; p. 337.
37. National Highway Authority. *General Specifications*; SAMPAK International (Pvt) Ltd.: Lahore, Pakistan, 1998; pp. 206–208.
38. *National Highway Authority National Highway Authority (NHA)*; SAMPAK International (Pvt.) Ltd.: Lahore, Pakistan, 2016. Available online: https://nha.gov.pk/wp-content/uploads/2016/08/CSR-2014-Punjab.pdf (accessed on 2 June 2022).
39. Kristjánsdóttir, Ó.; Muench, S.T.; Michael, L.; Burke, G. Assessing Potential for Warm-Mix Asphalt Technology Adoption. *Transp. Res. Rec.* **2022**, *2040*, 91–99. [CrossRef]
40. Jie, G.; Yuquan, Y.; Liang, S.; Jing, X.; Jiangang, Y. Determining the maximum permissible content of recycled asphalt pavement stockpile in plant hot-mix recycled asphalt mixtures considering homogeneity: A case study in China. *Case Stud. Constr. Mater.* **2022**, *16*, e00961.
41. Ahmad, G. The study of the feasibility of using recycled steel slag aggregate in hot mix asphalt. *Case Stud. Constr. Mater.* **2022**, *16*, e00861.

Disclaimer/Publisher's Note: The statements, opinions and data contained in all publications are solely those of the individual author(s) and contributor(s) and not of MDPI and/or the editor(s). MDPI and/or the editor(s) disclaim responsibility for any injury to people or property resulting from any ideas, methods, instructions or products referred to in the content.

Article

Influence of Glycerol on the Surface Morphology and Crystallinity of Polyvinyl Alcohol Films

Ganna Kovtun [1,2,*], David Casas [2] and Teresa Cuberes [2]

1. Institute of Magnetism NAS of Ukraine and MES of Ukraine, 03142 Kyiv, Ukraine
2. Group of Nanotechnology and Materials, Mining and Industrial Engineering School of Almaden, University of Castilla-La Mancha, 13400 Almaden, Spain; david.casas@uclm.es (D.C.); teresa.cuberes@uclm.es (T.C.)
* Correspondence: anna-kovtun@ukr.net; Tel.: +38-050-918-6960

Abstract: The structure and physicochemical properties of polyvinyl alcohol (PVA) and PVA/glycerol films have been investigated by Fourier transform infrared spectroscopy (FT-IR), X-ray diffraction (XRD), thermogravimetry/differential thermal analysis (TG/DTA), and advanced scanning probe microscopy (SPM). In the pure PVA films, SPM allowed us to observe ribbon-shaped domains with a different frictional and elastic contrast, which apparently originated from a correlated growth or assembly of PVA crystalline nuclei located within individual PVA clusters. The incorporation of 22% w/w glycerol led to modification in shape of those domains from ribbon-like in pure PVA to rounded in PVA/glycerol 22% w/w films; changes in the relative intensities of the XRD peaks and a decrease in the amorphous halo in the XRD pattern were also detected, while the DTA peak corresponding to the melting point remained at almost the same temperature. For higher glycerol content, FT-IR revealed additional glycerol-characteristic peaks presumably related to the formation of glycerol aggregates, and XRD, FT-IR, and DTA all indicated a reduction in crystallinity. For more than 36% w/w glycerol, the plasticization of the films complicated the acquisition of SPM images without tip-induced surface modification. Our study contributes to the understanding of crystallinity in PVA and how it is altered by a plasticizer such as glycerol.

Keywords: polyvinyl alcohol films; PVA/glycerol blends; crystallinity; infrared spectroscopy; X-ray diffraction; thermogravimetry; differential thermal analysis; atomic force microscopy; lateral force microscopy; ultrasonic force microscopy

Citation: Kovtun, G.; Casas, D.; Cuberes, T. Influence of Glycerol on the Surface Morphology and Crystallinity of Polyvinyl Alcohol Films. *Polymers* 2024, *16*, 2421. https://doi.org/10.3390/polym16172421

Academic Editor: Hsin-Lung Chen

Received: 7 June 2024
Revised: 19 August 2024
Accepted: 21 August 2024
Published: 27 August 2024

Copyright: © 2024 by the authors. Licensee MDPI, Basel, Switzerland. This article is an open access article distributed under the terms and conditions of the Creative Commons Attribution (CC BY) license (https://creativecommons.org/licenses/by/4.0/).

1. Introduction

Nowadays, the development of materials with film-forming capacity is of growing interest. Polyvinyl alcohol (PVA) is one of the most popular synthetic polymers due to its biodegradability, biocompatibility, high elasticity, tribological response, hydrophilicity, and acceptable thermal properties. PVA has excellent film-forming properties and is widely used in the development of sensor materials, drug delivery systems, wound dressings, artificial cartilages, implants, for packaging applications, recovery of organic and inorganic pollutants, etc. [1–6].

However, due to its poor ductility, in practical applications PVA is often used with the addition of plasticizers, which influence its mechanical and thermodynamic properties [7–9]. Plasticizers such as glycerol, sorbitol, sucrose, and some organic acids are well-known as low-molecular-weight compounds that reduce polymer–polymer chain bonding [10]. The breaking of hydrogen bonds increases the molecular mobility of the polymer, which results in increased flexibility of the polymer films. Being nontoxic to humans at low concentrations, glycerol is a widespread plasticizer incorporated in PVA films [11–19], as well as in its nanocomposites and blends.

PVA is a semicrystalline polymer composed of both amorphous and crystalline phases, with the percentage of crystallinity playing a crucial role in determining its performance.

Understanding and controlling the crystallinity of PVA is essential for applications such as drug delivery and wound dressings, as the amount of the crystalline phase will affect the release of active compounds and the material's mechanical properties. Higher crystallinity may enhance the durability and structural integrity of films for biomedical applications. Additionally, in films used as membranes for water treatment and the recovery of organic and inorganic pollutants, crystalline regions can improve mechanical strength and chemical resistance, leading to better performance and longer service life. For adhesives and coatings, the crystallinity of PVA affects adhesion properties, flexibility, and resistance to solvents and environmental factors. In packaging, it influences the film's strength, barrier properties, and stability. Controlling crystallinity can help optimize these films for specific industrial purposes.

In this work, we have conducted a comprehensive analysis of the crystallinity in pure PVA and PVA/glycerol films through a correlative study of data from fast Fourier transform infrared spectroscopy (FT-IR), X-ray diffraction (XRD), thermogravimetry (TG)/differential thermal analysis (DTA), and advanced scanning probe microscopy (SPM).

The use of atomic force microscopy (AFM), friction force microscopy (FFM), and ultrasonic force microscopy (UFM) enables us to map the topographical features of these films, as well as their frictional and mechanical responses, at the nanoscale. UFM is a relatively new AFM-based technique that provides contrast based on nanoscale differences in surface or near-subsurface sample stiffness and adhesive properties [20]. To the best of our knowledge, this is the first time an extensive scanning probe microscopy (SPM) study of this nature has been conducted on these films. SPM offers a unique platform for analyzing the surface properties of polymer films at the nanoscale. It not only allows for detailed examination of surface morphology but also facilitates the investigation of frictional and mechanical characteristics through direct interaction with surface atoms. Unlike scanning electron microscopy, which typically requires ultra-high vacuum conditions and involves the interaction of an electron beam with the polymer surface, SPM studies can be conducted in ambient conditions. This makes SPM a versatile and practical tool for these investigations. By additionally integrating data for structural, mechanical, physicochemical, and thermal characterization from the aforementioned experimental techniques, we seek to achieve a detailed understanding of how glycerol influences the crystallinity and overall behavior of PVA film.

Our study discloses valuable insights into how plasticizers, such as glycerol, impact the PVA nanostructure, thereby offering avenues for tailoring PVA properties to achieve optimal performance for specific applications.

2. Materials and Methods

2.1. Materials and Sample Preparation

PVA (Mw 31,000–50,000 g/mol, 98–99% hydrolyzed) and glycerol ($\geq$99.0%) were supplied by Merck, Darmstadt, Germany. The PVA was dissolved in distilled water with stirring at 90 °C, 400 rpm, when heated in a water bath for 2 h, to prepare a 7.0 wt.% stock solution. Solutions of PVA with 3.5 wt.% concentration and mixed PVA/glycerol solutions containing 3.5 wt.% PVA and 1.0 wt.%, 2.0 wt.%, and 4.4 wt.% glycerol were prepared from the 7.0 wt.% PVA stock solution, glycerol, and distilled water, stirred at 25 °C, 400 rpm, 20 min, poured on polystyrene Petri dishes, and evaporated at room temperature (20–25 °C) and ambient humidity ~50% R.H. The glycerol content in the prepared films in relation to the amount of dry PVA was 22% w/w, 36% w/w, and 55% w/w. Films ~200 μm thick were obtained by this method (solution casting). The film preparation procedure is illustrated in Figure 1.

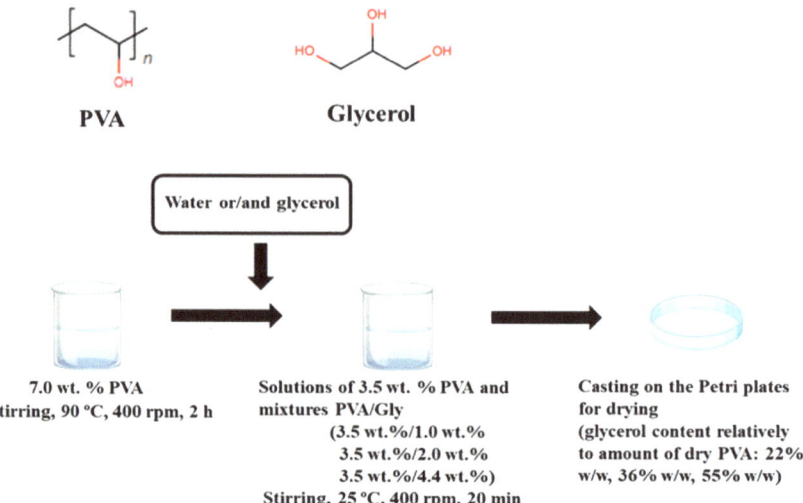

Figure 1. Structural formulas of PVA and glycerol; film preparation scheme.

2.2. Fourier-Transformed Infrared Spectroscopy (FT-IR)

FTIR spectra (4 cm^{-1} resolution, wavenumber range 500–4000 cm^{-1}) were recorded with a Shimadzu IRPrestige-21 spectrometer (Tokio, Japan), using the ATR method. Small pieces of the PVA and PVA/glycerol films were cut and placed in the instrument sample holder. The data were acquired using the software Shimadzu IR Solution 1.21 (Tokio, Japan) and analyzed using Originpro 2024 (Northampton, MA, USA) software.

2.3. X-ray Diffraction (XRD)

XRD measurements were performed in an equipment Philips X'Pert MPD (Eindhoven, Holland) using CuK α radiation (1.54056 Å) with 40 KV and 40 mA. It incorporates 0.04 rad soller slits for both incident and diffracted beams, an automatic 12.5 mm programmable divergence slit, and a Xe gas-sealed proportional detector. Data were collected in an angular range between 1° and 50° (2θ) with a step size of 0.02° and a counting time of 0.70 s per step, and analyzed using Originpro 2024 (Northampton, MA, USA) software.

2.4. Thermogravimetric Analysis (TGA) and Differential Thermal Analysis (DTA)

The thermal behavior of the samples was examined with a Setaram model TG/DTA92 equipment (Caluire-et-Cuire, France) using a Pt crucible. Typically, thermograms were recorded in air atmosphere, within a temperature range from 20 °C to 500 °C, with a heating rate of 25 °C/min. The data were analyzed using Originpro 2024 (Northampton, MA, USA) software.

2.5. Scanning Probe Microscopy

Contact-mode atomic force microscopy (AFM), lateral force microscopy (LFM), and ultrasonic force microscopy (UFM) were performed using a NANOTEC (Madrid, Spain) instrument. The modification of the AFM equipment for the incorporation of UFM facilities is described in [20]. For UFM, ultrasonic frequencies of ~3.8 MHz and modulation frequencies of 2.4 KHz were applied from a piezoelectric element placed under the sample. Typically, Olympus silicon nitride cantilevers with a nominal spring constant of 0.06 N/m and a nominal tip radius of 20 nm were used. The measurements were performed in air at ambient conditions (20–25 °C, ~50% R.H.). Data were analyzed using the WSxM 5.0 software (Madrid, Spain) [21].

3. Results and Discussion

FTIR spectra for the pure PVA and PVA/glycerol blended films with different glycerol contents (22, 36, and 55% w/w) are displayed in Figure 2a.

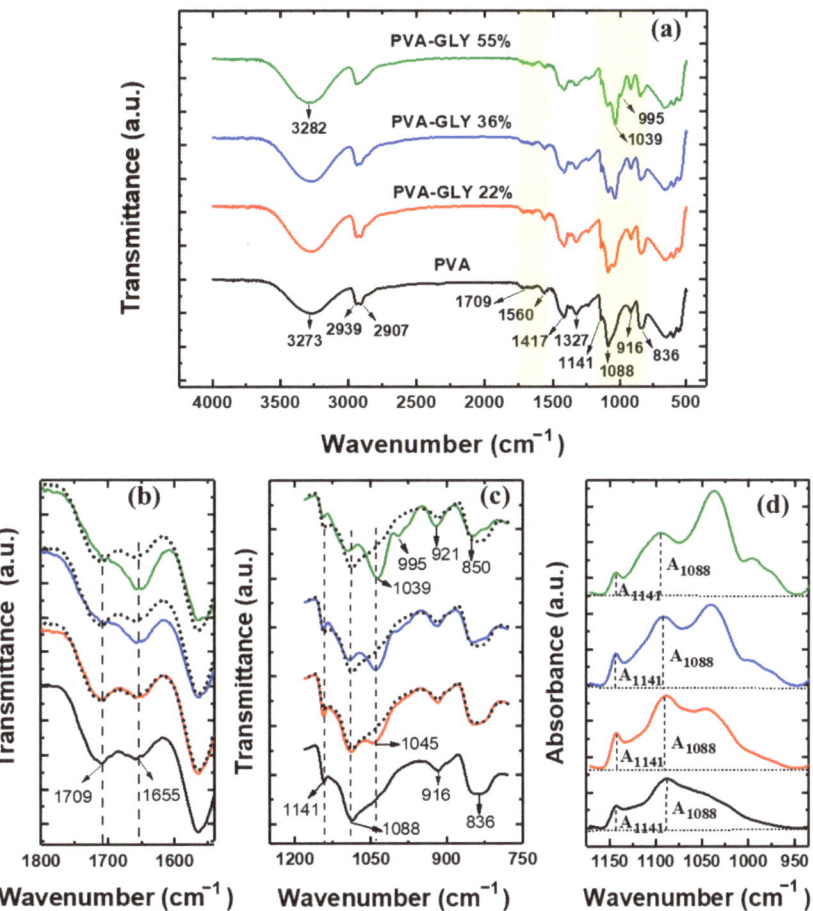

Figure 2. (a) Normalized FT-IR spectra of the pure PVA (black curve) and PVA/glycerol films with different concentrations of glycerol: 22 (red curve), 36 (blue curve), 55 (green curve) % w/w. (b,c) Zoom of regions highlighted in yellow in (a). In (b,c), the dotted black curves correspond to the pure PVA film, which has been shifted and superimposed on each of the other curves to facilitate comparison. (d) Absorbance spectra (inverse from the transmittance) in the selected region to illustrate the procedure for constructing the baseline to determine A_{1141} and A_{1088} in Equation (1) (see text).

For the pure PVA film, the band at 3273 cm^{-1} corresponds to (O-H) stretching vibration from the intermolecular and intramolecular hydrogen bonds [18]. The two bands at 2939 and 2907 cm^{-1} correspond to the asymmetric and symmetric stretching vibrations of methylene (–CH$_2$–), respectively [22]. The peaks at 1709 cm^{-1}, 1655 cm^{-1}, and 1560 cm^{-1} have been related to the stretching vibrations of the (C=O) and (C–O) bonds present in the remaining acetate units [23], which in our case must be very few, as we are using highly hydrolyzed PVA. The 1655 cm^{-1} band has also been assigned to absorbed water [24,25]. The peaks at 1417 cm^{-1} and 1327 cm^{-1} have been attributed to bending vibrations of hydroxyl (–OH) and wagging of (C–H), respectively [26]. The peak at 1141 cm^{-1} has been assigned to (C–O) stretching vibrations in C–OH groups of the crystalline polymer phase [27] and also to (C–C) stretching vibrations of the carbon framework of the polymer chain in the crystalline phase [28]. The peak at 1088 cm^{-1} corresponds to the (C–O) stretching vibrations [25], the band at 916 cm^{-1} to CH$_2$ rocking vibration, and this at 836 cm^{-1} to (C–C) stretching and (C–H) out-of-plane vibrations.

For the PVA-glycerol films, (Figure 2a) a shift of the (O–H) stretching vibration band from 3273 cm^{-1} to higher wavenumbers is observed when increasing the glycerol content (up to 3283 cm^{-1} for 55% w/w of glycerol). A possible reason for this effect is the dissociation of hydrogen bonds between the PVA chains and the formation of new hydrogen bonds between the PVA and glycerol molecules [18,29]. Similar results were obtained for films of PVA/chitosan with the addition of glycerol [29] and films of wheat starch with glycerol [30].

For the PVA/glycerol 55% w/w film, the band at 2939 cm^{-1} experiences a slight increase compared to the 2907 cm^{-1} band. For pure glycerol, the aliphatic (C–H) group bands appear at 2931 cm^{-1} and 2879 cm^{-1}. Hence, the aggregation of glycerol molecules within the PVA matrix might possibly account for this effect [31,32].

The peak at 1655 cm^{-1} increases with increasing glycerol concentration (see Figure 2b). A similar effect is noticeable from the FT-IR spectra for PVA/glycerol films in [13], although the origin of this effect was not discussed in this reference. According to [31] an FT-IR band at ~1653 cm^{-1} appears for commercial glycerol. This band may be assigned to (H–O–H) bending vibrations of water. Hence, the observed increase in the peak at ~1653 cm^{-1} with increasing glycerol content (see Figure 2b) may be indicative of an increase in the amount of water in the film.

The most noticeable change in the glycerol-modified PVA films is the appearance of a new peak at 1039 cm^{-1} (see Figure 2a,c).

In the FT-IR spectrum of pure glycerol, five characteristic bands have been observed, located at 800 up to 1150 cm^{-1} corresponding to vibrations of (C–C) and (C–O) linkages: three broad bands at 850, 925, and 995 cm^{-1} that have been attributed to the vibration of the skeleton (C–C); a peak at 1045 cm^{-1} associated to the stretching of the (C–O) linkage in the primary alcohol groups and a band at 1117 cm^{-1} that corresponds to the stretching of (C–O) in the secondary alcohol group [32–34].

In Figure 2c, a peak at 1045 cm^{-1} appears for the film with 22% w/w glycerol content (red curve) and increases in intensity while slightly displacing to lower wavelengths, up to 1039 cm^{-1}, as the amount of glycerol incorporated into PVA is increased (blue and green curves). Such peak corresponds to (C–O) stretching of the primary alcohol groups of glycerol, and its shifting to lower wavelengths is indicative of increased glycerol/PVA interactions—presumably of the hydrogen bonding type—for higher glycerol contents. For PVA films with 36 and 55% w/w glycerol (blue and green curves in Figure 2c,d), the PVA band at 1088 cm^{-1} shows a clear modification of its shape due to an increase in intensity at ~1117 cm^{-1} (corresponding to stretching of (C–O) in the secondary alcohol group of glycerol), and slightly displaces to higher wavelengths (1091 and 1093 cm^{-1} for 36 and 55% w/w glycerol, respectively).

The fact that for 22% w/w glycerol the characteristic peak of glycerol for the primary alcohol group appears at 1045 cm^{-1}, as in pure glycerol, suggests that for that film (22% w/w glycerol), it is the secondary alcohol group that preferentially forms hydrogen bonds with the PVA molecules. As the amount of glycerol is increased, additional types of glycerol interactions within the PVA matrix may take place, and the peak corresponding to the primary alcohol group displaces to lower wavelengths, up to 1039 cm^{-1}.

For the 36% w/w glycerol films, the emergence of a band at ~995 cm^{-1} is also apparent, indicative of the excitation of vibrations of the glycerol (C–C) skeleton. For the film with 55% w/w glycerol, four glycerol bands are apparent in the covered wavelength range in Figure 2c. In addition to bands at 1039 and 995 cm^{-1}, the intensity of the band at 921 cm^{-1}

increases, and a clear peak at 850 cm^{-1} appears, while for pure PVA a relatively broad band at 836 cm^{-1} is observed (Figure 2c).

According to the literature [25,35], the crystallinity of various PVA films can be calculated from the FT-IR data by quantitative correlation between the intensities of the bands corresponding to the crystalline and amorphous phases. The positions of the PVA crystallinity absorption maximum given in the literature vary in the range 1141–1145 cm^{-1}. We have used the value 1141 cm^{-1} that corresponds to the absorption maximum obtained in our experiments to calculate the crystallinity α of our PVA and PVA/glycerol films and considered the peak at 1088 cm^{-1} attributed to (C–O) stretching vibration as representative of the amorphous phase, according to the equation [25]:

$$\alpha = -13.1 + 89.5 \times \left(\frac{A_{1141}}{A_{1088}}\right) \quad (1)$$

where A_{1141} and A_{1088} are the intensities of the absorption peaks in the crystalline and amorphous phases, respectively. According to [25], this equation is applicable for the calculation of α values in the range 18–60% with a correlation coefficient of 0.999 and an absolute error of determining $\alpha \pm 0.3\%$.

The detailed procedure for constructing the baseline relative to which the intensities of the peaks A_{1141} and A_{1088} are to be measured is described in [25], and it is illustrated in Figure 2d. The baseline must pass through the minima in the absorption spectrum located at the edges of the examined spectral range, as depicted in Figure 2d. The position of the (C–O) band maximum at 1088 cm^{-1} varied slightly with increasing glycerol concentration, 1091 and 1093 cm^{-1} for PVA/glycerol 36 and 55% w/w, respectively (for sake of clarity, we refer to it as the band at 1088 cm^{-1} in all cases). The crystallinity band maximum is almost at the same position for the different films: 1141 cm^{-1} for the pure PVA film and 1142 cm^{-1} for PVA/glycerol 55% w/w.

The crystallinity of the PVA films with different glycerol content calculated by FT-IR is shown in Table 1. It is apparent that the crystallinity first increases with the addition of glycerol and then decreases to a level below that of the pure PVA film. Possibly, a small amount of glycerol, with their secondary alcohol groups linked by hydrogen bonding to the PVA chains, may even locally enhance their regular arrangements, resulting in an increase in PVA crystallinity. As the glycerol content increases, more glycerol molecules are available to interact with the PVA or with each other, destabilizing the interactions between the PVA chains and thus reducing the crystallinity.

Table 1. Crystallinity percentage (α) of PVA films with different glycerol content calculated by FT-IR (Figure 2) and XRD (Figure 3, Debye–Scherrer formula applied to peak at 19.7°); melting points (T_M) determined from DTA (Figure 4c).

Glycerol (% w/w)	α(FT-IR)	α(XRD)	Crystal Size (nm)	T_M (°C)
0	27.1	27	4.25	255
22	30.4	30	5.08	253
36	27.3	25	4.90	231
55	19.5	-	-	-

The influence of glycerol on the crystalline structure of PVA films prepared via solution casting was also investigated by means of XRD. According to [36,37], the XRD diffractogram of pure PVA presents characteristic peaks at 2θ ~11.2° (100), 16.1° (001), 19.4° and 20.1° $(10\bar{1})/(101)$, and 22.5° (200). A crystalline peak at ~40–42° related to diffraction from $(1\bar{1}1)/(210)$ planes has also been reported [37–39].

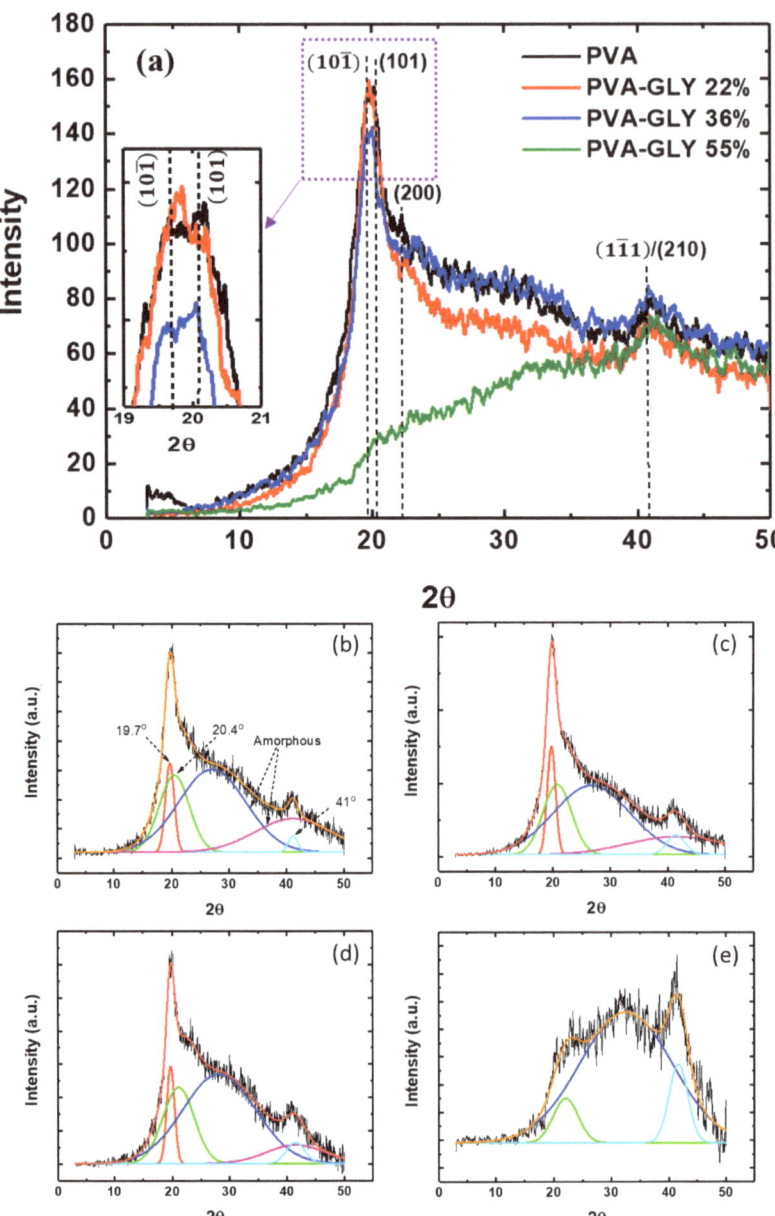

Figure 3. (**a**) XRD patterns of pure PVA (black curve) and PVA/glycerol films with different glycerol content: 22 (red curve), 36 (blue curve), 55 (green curve) % w/w. The inset displays an enlarged region of the peak at approximately 20° for pure PVA (black curve) and PVA/glycerol films with different glycerol content: 22 (red curve), 36 (blue curve) % w/w. (**b**–**e**) Fitting of XRD curve for PVA (**b**), PVA/glycerol 22% w/w (**c**), PVA/glycerol 36% w/w (**d**), PVA/glycerol 55% w/w (**e**).

Figure 3 shows the XRD diffractograms of the pure PVA and PVA/glycerol films with different glycerol content. The expected positions for the (100), (001), $(10\bar{1})/(101)$, and (200) diffraction peaks are indicated by dashed lines. For the films of pure PVA, and those with 22 and 36% w/w glycerol content, a diffraction peak about $2\theta = 19.7°$ is clearly distinguished, as well as a small hump at ~40.7°. The obtained XRD patterns are typical for a semicrystalline polymer, with a broad amorphous halo originating from disordered (noncrystalline) regions. We understand that broad amorphous peaks (halos) in the XRD diffractogram reflect preferential interatomic distances, corresponding to the maximum of the amorphous halo, within the disordered regions of the polymer. From Figure 3, it can be seen that, as a result of the addition of 22% w/w glycerol (red curve in the figure), the diffuse scattering associated with the amorphous phase is reduced, which may be due to a structural rearrangement of the polymer chains. In [17], it was suggested that the introduction of glycerol in PVA films could significantly influence the structure of the amorphous domains. According to the FT-IR data, the crystallinity of this film experiences a slight increase (see Table 1). For a glycerol concentration of 36% w/w (blue curve), the XRD diffractogram shows now a larger contribution of the amorphous phase, similar to that of the pure PVA film, although the intensity of the peak at $2\theta = 19.7°$ is lower. For films with 55% w/w glycerol (red curve), this peak does not appear in the diffractogram. The small peak at 22.5° decreases for films with glycerol compared to pure PVA. Interestingly, the peak at $2\theta = 40.7°$ does not change substantially for the different samples studied.

It is worth noticing that for the diffraction peak at $2\theta = 19.7°$ in the Figure 3, apparently, the ratio between the intensities corresponding to the diffraction of $(10\bar{1})/(101)$ planes changes with the addition of glycerol (see the inset in Figure 3). For the film with 22% w/w glycerol, the diffraction from $(10\bar{1})$ planes slightly increases compared to that of the pure PVA films. A similar effect was reported in [36,39] for pure PVA films with different water content. The incorporation of water molecules in the PVA matrix induced an increase in the diffraction signal of the $(10\bar{1})$ planes compared to that of the (101) planes. As this could be an effect of water plasticization, it is understandable that the incorporation of a small amount of glycerol also induces a similar effect. In addition, the incorporation of glycerol is expected to increase the water content in the PVA film. As more glycerol is incorporated into the film (36% w/w), the effect is reversed. The ratio of the $(10\bar{1})/(101)$ signals and the shape of the amorphous halo in the diffractogram of the PVA/glycerol 36% w/w film become similar to those of the pure PVA film.

For the quantitative description and understanding of glycerol impact on crystallinity of PVA, the diffraction scattering peaks in Figure 3 were deconvoluted into fundamental Gaussian curves with $R^2 = 0.99; 0.95; 0.98; 0.95$ for 0, 22, 36, and 55% w/w glycerol, respectively [40] (Figure 3b–e), and the degree of crystallinity of the PVA films was calculated from the XRD data using the following formula [38]:

$$\alpha = \frac{A_c}{A_c + A_a} \times 100\% \tag{2}$$

where A_c and A_a are the areas of the crystalline and amorphous phases, respectively.

The crystallinity for PVA/glycerol 55% w/w was not calculated, as the main PVA crystalline peak at 19.7° did not appear in the diffractogram, and the most prominent feature was the broad amorphous halo.

The results obtained for the different films are indicated in Table 1. As it is noticeable from Table 1, the results derived from analysis of the XRD data support those obtained from the FT-IR spectra.

Interestingly, the deconvolutions in Figure 3b–d show that the characteristic broad amorphous halo in the films of pure PVA and 22% w/w glycerol is centered at $2\theta = 27°$ (Figure 3b,c). As discussed above, for the film with 22% w/w glycerol, the amorphous halo decreases compared to that of the pure PVA film. According to the FT-IR results, the glycerol molecules in this film are expected to be preferentially attached to the PVA chains through H bonding of their secondary alcohol. These molecules may cause the PVA chains to restructure, altering the organization of the amorphous phase and resulting in a slight increase in crystallinity. For the film with 33% w/w glycerol, the amorphous halo in the XRD is now centered at $2\theta = 28°$, and its intensity increases (Figure 3d). FT-IR indicates that as the amount of glycerol in the films increases, additional types of glycerol/PVA interactions occur. This may again lead to a restructuring of the amorphous phase and possibly the emergence of a new amorphous phase formed by aggregates of glycerol or PVA-glycerol complexes. Eventually, for the film with 55% w/w glycerol, the amorphous halo appears centered at $2\theta = 33°$ and is prominent in the diffractogram (Figure 3e). The displacement of the amorphous halo to higher scattering angles, indicative of preferentially smaller interatomic distances in the amorphous phase, strongly supports the hypothesis that a different amorphous phase than this in pure PVA is responsible for the X-ray scattering in this case. The FT-IR data discussed above are consistent with the presence of a new glycerol-related phase in the films with 55% w/w glycerol.

In order to further study the relation between the XRD pattern and the crystallite size, the well-known Debye–Scherrer formula was used:

$$D = K\lambda/\beta\cos\theta \qquad (3)$$

where D is the mean crystallite size in the direction perpendicular to the lattice planes of the selected diffraction peak, λ is the X-ray wavelength, β is the full width at half maximum (FWHM) of the peak, K is the Scherrer constant with a value of about 0.9, and θ is the diffraction angle corresponding to the selected peak in the XRD pattern. The values for D estimated from the peak at $2\theta = 19.7°$ are displayed in Table 1.

From Table 1, it is also observed that both the degree of crystallinity and the crystallite size are slightly higher for the film of PVA with 22% w/w glycerol than for this of pure PVA. For 36% w/w glycerol, the degree of crystallinity is smaller, even though the size of the existing crystals does not experience significant modifications.

A decrease in crystallinity as a result of the addition of glycerol in PVA and PVA/starch films has been previously reported [17,41]. Also, an increase in crystallinity in PVA for low glycerol concentrations was observed by XRD [42].

The effect of glycerol on PVA crystallinity may be explained in terms of its influence on the interactions between the PVA chains. A small amount of plasticizer may increase the ability of the molecular chains to adopt different conformations and promote their ordering [43]. In order to prepare the PVA/glycerol films, we first prepare a 7.0 wt.% PVA stock solution by dissolving the purchased PVA granules in distilled water at 90 °C and then cooling to room temperature (see Section 2.1). Crystalline PVA nuclei are expected to form at this stage. Subsequently, distilled water and glycerol are added in different concentrations to prepare the final solutions, which are poured into Petry dishes and dried at room temperature. Increasing the amount of water may facilitate the ordering and assembly of existing PVA nuclei, and the presence of a small amount of glycerol in the solution may play a similar role. If glycerol is incorporated in higher quantities, glycerol aggregates may form; the glycerol molecules may selectively interact with the PVA molecular chains at the crystalline interface and even penetrate into the crystals, disrupting the molecular order [9,18].

TGA/DTA was performed to determine the temperature-dependent water/glycerol mass loss, the thermal stability, the glass transition temperature, and the melting point and enthalpy of fusion of the pure PVA and PVA/glycerol films.

Figure 4 shows the thermogravimetric curves (mass loss versus temperature) (TG) (Figure 4a) and DTA curves (Figure 4b,c) for the PVA and PVA/glycerol films. Thermal analysis was carried out in a static air atmosphere at a 25 °C/min heating rate. The inset in Figure 4a represents the determination of the onset of mass loss for the studied samples (the curves at the inset have been arbitrarily shifted on the y-axis for ease of observation). The temperature at the onset decreases with an increase in glycerol content from 140.5 °C for PVA to 112.2 °C for 36% w/w glycerol.

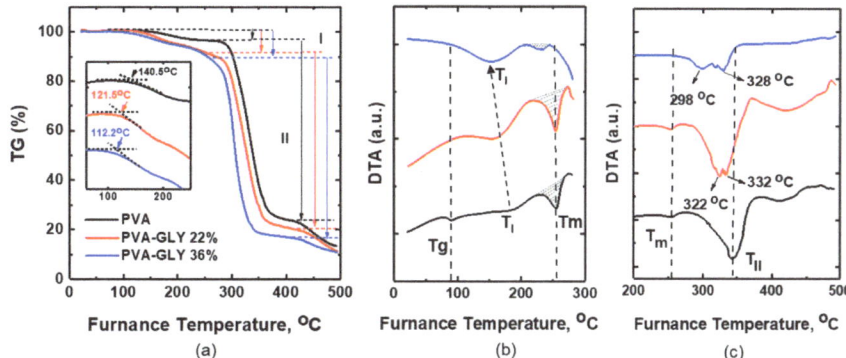

Figure 4. (**a**)—TG, (**b**,**c**) DTA curves of PVA (black curve) and PVA/glycerol films with 22% w/w (red curve), 36% w/w (blue curve) glycerol. The inset on (**a**) represents the onset of mass loss.

The thermal transformations of pure PVA films typically occur in 4 main stages of mass loss [44,45]. The first stage corresponds to the loss of physically absorbed water, which is related to an endothermic peak in the DTA curve labeled as T_I in Figure 4b. The second corresponds to the partial dehydration of PVA, accompanied by polyene formation, and is related to the endothermic peak in the DTA curve labeled as T_{II} in Figure 4c. The third and fourth degradation stages are related to polyene decomposition and thermo-oxidation of carbonized residues. The degradation mechanism of PVA is explained in [46].

Table 2 lists the weight loss percentage and temperatures of the DTA peaks (T_I and T_{II}) for the first two stages, as derived from Figure 4a–c for each sample.

Table 2. Weight loss and temperatures corresponding to maximum rate of weight loss at first two stages (determined in Figure 4a–c).

Sample	Degradation Step	Weight Loss, %	$T_{DTA\ peak}$, °C
PVA	I	3.6	186
	II	72.9	342
PVA-GLY 22%	I	8.6	161
	II	71.4	332/322
PVA-GLY 36%	I	10.9	150
	II	72.4	328/298

The first degradation stage (Figure 4) is characterized by a weight loss of about 3.6–10.9% (Table 2), presumably related to the removal of water. According to FT-IR results, the water amount increases with an increase in glycerol content (see Figure 2b, the intensity of the peak at 1655 cm^{-1} increases). In our case, the weight loss on the first stage for PVA/glycerol films is higher than for pure PVA. However, the weight loss in this case cannot be straightforwardly attributed solely to water loss. According to [47], the thermal degradation of pure glycerol in air is characterized by a single event that takes place at 194–246 °C for a heating rate of 10 °C/min. As the first stage for PVA/glycerol 22 and 36% w/w finishes at about 270–250 °C, respectively, already above the degradation temperature of glycerol, we understand that the observed weight loss in the first stage should be due not only to water loss but also due to degradation (or initial degradation) of glycerol. As per the studies in [48–50], the weight loss observed in the temperature range of 125–290 °C for glycerol plasticized polymer films can be attributed to the degradation of glycerol and the loss of chemisorbed water.

The second degradation stage for the pure PVA and PVA/glycerol films is characterized by the highest weight loss. It is noticeable from Figure 4c that for the PVA/glycerol films, T_{II} shifts to lower temperatures as the glycerol content is increased. For glycerol-containing films, this peak splits into two peaks, which become more distinct for 36% w/w glycerol (298 °C and 328 °C, Figure 4c). This may be related to the degradation of a glycerol-based phase in the PVA matrix, possibly consisting of glycerol/PVA complexes, which degrade at a different temperature than pure PVA in this degradation stage.

Considering the trend for the various PVA/glycerol samples in Figure 4 and Table 2, the temperatures corresponding to the onset of mass loss and the endothermic minima of the DTA peaks T_I and T_{II} shift to lower values as the glycerol concentration increases. This indicates that the increase in glycerol content decreases the thermal stability of the PVA/glycerol films.

Focusing on Figure 4b, an endothermic minimum at 90 °C can be observed on the DTA curve for pure PVA. This peak, labeled Tg, is attributed to the glass transition temperature of PVA [51,52]. This peak was not observed in the DTA curves of the PVA/glycerol films.

The endothermic peak labeled T_m in Figure 4b can be related to the melting points of pure PVA and PVA/glycerol films [13,53,54]. Interestingly, the incorporation of 22% w/w glycerol did not lead to a significant reduction in the PVA melting point. This peak shifts to lower temperatures as the glycerol content in the film is increased to 36% w/w [13,54], indicating that the ordered association of the PVA molecules decreases with higher amounts of glycerol. The enthalpy of melting, represented by the area under the curve corresponding to the melting peak (Figure 4b), increases slightly (by ~1.6 times) for the film with 22% w/w glycerol content compared to that of the pure PVA films and decreases (by ~3.5 times) for 36% w/w. This provides further evidence of the increase and decrease in crystallinity in those samples, consistent with the FT-IR and XRD studies.

SPM images of the pure PVA films are displayed in Figures 5 and 6. They were obtained from different pieces cut from the same sample cast on a 90 mm Petri dish, and, as will be explained in detail below, we understand that they illustrate different stages—possibly determined by the rate of water evaporation—of the development of crystalline domains in the semicrystalline PVA film.

In Figure 5a, the topography is characterized by clusters of aggregates, some of which appear aligned along a well-defined direction, like those enclosed within the white dashed lines. In the UFM image, recorded over the same surface area (Figure 5b), the shape of the clusters that form the topographic aggregates is much better resolved. A close look into the area outlined by the white dashed ellipse in Figure 5b reveals the presence of clusters ~45 nm in diameter, which yield a higher (stiffer) UFM contrast, forming a zig-zag chain. In the topographic image (Figure 5a), the cluster size appears to be ~80–100 nm, which may be due to UFM detecting the stiffer zones within the individual clusters. Most of the clusters within the white dashed lines in Figure 5b exhibit a stiffer contrast. In LFM (Figure 5c,d), aggregates within those dashed lines are imaged with a lower frictional contrast. Notice that in Figure 5c,d, other areas, like, for instance, that indicated by the white slanted arrows, give rise to a similar lower frictional contrast, which now does not directly correlate with any particular features of the topographic and/or UFM contrast.

We conjecture that the alignment of PVA clusters along well-defined orientations observed in Figure 5 originates from a correlated growth or assembly of PVA crystallites within the amorphous PVA matrix. Semicrystalline polymers typically crystallize into lamellar structures where molecular chains fold into layers. Spherulitic crystal growth begins with a primary nucleus at the center, from which radial fibrils consisting of individual lamellar structures separated by amorphous material radiate outwards [55]. The stiffer contrast observed in UFM for the aligned clusters is consistent with crystalline order, leading to a locally higher density in those regions. Previous studies of crystallite nucleation in PVA films with sodium montmorillonite fillers by AFM techniques support this interpretation [56]. The crystalline areas of polymer films are expected to yield a lower friction contrast, as in the case of the aligned clusters in Figure 5c,d [56]. A higher amount of water is actually expected at the PVA amorphous/crystalline interface regions [57], which may act as a lubricant, reducing friction. The areas with a similar lower frictional contrast in Figure 5c,d not correlated to specific features in the topographic or UFM image (e.g., area pointed by the white slanted arrow) may be due to an inhomogeneous distribution of water at the surface or near-surface area of the polymer films.

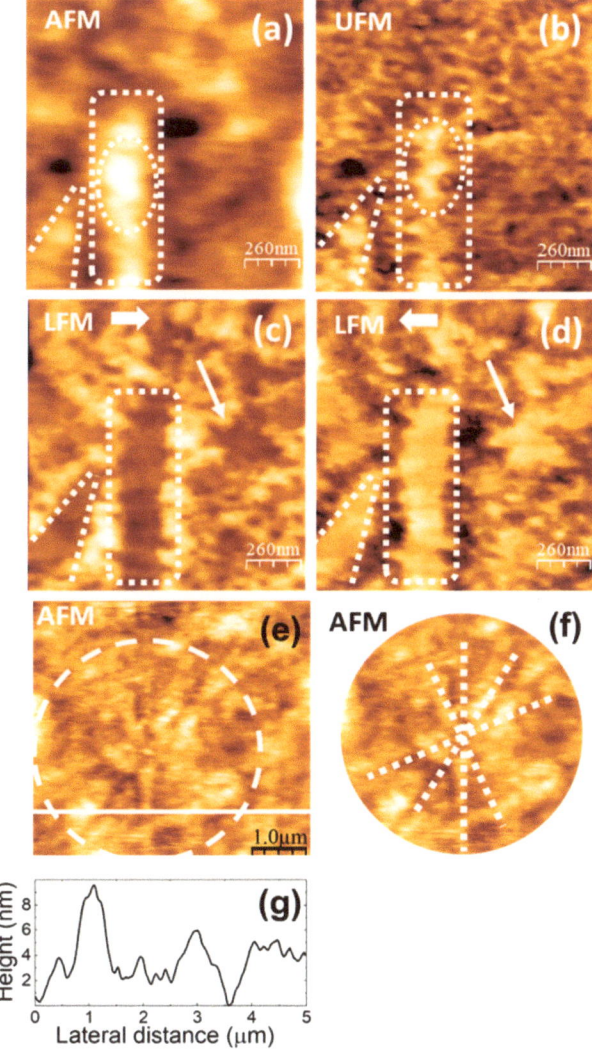

Figure 5. SPM images recorded on pure PVA. (**a**) Contact-mode AFM topography. Color-scale range: 8 nm. (**b**) UFM image recorded simultaneously with (**a**). (**c**,**d**) LFM images recorded over the same surface region than (**a**) scanning from left to right ➡ (**b**) and from right to left ⬅ (**c**). (**e**) Contact-mode topographic image. Color-scale range: 16 nm. (**f**) Region enclosed by the dashed white circle in (**e**). (**g**) Height-contour profile over the continuous white line in (**e**).

Figure 5e shows a larger topographic image of the PVA surface. Intersecting rows of PVA cluster aggregates are noticeable in this image. In the area within the white dashed circle in the image, some of the rows appear to stem from a primary nucleus at the center. In Figure 5f, white dotted straight lines have been drawn following the paths of some of the intersecting aligned-aggregate rows in this region to facilitate their visualization. Figure 5g corresponds to a height contour profile along the white continuous line, showing that the surface features are maintained at about 8 nm in height. The observed radial structures in the surface morphology strongly suggest incipient dendritic or spherulitic crystal growth.

Images in Figure 6 were obtained on a different piece cut from the same PVA sample cast on a 90 mm Petri dish. Here, the surface topography (Figure 6a,e) exhibits higher terraces consisting of straight ribbons ~40 nm wide, similar to the "ribbons" enclosed within the white dashed lines in Figure 5a, presumably also formed by aligned clusters as in Figure 5b, even though in Figure 6a,e the clusters cannot be distinguished. Figure 6a,e strongly resembles the images reported in [56,58,59], from which we confidently infer that the ribbon-like regions correspond to PVA crystallites. According to [58,59], the longer sides of the parallelogrammic platelets in Figure 6a should correspond to (101) planes and the shorter sides to (100) planes, being the acute angle measured from Figure 6a indeed consistent with the reported value of 55°.

The LFM images (Figure 6c,d) indicate that the ribbon-like areas yield lower frictional contrast, as expected for crystalline regions [56].

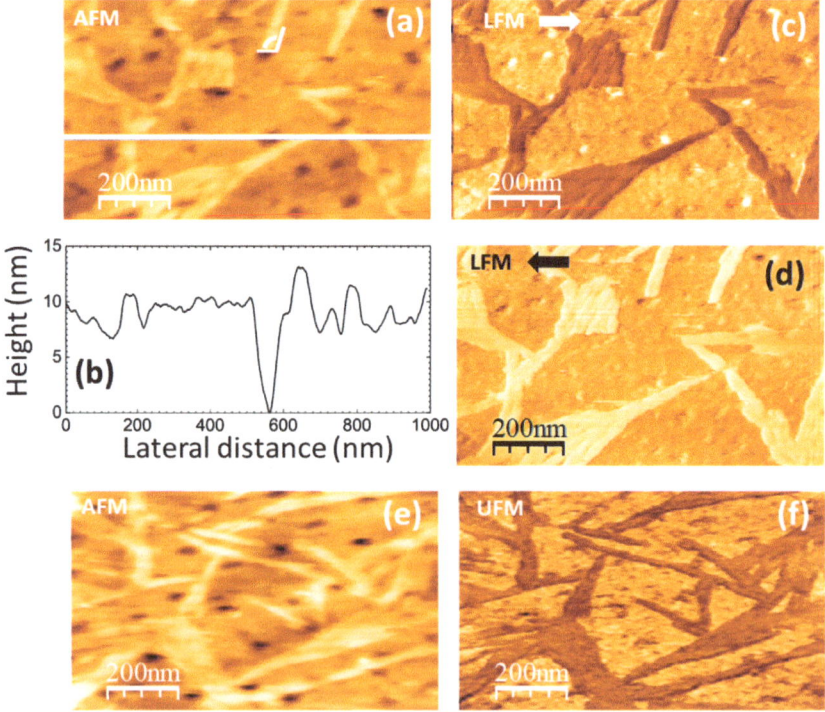

Figure 6. SPM images recorded on pure PVA. (**a**) Contact-mode AFM topography. Color-scale range: 17 nm. (**b**) Height-contour profile over the continuous white line in (**a**). (**c**,**d**) LFM images recorded over the same surface region than (**a**) scanning from left to right ➡ (**c**) and from right to left ⬅ (**d**). (**e**) Contact-mode AFM topographic image. Color-scale range: 17 nm. (**f**) UFM image recorded simultaneously with (**e**) over the same surface region.

Figure 6e,f was recorded over a region similar to that of Figure 6a–c. The topography in Figure 6e shows higher ribbon-like features similar to those in Figure 6a. Nevertheless, in contrast with the results in Figure 5a,b, UFM now yields a lower (softer) contrast at the ribbon areas, and individual PVA clusters cannot be resolved.

The topographical differences between the PVA surfaces in Figures 5 and 6 might be explained by a slightly different water content in the two regions. According to the XRD data, the size of the PVA crystallites is expected to be ~5 nm. Hence, several small crystalline nuclei may coexist within the clusters observed in Figure 5a,b. In Figure 5, we observed that PVA clusters align, forming rows and ribbon-like structures. In Figure 6, the ribbon-like structures appear much better defined. In view of our results, we attribute the differences between Figures 5 and 6 to the fact that the sample areas on which the images were taken are at different stages of the evolution of the surface PVA crystalline structures that result from the assembly of the crystalline nuclei. The presence of water is expected to facilitate the reorganization and assembly of the PVA clusters. Hence, the formation of better-defined ribbon-like structures—presumably related to PVA crystalline domains—in the regions where Figure 6 was recorded can be explained if the local content of water in this region is slightly higher. In PVA solutions, water may be bonded to the PVA molecular chains through hydrogen bonds at hydroxyl sites (bound water) or remain as free water within the polymer matrix [60]. As PVA crystallization evolves, it is expected that bound water remains at the amorphous/crystalline interface. This may exert a plasticizing effect and lead to a softer (lower) UFM contrast at the PVA crystalline ribbons, as well as hinder the identification of individual PVA clusters at those sites.

Figure 6b depicts a height-contour profile along the line in Figure 6a, which is consistent with a multilayer structure of the surface topography with steps of ~2 nm defining the different height levels. In both Figure 5a,d and Figure 6a,e, pores are apparent. These pores, with a diameter similar to the width of the ribbons, might result from the reorganization of surface clusters during the assembly of the ribbon-like structures.

Figure 7 shows SPM images of the PVA/glycerol 22% w/w film. The size of the AFM topographic image in Figure 7a is the same size as this in Figure 5e. In Figure 7a, the topography is also characterized by cluster aggregates with a size similar to those on the pure PVA film surface (~100 nm as measured in the contact-mode AFM topographic image). Also, in this image some of the clusters are aligned (e.g., those beside the dashed line on the image), even though the length of the clusters row is considerably shorter.

The LFM images (Figure 7b,c) indicate here the presence of areas with a lower frictional contrast, similar to those in Figures 5c,d and 6c,d, although now those are not ribbon-like or parellogrammic in shape but rather rounded. Such areas do not appear straightforwardly correlated to specific topographic features.

Figure 7d corresponds to an enlargement of the topographic area enclosed by the rectangle in Figure 7a. Figure 7e displays a height-contour profile along the straight line in Figure 7d. Figure 7f,g shows enlargements of the LFM images of the same surface area, also enclosed by rectangles in Figure 7b,c, respectively. The rounded areas with a lower frictional contrast have been enclosed by circles in Figure 7f,g, and circles over the same area have also been drawn in Figure 7d to facilitate the analysis of the topography on those areas. In both the topography and the LFM images, clusters may be clearly distinguished in the regions enclosed by the circles. However, clusters that appear topographically identical do not produce the same frictional contrast. In Figure 7d, we observe clusters located at different layers. A detailed examination of Figure 7a–c allows us to infer that at the topographic regions around those that yield a lower frictional contrast, the clusters tend to be located at higher sites. It is also interesting to remark that in Figure 7f,g, the boundary regions of the lower friction domains appear defined by individual clusters, as is also the case in Figure 5c,d. This supports the interpretation that the assembly of the PVA crystalline domains proceeds through the assembly of the PVA clusters themselves.

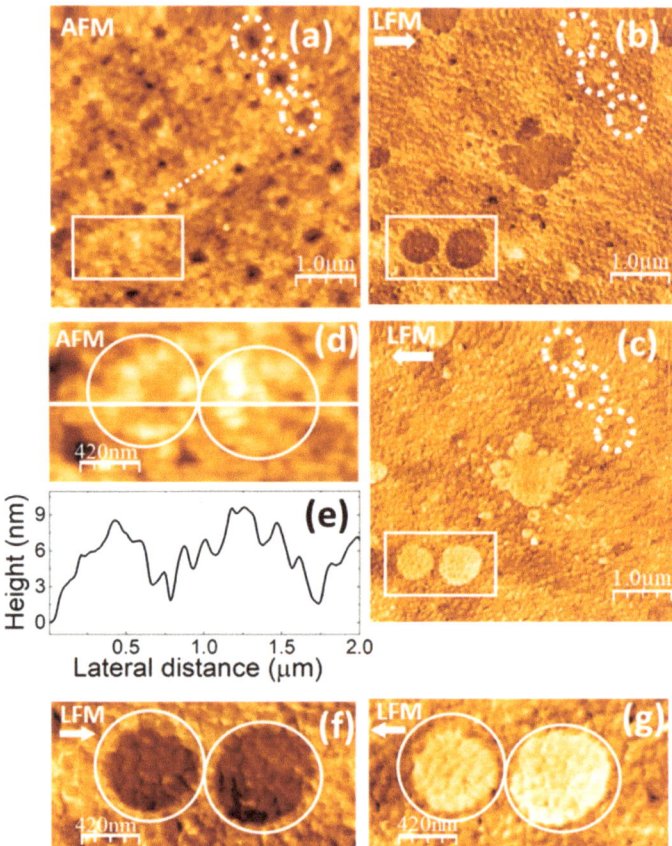

Figure 7. SPM images recorded on PVA/glycerol 22% w/w. (**a**) Contact-mode AFM topography. Color-scale range: 26 nm. (**b,c**) LFM images recorded over the same surface region than (**a**) scanning from left to right ➡ (**b**) and from right to left ⬅ (**c**). (**d**) Enlargement of the region enclosed by the white rectangle in (**a**). (**e**) Height-contour profile over the while line in (**d**). (**f,g**) Enlargements of the regions enclosed by the white rectangles in (**b,c**).

We hypothesize that the areas with distinct frictional contrast in Figure 7b,c,f,g correspond to crystalline PVA domains, whose shape has been modified by the incorporation of 22% w/w glycerol in the PVA matrix. According to our preparation method, we first prepare a 7.0 wt.% PVA stock solution, which we then dilute to 3.5 wt.% PVA by adding a glycerol/water solution at the appropriate concentration. The addition of the glycerol/water solution is carried out at room temperature (25 °C). We presume that the initial PVA crystalline nuclei are already present in the 7.0 wt.% PVA stock solution. These nuclei would have formed as the solution was slowly cooled from 90 °C to room temperature. The growth or assembly of these nuclei is expected to take place as the resulting solution is eventually dried. Glycerol will incorporate into the PVA chains through hydrogen bonds, but if only in a small quantity, it will most likely not penetrate inside the PVA crystallites. Instead, it will preferentially accumulate at the amorphous/crystalline interface, as it happens with water. The incorporation of a small amount of glycerol into the PVA molecules may help the assembly of the crystalline domains, fostering the reorganization of the amorphous PVA regions. This would explain the fact that the amorphous halo observed in the XRD diffractogram (Figure 3) diminishes for the PVA/glycerol 22% w/w films and the fact that the crystallinity slightly increases as inferred from the XRD, FT-IR, and DTA data. In addition, the modification of the shape of the PVA crystalline domains allows us to explain the changes in the relative intensities of the XRD peaks.

The presence of numerous pores is also noticeable in Figure 7a. As in Figure 6a, many of those pores appear to be originated by a missing PVA cluster. Pores with a larger diameter, which might correspond to the displacement of various neighboring clusters, may also be seen in the topographic image. For instance, those within the white dotted circles in Figure 7a have a diameter of ~300 nm. Notice that those pores produce a higher frictional contrast in the LFM images, in which the same locations have also been marked with white dotted circles to facilitate their localization. The origin of this contrast will be discussed in detail below.

In Figure 8, also referred to as the PVA/glycerol 22% w/w film, the topography (Figure 8a) and the simultaneous UFM image (Figure 8b) were recorded over the same surface region as Figure 7a–c. The three pores are enclosed by the white dotted circles in the upper right corner in both Figures 7a and 8a and allow the identification of the regions slightly displaced from each other due to drift. Figure 8a,b was recorded with the tip scanning forwards (from the left to right), and Figure 8c was recorded together with Figure 8b with the tip scanning backwards (from right to left). As it is apparent from Figure 8b,c, while scanning from left to right in the presence of ultrasonic vibration, some of the PVA surface clusters from the right-hand side were dragged to the left by the tip and remained there as the tip scanned back to produce Figure 8c. A close look at Figure 8a shows straight features resulting from the dragging of the clusters, even though the topography does not seem to be severely affected by the tip-induced modifications. In Figure 8c, we find areas with a lower (softer) UFM contrast in similar locations as those that in Figure 7b,c yielded lower frictional contrast. As it is apparent, the shape of domain located at the center in Figure 8c is markedly different from this in Figure 7b,c, presumably as a result of the tip action. At the lower right-hand border of Figure 8c, the image shows a heap with darker (softer) contrast. Figure 8d corresponds to a larger UFM image recorded over a region that included the area of Figure 8a–c in its central part, recorded following the latter. The heap with lower UFM contrast induced by the tip while producing Figure 8b can be clearly distinguished here. Such heap presumably consists of clusters drugged by the tip while scanning forwards and piled up at the end of the scan. Interestingly, some of the other domains in Figure 8a–c, like those with a rounded shape at the lower right-hand corner of the image, remained unaffected.

When discussing Figure 7a–c we identified the lower friction domains as PVA crystalline regions. The fact that those domains yield a lower (softer) UFM contrast may be explained because of the presence of glycerol (and possibly also water) at the amorphous/crystalline interface areas. UFM images in Figure 6f also yielded a lower UFM contrast in the crystalline domains of the pure PVA films. However, we cannot rule out that in PVA/glycerol films, clusters incorporating glycerol molecules produce a softer and lower frictional contrast even if they do not contain crystalline nuclei.

Glycerol molecules located at PVA cluster surfaces may facilitate the displacement of those clusters relative to each other, and this effect may certainly be enhanced by the presence of surface ultrasonic vibration, promoting tip-induced modifications while scanning.

In addition, the glycerol molecules might favor the detachment of surface clusters and induce surface pores. The simultaneous topographic and UFM images in Figure 8e,f show an area in which various surface pores of different sizes up to ~300 nm are apparent. As can be seen in the images, the UFM contrast is stiffer (brighter) within the pores. Moreover, as it was mentioned above, a higher frictional contrast is also found within pores (see Figure 8a–c, pores located within the white dashed circles). The size of the pores rules out that this contrast is an artifact. In fact, such contrast is not observed in the pores found in the pure PVA films (see Figure 5). The presence of pores on the PVA/glycerol 22% w/w films and the UFM and LFM contrast at the pores' location may indicate that PVA clusters located at the bottom of the pores have incorporated glycerol molecules on their surfaces.

On the one hand, this may facilitate the detachment of the clusters at the top, explaining the formation of the pore itself. On the other hand, glycerol molecules in a cluster bonded with their primary alcohol groups to a PVA molecular chain and with their secondary alcohol groups to a neighboring PVA molecular chain could exert a hydrogen bond-mediated crosslinking effect on PVA, increasing the stiffness of the confined cluster [11].

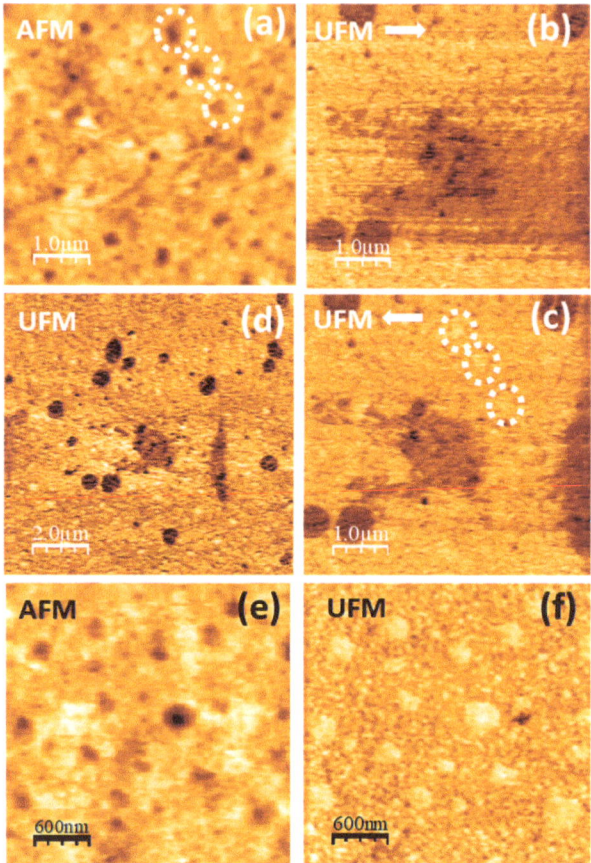

Figure 8. SPM images recorded on PVA/glycerol 22% w/w. (**a**) Contact-mode AFM topography over the same surface region as Figure 7a. (**b**) UFM image recorded simultaneously with (**a**), scanning from left to right ➡. (**c**) UFM image recorded together with (**a**) but scanning from right to left ⬅. (**d**) UFM image over a larger area than (**c**), centered at the same position. (**e**) Contact-mode AFM topography at nearby region. Color-scale range: 24 nm. (**f**) UFM image recorded simultaneously with (**e**).

Figure 9 shows SPM images of the PVA/glycerol 36% w/w film. Figure 9a corresponds to a contact-mode AFM topographic image, recorded over an area of the same size as Figures 5e and 7a. As in the case of the pure PVA and the PVA/glycerol 22% w/w film, the topography is characterized by cluster aggregates, some of which are aligned into rows. Figure 9b displays a contour-line profile along the straight line in Figure 9a.

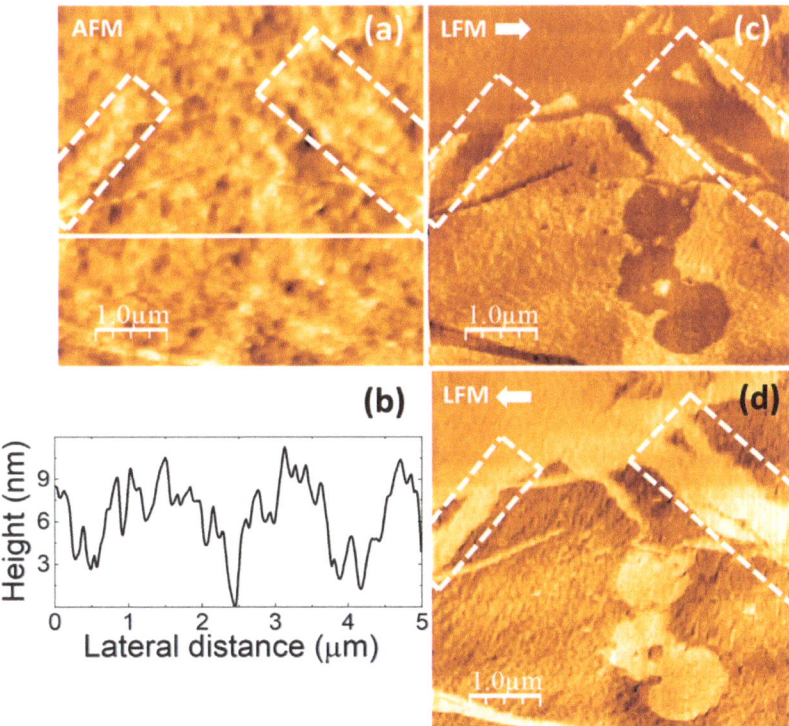

Figure 9. SPM images recorded on PVA/glycerol 36% w/w. (**a**) Contact-mode topography. Color-scale range: 28 nm. (**b**) Height-contour profile recorded along the continuous white line in (**a**). (**c,d**) LFM images recorded over the same surface area as (**a**), scanning from left to right ➡ (**c**) and from right to left ⬅ (**d**).

Figure 9c,d corresponds to lateral force microscopy images scanning from left to right (Figure 9c) and from right to left (Figure 9d) recorded over the same surface region as (Figure 9a). As it is clearly seen in Figure 9c,d, we find areas with a lower frictional contrast, similar to those in Figure 5c,d, Figures 6c,d, and 7b,c. While some of those areas exhibit rounded interfaces similar to those in Figure 6c,d, others rather resemble the ribbon-like shape domains found in the pure PVA film surface (Figures 5c,d and 6c,d). Nevertheless, here, the extension of the lower contrast domains is apparently larger, as if some of the areas linking the domains in Figure 6 that for the pure PVA produced a higher friction contrast, were now also yielding a low frictional response. Interestingly, some of the ribbon-like low friction domains in Figure 9c,d (like, for instance, those within the white dashed rectangles) seem to be correlated to higher topographic regions, but at other areas of the low friction domains, no correlation with the topographic image could be appreciated.

According to the FT-IR and XRD data, the crystallinity of the PVA/glycerol 36% w/w films is reduced compared to this for the pure PVA and 22% w/w glycerol films. This may be explained by the formation of glycerol aggregates for the higher glycerol concentrations, which may penetrate within the crystalline PVA domains and partially destroy them. For these films, in the DTA curves, the endothermic peak corresponding to the maximum rate of degradation for PVA appears split into two well-resolved peaks, which might be indicative of the formation of a new phase formed by glycerol aggregates or glycerol/PVA complexes. The lower friction contrast in Figure 9b,c might originate either because of the presence of crystalline PVA domains, or of the presence of a distinct phase formed by glycerol/PVA complexes, even though the LFM image provides no clue to justify the presence of lower frictional domains of a different nature.

It should be pointed out that on the PVA/glycerol 36% w/w films, we did not succeed in recording good quality UFM images, as the film was severely affected by tip actuation in the presence of ultrasound. And for the case of PVA/glycerol (55% w/w), nor even contact AFM and LFM images could be recorded without surface modification.

According to FT-IR, XRD, and DTA data, the incorporation of 55% w/w glycerol leads to a significant reduction in the PVA crystallinity. The effect may be similar to this observed with water, which in a small amount enhances the PVA crystallinity, but excess water leads to destruction of the PVA crystallites.

4. Conclusions

The study presents a thorough investigation of the structural properties, crystallinity, and thermal response of pure PVA and PVA/glycerol films using FT-IR, XRD, TG/DTA, and advanced SPM. The combination of different types of SPM techniques (contact mode AFM, LFM, and UFM) allowed us to obtain information on the nanoscale topography and frictional and mechanical response of the films' surface.

In the pure PVA films, SPM allowed us to observe ribbon-shaped domains and clusters aligning along a well-defined direction and forming intersecting rows with a different frictional and elastic contrast, which apparently originated from a correlated growth or assembly of PVA crystalline nuclei (~5 nm in diameter according to XRD) located within individual PVA clusters (~80–100 nm in diameter). The incorporation of 22% w/w glycerol leads to a modification in shape of those domains from ribbon-like in pure PVA to rounded in the 22% w/w PVA/glycerol films. In correlation with this, the XRD patterns show changes in the relative intensities of the XRD crystalline peaks and a decrease in the amorphous halo, while the melting point remained almost the same. In addition, pores of various sizes up to 300 nm in diameter were observed, which produced a characteristically higher frictional and higher (stiffer) UFM contrast. At the film with 36% w/w glycerol content, domains with lower frictional contrast were also observed, some with rounded interfaces like those seen in PVA/glycerol 22% w/w films, and others that rather resembled the ribbon-like domains found in the pure PVA film surface but were wider and joined together. For these films, the amorphous halo in the XRD diffractogram increases and slightly displaces to higher 2θ angles. The observation of larger lower friction domains in this case may also originate from a new phase formed by glycerol aggregates or glycerol/PVA complexes, even though LFM does not provide any clue to justify a different nature of coexisting lower friction domains. The aforementioned changes are accompanied by a decrease in crystallinity to values of pure PVA according to FT-IR and XRD and a decrease in melting point. For glycerol contents higher than 36% w/w, tip-induced modifications severely complicated the acquisition and analysis of SPM images. FT-IR revealed additional glycerol-characteristic peaks presumably related to the formation of glycerol aggregates; the amorphous halo was prominent in the XRD diffraction pattern and was clearly displaced to a higher 2θ, and both XRD and FT-IR confirmed a drastic reduction in crystallinity.

According to the obtained results, glycerol molecules in a small quantity do not penetrate inside the PVA crystallites but foster the reorganization of the amorphous PVA regions, alter the shape of the PVA crystalline domains, and preferentially accumulate at the amorphous/crystalline interface. In the PVA/glycerol films, the domains could only be clearly identified from LFM and UFM; no topographical correlation was apparent. As the glycerol content increases, glycerol begins to interact with PVA within the crystalline structures, being those nearly completely disrupted for high glycerol contents. The mobility of PVA chains increases substantially, and tip-induced surface modification effects were observed, which were more significant in the presence of ultrasonic vibration.

The presented research contributes to the understanding of PVA crystallinity and its modification by plasticizers such as glycerol. Our findings deliver crucial insights into how glycerol impacts that nanostructure of PVA, thereby paving the way for tailoring PVA properties to achieve optimal performance.

Author Contributions: Conceptualization, G.K. and T.C.; methodology, G.K. and T.C.; formal analysis, G.K., D.C. and T.C.; investigation, G.K., D.C. and T.C.; resources, T.C.; data curation, G.K., D.C. and T.C.; writing—original draft preparation, G.K. and T.C.; writing—review and editing, G.K. and T.C.; visualization, G.K. and T.C.; supervision, T.C.; funding acquisition, T.C. All authors have read and agreed to the published version of the manuscript.

Funding: We acknowledge financial support under project Ref. 2022-GRIN-34226 (Plan Propio UCLM cofunded 85% by FEDER). G.K. acknowledges financial support from the UCLM for her stay in Almadén, Spain, under contract 2022-UNIVERS-11036, ref. 2022-POST-20987.

Institutional Review Board Statement: Not applicable.

Data Availability Statement: The original contributions presented in the study are included in the article, further inquiries can be directed to the corresponding author.

Acknowledgments: Carlos Rivera Cavanillas (UCLM IRICA Instrumentation Service) is gratefully acknowledged for technical aid in XRD, FT-IR, and TGA/DTA data acquisition.

Conflicts of Interest: The authors declare no conflicts of interest. The funders had no role in the design of the study; in the collection, analyses, or interpretation of data; in the writing of the manuscript, or in the decision to publish the results.

References

1. Abdullah, Z.W.; Dong, Y.; Davies, I.J.; Barbhuiya, S. PVA, PVA Blends, and Their Nanocomposites for Biodegradable Packaging Application. *Polym.-Plast. Technol. Eng.* **2017**, *56*, 1307–1344. [CrossRef]
2. Aslam, M.; Kalyar, M.A.; Raza, Z.A. Polyvinyl alcohol: A review of research status and use of polyvinyl alcohol based nanocomposites. *Polym. Eng. Sci.* **2018**, *58*, 2119–2132. [CrossRef]
3. Panda, P.K.; Sadeghi, K.; Seo, J. Recent advances in poly (vinyl alcohol)/natural polymer based films for food packaging applications: A review. *Food Packag. Shelf Life* **2022**, *33*, 100904. [CrossRef]
4. Oliveira, A.S.; Seidi, O.; Ribeiro, N.; Colaço, R.; Serro, A.P. Tribomechanical Comparison between PVA Hydrogels Obtained Using Different Processing Conditions and Human Cartilage. *Materials* **2019**, *12*, 3413. [CrossRef]
5. Rahmadiawan, D.; Abral, H.; Shi, S.-C.; Huang, T.-T.; Zainul, R.; Nurdin, H. Tribological Properties of Polyvinyl Alcohol/Uncaria Gambir Extract Composite as Potential Green Protective Film. *Tribol. Ind.* **2023**, *45*, 367–374. [CrossRef]
6. Shekaryar, H.; Norouzbahari, S. A review on versatile applications of polyvinyl alcohol thin films, specifically as sensor devices. *Polym. Eng. Sci.* **2024**, *64*, 455–468. [CrossRef]
7. Lim, L.Y.; Wan, L.S.C. The Effect of Plasticizers on the Properties of Polyvinyl Alcohol Films. *Drug Dev. Ind. Pharm.* **1994**, *20*, 1007–1020. [CrossRef]
8. Jang, J.; Lee, D.K. Plasticizer effect on the melting and crystallization behavior of polyvinyl alcohol. *Polymer* **2003**, *44*, 8139–8146. [CrossRef]
9. Cortés-Morales, E.C.; Rathee, V.S.; Ghobadi, A.; Whitmer, J.K. A molecular view of plasticization of polyvinyl alcohol. *J. Chem. Phys.* **2021**, *155*, 174903. [CrossRef]
10. Chen, M.; Runge, T.; Wang, L.; Li, R.; Feng, J.; Shu, X.-L.; Shi, Q.-S. Hydrogen bonding impact on chitosan plasticization. *Carbohydr. Polym.* **2018**, *200*, 115–121. [CrossRef]
11. Shi, S.; Peng, X.; Liu, T.; Chen, Y.-N.; He, C.; Wang, H. Facile preparation of hydrogen-bonded supramolecular polyvinyl alcohol-glycerol gels with excellent thermoplasticity and mechanical properties. *Polymer* **2017**, *111*, 168–176. [CrossRef]
12. Sin, L.T.; Aizan, W.; Abdul, W.; Rahmat, A. Specific heats of neat and glycerol plasticized polyvinyl alcohol. *Pertanika J. Sci. Technol.* **2010**, *18*, 387–391.
13. Mohsin, M.; Hossin, A.; Haik, Y. Thermal and mechanical properties of poly(vinyl alcohol) plasticized with glycerol. *J. Appl. Polym. Sci.* **2011**, *122*, 3102–3109. [CrossRef]
14. Pu-You, J.; Cai-ying, B.; Li-hong, H.; Yong-hong, Z. Properties of Poly (vinyl alcohol) Plasticized by Glycerin. *J. For. Prod. Ind* **2014**, *3*, 151–153.
15. Tian, H.; Yuan, L.; Zhou, D.; Niu, J.; Cong, H.; Xiang, A. Improved mechanical properties of poly (vinyl alcohol) films with glycerol plasticizer by uniaxial drawing. *Polym. Adv. Technol.* **2018**, *29*, 2612–2618. [CrossRef]
16. Lv, C.; Liu, D.; Tian, H.; Xiang, A. Non-isothermal crystallization kinetics of polyvinyl alcohol plasticized with glycerol and pentaerythritol. *J. Polym. Res.* **2020**, *27*, 66. [CrossRef]
17. Fei, W.; Wu, Z.; Cheng, H.; Xiong, Y.; Chen, W.; Meng, L. Molecular mobility and morphology change of poly(vinyl alcohol) (PVA) film as induced by plasticizer glycerol. *J. Polym. Sci.* **2023**, *61*, 1959–1970. [CrossRef]

18. Fu, Z.Z.; Yao, Y.H.; Guo, S.J.; Wang, K.; Zhang, Q.; Fu, Q. Effect of Plasticization on Stretching Stability of Poly(Vinyl Alcohol) Films: A Case Study Using Glycerol and Water. *Macromol. Rapid Commun.* **2023**, *44*, 2200296. [CrossRef]
19. Gassab, M.; Papanastasiou, D.T.; Sylvestre, A.; Bellet, D.; Dridi, C.; Basrour, S. Dielectric Study of Cost-Effective, Eco-Friendly PVA-Glycerol Matrices with AgNW Electrodes for Transparent Flexible Humidity Sensors. *Adv. Mater. Interfaces* **2023**, *10*, 2201652. [CrossRef]
20. Cuberes, M.T. Mechanical Diode-Based Ultrasonic Atomic Force Microscopies. In *Applied Scanning Probe Methods Xi: Scanning Probe Microscopy Techniques*; Bhushan, B., Fuchs, H., Eds.; Nanoscience and Technology; Springer: Berlin/Heidelberg, Germany, 2009; pp. 39–71.
21. Horcas, I.; Fernández, R.; Gomez-Rodriguez, J.; Colchero, J.; Gómez-Herrero, J.; Baro, A. WSXM: A software for scanning probe microscopy and a tool for nanotechnology. *Rev. Sci. Instrum.* **2007**, *78*, 013705. [CrossRef]
22. Vu, T.H.N.; Morozkina, S.N.; Uspenskaya, M.V. Study of the Nanofibers Fabrication Conditions from the Mixture of Poly(vinyl alcohol) and Chitosan by Electrospinning Method. *Polymers* **2022**, *14*, 811. [CrossRef] [PubMed]
23. Chen, T.; Wu, Z.; Wei, W.; Xie, Y.; Wang, X.A.; Niu, M.; Wei, Q.; Rao, J. Hybrid composites of polyvinyl alcohol (PVA)/Si–Al for improving the properties of ultra-low density fiberboard (ULDF). *RSC Adv.* **2016**, *6*, 20706–20712. [CrossRef]
24. Anicuta, S.-G.; Dobre, L.; Stroescu, M.; Jipa, I. Fourier transform infrared (FTIR) spectroscopy for characterization of antimicrobial films containing chitosan. *Analele Univ. Ńii Din Oradea Fasc. Ecotoxicol. Zooteh. Şi Tehnol. Ind. Aliment.* **2010**, *2010*, 1234–1240.
25. Tretinnikov, O.N.; Zagorskaya, S.A. Determination of the degree of crystallinity of poly(vinyl alcohol) by FTIR spectroscopy. *J. Appl. Spectrosc.* **2012**, *79*, 521–526. [CrossRef]
26. Jaipakdee, N.; Pongjanyakul, T.; Limpongsa, E. Preparation and characterization of poly (vinyl alcohol)-poly (vinyl pyrrolidone) mucoadhesive buccal patches for delivery of lidocaine HCL. *Int. J. Appl. Pharm* **2018**, *10*, 115–123. [CrossRef]
27. Mallapragada, S.K.; Peppas, N.A.; Colombo, P. Crystal dissolution-controlled release systems. II. Metronidazole release from semicrystalline poly(vinyl alcohol) systems. *J. Biomed. Mater. Res.* **1997**, *36*, 125–130. [CrossRef]
28. Deshmukh, K.; Hägg, M.B. Influence of TiO_2 on the Chemical, Mechanical, and Gas Separation Properties of Polyvinyl Alcohol-Titanium Dioxide (PVA-TiO_2) Nanocomposite Membranes. *Int. J. Polym. Anal. Charact.* **2013**, *18*, 287–296.
29. Stachowiak, N.; Kowalonek, J.; Kozlowska, J. Effect of plasticizer and surfactant on the properties of poly(vinyl alcohol)/chitosan films. *Int. J. Biol. Macromol.* **2020**, *164*, 2100–2107. [CrossRef]
30. Farahnaky, A.; Saberi, B.; Majzoobi, M. Effect of Glycerol on Physical and Mechanical Properties of Wheat Starch Edible Films. *J. Texture Stud.* **2013**, *44*, 176–186. [CrossRef]
31. Syahputra, R.A.; Rani, Z.; Ridwanto, R.; Miswanda, D.; Pulungan, A. Isolation and characterization of glycerol by transesterification of used cooking oil. *Rasayan J. Chem.* **2023**, *16*, 648–652. [CrossRef]
32. Habuka, A.; Yamada, T.; Nakashima, S. Interactions of Glycerol, Diglycerol, and Water Studied Using Attenuated Total Reflection Infrared Spectroscopy. *Appl. Spectrosc.* **2020**, *74*, 767–779. [CrossRef] [PubMed]
33. Basiak, E.; Lenart, A.; Debeaufort, F. How Glycerol and Water Contents Affect the Structural and Functional Properties of Starch-Based Edible Films. *Polymers* **2018**, *10*, 412. [CrossRef] [PubMed]
34. Guerrero, P.; Retegi, A.; Gabilondo, N.; De La Caba, K. Mechanical and thermal properties of soy protein films processed by casting and compression. *J. Food Eng.* **2010**, *100*, 145–151. [CrossRef]
35. Xiang, A.; Lv, C.; Zhou, H. Changes in Crystallization Behaviors of Poly(Vinyl Alcohol) Induced by Water Content. *J. Vinyl Addit. Technol.* **2020**, *26*, 613–622. [CrossRef]
36. Assender, H.E.; Windle, A.H. Crystallinity in poly(vinyl alcohol). 1. An X-ray diffraction study of atactic PVOH. *Polymer* **1998**, *39*, 4295–4302. [CrossRef]
37. Kumar, A.; Negi, Y.S.; Choudhary, V.; Bhardwaj, N.K. Microstructural and mechanical properties of porous biocomposite scaffolds based on polyvinyl alcohol, nano-hydroxyapatite and cellulose nanocrystals. *Cellulose* **2014**, *21*, 3409–3426. [CrossRef]
38. Sreekumar, P.A.; Al-Harthi, M.A.; De, S.K. Studies on compatibility of biodegradable starch/polyvinyl alcohol blends. *Polym. Eng. Sci.* **2012**, *52*, 2167–2172. [CrossRef]
39. Assender, H.E.; Windle, A.H. Crystallinity in poly(vinyl alcohol) 2. Computer modelling of crystal structure over a range of tacticities. *Polymer* **1998**, *39*, 4303–4312. [CrossRef]
40. Murthy, N.S.; Minor, H. General procedure for evaluating amorphous scattering and crystallinity from X-ray diffraction scans of semicrystalline polymers. *Polymer* **1990**, *31*, 996–1002. [CrossRef]
41. Sreekumar, P.A.; Al-Harthi, M.A.; De, S.K. Effect of glycerol on thermal and mechanical properties of polyvinyl alcohol/starch blends. *J. Appl. Polym. Sci.* **2012**, *123*, 135–142. [CrossRef]
42. Liu, T.; Peng, X.; Chen, Y.-N.; Bai, Q.-W.; Shang, C.; Zhang, L.; Wang, H. Hydrogen-Bonded Polymer–Small Molecule Complexes with Tunable Mechanical Properties. *Macromol. Rapid Commun.* **2018**, *39*, 1800050. [CrossRef] [PubMed]
43. Yin, D.; Xiang, A.; Li, Y.; Qi, H.; Tian, H.; Fan, G. Effect of Plasticizer on the Morphology and Foaming Properties of Poly(vinyl alcohol) Foams by Supercritical CO_2 Foaming Agents. *J. Polym. Environ.* **2019**, *27*, 2878–2885. [CrossRef]
44. Budrugeac, P. Kinetics of the complex process of thermo-oxidative degradation of poly(vinyl alcohol). *J. Therm. Anal. Calorim.* **2008**, *92*, 291–296. [CrossRef]
45. Wiśniewska, M.; Chibowski, S.; Urban, T.; Sternik, D. Investigation of the alumina properties with adsorbed polyvinyl alcohol. *J. Therm. Anal. Calorim.* **2011**, *103*, 329–337. [CrossRef]

46. Pandey, S.; Pandey, S.K.; Parashar, V.; Mehrotra, G.K.; Pandey, A.C. Ag/PVA nanocomposites: Optical and thermal dimensions. *J. Mater. Chem.* **2011**, *21*, 17154. [CrossRef]
47. Castelló, M.; Dweck, J.; Aranda, D. Thermal stability and water content determination of glycerol by thermogravimetry. *J. Therm. Anal. Calorim.* **2009**, *97*, 627–630. [CrossRef]
48. Tarique, J.; Sapuan, S.M.; Khalina, A. Effect of glycerol plasticizer loading on the physical, mechanical, thermal, and barrier properties of arrowroot (*Maranta arundinacea*) starch biopolymers. *Sci. Rep.* **2021**, *11*, 13900. [CrossRef]
49. Ilyas, R.A.; Sapuan, S.M.; Ishak, M.R.; Zainudin, E.S. Development and characterization of sugar palm nanocrystalline cellulose reinforced sugar palm starch bionanocomposites. *Carbohydr. Polym.* **2018**, *202*, 186–202. [CrossRef]
50. Sanyang, M.; Sapuan, S.; Jawaid, M.; Ishak, M.; Sahari, J. Effect of Plasticizer Type and Concentration on Tensile, Thermal and Barrier Properties of Biodegradable Films Based on Sugar Palm (*Arenga pinnata*) Starch. *Polymers* **2015**, *7*, 1106–1124. [CrossRef]
51. Guirguis, O.W.; Moselhey, M.T.H. Thermal and structural studies of poly (vinyl alcohol) and hydroxypropyl cellulose blends. *Nat. Sci.* **2012**, *04*, 57–67. [CrossRef]
52. Tsioptsias, C.; Fardis, D.; Ntampou, X.; Tsivintzelis, I.; Panayiotou, C. Thermal Behavior of Poly(vinyl alcohol) in the Form of Physically Crosslinked Film. *Polymers* **2023**, *15*, 1843. [CrossRef]
53. Abd-Elrahman, M.I. Enhancement of thermal stability and degradation kinetics study of poly(vinyl alcohol)/zinc oxide nanoparticles composite. *J. Thermoplast. Compos. Mater.* **2014**, *27*, 160–166. [CrossRef]
54. Tubbs, R.K. Melting point and heat of fusion of poly(vinyl alcohol). *J. Polym. Sci. Part A* **1965**, *3*, 4181–4189. [CrossRef]
55. Sperling, L.H. *Introduction to Physical Polymer Science*; John Wiley & Sons: Hoboken, NJ, USA, 2005.
56. Strawhecker, K.E.; Manias, E. AFM of Poly(vinyl alcohol) Crystals Next to an Inorganic Surface. *Macromolecules* **2001**, *34*, 8475–8482. [CrossRef]
57. Hodge, R.M.; Edward, G.H.; Simon, G.P. Water absorption and states of water in semicrystalline poly(vinyl alcohol) films. *Polymer* **1996**, *37*, 1371–1376. [CrossRef]
58. Tsuboi, K. Solution-grown crystals of poly(vinyl alcohol). *J. Macromol. Sci. Part B* **1968**, *2*, 603–622. [CrossRef]
59. Tsuboi, K.; Mochizuki, T. Single crystals of polyvinyl alcohol. *J. Polym. Sci. Part B Polym. Lett.* **1963**, *1*, 531–534. [CrossRef]
60. Song, Y.; Zhang, S.; Kang, J.; Chen, J.; Cao, Y. Water absorption dependence of the formation of poly(vinyl alcohol)-iodine complexes for poly(vinyl alcohol) films. *RSC Adv.* **2021**, *11*, 28785–28796. [CrossRef]

Disclaimer/Publisher's Note: The statements, opinions and data contained in all publications are solely those of the individual author(s) and contributor(s) and not of MDPI and/or the editor(s). MDPI and/or the editor(s) disclaim responsibility for any injury to people or property resulting from any ideas, methods, instructions or products referred to in the content.

Article

Poly(1,3-Propylene Glycol Citrate) as a Plasticizer for Toughness Enhancement of Poly-L-Lactic Acid

Dengbang Jiang [1,†], Junchao Chen [1,†], Minna Ma [1], Xiushuang Song [1], Huaying A [1], Jingmei Lu [1], Conglie Zi [1], Wan Zhao [1], Yaozhong Lan [2] and Mingwei Yuan [1,*]

[1] Green Preparation Technology of Biobased Materials National &Local Joint Engineering Research Center, Yunnan Minzu University, Kunming 650500, China; 041814@ymu.edu.cn (D.J.); cjc9931@163.com (J.C.);
[2] School of Ecology and Environmental Science, Yunnan University, Kunming 650091, China; ynulanyaozhong@163.com
* Correspondence: ymujiang@163.com
† These authors contributed equally to this work.

Abstract: Despite the unique features of poly-L-lactic acid (PLLA), its mechanical properties, such as the elongation at break, need improvement to broaden its application scope. Herein, poly(1,3-propylene glycol citrate) (PO3GCA) was synthesized via a one-step reaction and evaluated as a plasticizer for PLLA films. Thin-film characterization of PLLA/PO3GCA films prepared via solution casting revealed that PO3GCA shows good compatibility with PLLA. The addition of PO3GCA slightly improves the thermal stability and enhances the toughness of PLLA films. In particular, the elongation at break of the PLLA/PO3GCA films with PO3GCA mass contents of 5%, 10%, 15%, and 20% increases to 172%, 209%, 230%, and 218%, respectively. Therefore, PO3GCA is promising as a plasticizer for PLLA.

Keywords: polylactic acid; citric acid; poly(1,3-propylene glycol); toughening agent

Citation: Jiang, D.; Chen, J.; Ma, M.; Song, X.; A, H.; Lu, J.; Zi, C.; Zhao, W.; Lan, Y.; Yuan, M. Poly(1,3-Propylene Glycol Citrate) as a Plasticizer for Toughness Enhancement of Poly-L-Lactic Acid. Polymers 2023, 15, 2334. https://doi.org/10.3390/polym15102334

Academic Editors: Beata Kaczmarek and Marcin Wekwejt

Received: 22 March 2023
Revised: 21 April 2023
Accepted: 23 April 2023
Published: 17 May 2023

Copyright: © 2023 by the authors. Licensee MDPI, Basel, Switzerland. This article is an open access article distributed under the terms and conditions of the Creative Commons Attribution (CC BY) license (https://creativecommons.org/licenses/by/4.0/).

1. Introduction

Polylactic acid (PLA) is a biodegradable and eco-friendly polymer that has garnered significant attention in recent years due to its numerous advantages. PLA is derived from renewable plant resources, such as corn, potato, and jackfruit, through a fermentation process that produces high-purity lactic acid. The lactic acid is then used to synthesize PLA of the desired molecular weight through chemical processes [1–3].

PLA is biocompatible and is widely used in the medical field, including as implant materials, drug delivery systems, and tissue engineering scaffolds [4]. It is also used in commodity packaging, where its biodegradability and low environmental impact make it an attractive alternative to traditional plastics [5]. Additionally, PLA has been explored for use in textiles due to its biodegradability and potential to reduce the environmental impact of textile production [6].

Despite its numerous advantages, PLA has limitations that have hindered its widespread application. One of the major drawbacks of PLA is its high glass transition temperature (Tg) which ranges from 55 °C to 65 °C, making it brittle and fragile at room temperature. Additionally, the elongation at the break of PLA is limited to only ~5%, which further limits its practical use [7]. Efforts have been made to improve the mechanical properties of PLA through various methods, including blending it with other polymers, copolymerization, and the addition of nanoparticles [7–9].

To improve the properties and performance of Polylactic acid (PLA), two commonly used methods include the copolymerization of lactic acid with other polymer monomers, such as trimethylene carbonate and caprolactone [8,9], and the preparation of physical blends of PLA with different polymers, small molecular compounds, or even inorganic materials. These approaches have shown great potential in enhancing the thermal stability, mechanical strength, and elongation at the break of PLA. Furthermore, they have opened up new avenues for the diversified application of PLA in various fields, including automotive, construction, and electronics.

However, there are some challenges associated with these methods that need to be addressed. For instance, the compatibility of the blended systems needs to be optimized to achieve the desired properties of PLA-based materials. The processing conditions, such as temperature, pressure, and blending time, also need to be carefully controlled to obtain the desired properties of PLA-based materials [10–12].

Despite these challenges, research efforts have been ongoing to overcome these limitations and optimize the properties of PLA-based materials. A better understanding of the fundamental properties of PLA and the mechanisms underlying the modification of its properties will enable researchers to design new strategies for enhancing the performance of PLA-based materials. Moreover, the development of sustainable and cost-effective methods for the production of PLA will further promote the widespread application of this eco-friendly polymer.

In the process of blending modification, a plasticizer is commonly used to enhance the properties of Polylactic acid (PLA). The ideal plasticizer should exhibit non-toxicity, excellent compatibility with PLA, and high thermal stability [11–13]. Citrate is a plasticizer that meets these requirements, and various citrate derivatives, such as triethyl citrate (TEC), acetyl tributyl citrate (ATBC), tributyl citrate (TBC), and polyethylene glycol citrate, have been utilized as plasticizers for PLA [14]. For example, Ljungberg N et al. toughened PLA by incorporating TBC and ethyl triacetate (TAC). Although the addition of TBC did not significantly affect the crystallinity of PLA, it increased the elongation at the break of PLA/TBC blends up to 350%, considerably improving the toughness of PLA [13]. Mounira Maiza prepared and characterized PLA/TEC and PLA/ATBC blends using TEC and ATBC as plasticizers for PLA, respectively. TEC and ATBC effectively reduced the Tg and enhanced the crystallinity of the blends, while having no substantial effect on their transparency [14]. Gui Zongyan demonstrated that the incorporation of polyethylene glycol (PEG) and carboxylic acid copolymers could effectively enhance the toughness of PLA [12]. Further research is required to optimize the processing conditions and to investigate the compatibility of the blended systems, to achieve the desired properties of PLA-based materials. Poly(1,3-propylene glycol) (PO3G) is a hydrophilic polymer synthesized using 1,3-propylene glycol via an acid-catalyzed dehydration process. Although its structure is highly similar to that of PEG, PO3G has better compliance, mechanical properties, and thermal stability, which stem from its longer molecular chain [15,16]. Therefore, we envisioned that PO3G or its combination with other carboxylic acids could serve as a plasticizer for PLA.

This study utilized a novel approach to improve the properties of Polylactic-L-acid (PLLA) by physically blending it with poly(1,3-propylene glycol citrate) (PO3GCA), a copolymer obtained from biologically derived citric acid (CA) and PO3G. The aim of this study was to investigate the plasticizing effect of PO3GCA on PLLA. This approach is unique, as it involves the use of a copolymer obtained from natural sources to enhance the properties of PLLA.

The results of this study have significant implications for the development of biodegradable and eco-friendly plasticizers for PLLA, as well as for the sustainable utilization of renewable resources. By utilizing natural sources to synthesize copolymers, it is possible to reduce the environmental impact of plasticizers and promote the use of biodegradable materials. Moreover, the use of PO3GCA as a plasticizer for PLLA may offer enhanced mechanical properties, thermal stability, and elongation at break, making it a promising alternative to traditional plasticizers.

Overall, this study highlights the importance of developing sustainable and eco-friendly materials for the plastic industry. The use of renewable resources and biodegradable materials can significantly reduce the environmental impact of plastic production and promote the development of a more sustainable future.

2. Materials and Methods

2.1. Raw Materials

PLLA (4032D, Nature Works, Minnetonka, MN, USA) [17], PO3G (Mn = 2300, DuPont Company, Wilmington, DE, USA), Citric acid monohydrate (McLean Biochemical Technology Co., Ltd., Shanghai, China, 99.5%), 1,4-dioxane (Adamas Reagent Company, Shanghai, China), molybdenum trioxide (Adamas Reagent Company, Shanghai, China, 99.9%), and methylene chloride (DCM, Damao Chemical Reagent Factory, Tianjin, China) were used as received.

2.2. Experimental Equipment

The following instruments were used for the experiments: a heat-collection constant temperature heating magnetic agitator (DF-101S, Yuhua Instrument Co., Ltd., Gongyi, China), an electric blast drying oven (GZX-9240MBE, Medical Equipment Factory of Boxun Industrial Co., Ltd., Shanghai, China), an electronic balance (FA2004, Yueping Scientific Instrument Co., Ltd., Shanghai, China), a nuclear magnetic resonance (NMR) spectrometer (400 MHz, BrukerAvance-II, Bruker, Bremen, Germany), a Fourier transform infrared (FTIR) spectrometer (Nicolet IS10, Thermo Fischer Scientific, Waltham, MA, USA), a scanning electron microscope(EvoMA10, ZEISS, Oerkochen, Germany),a microcomputer-controlled universal tensile testing machine (CMT4104, Chuangcheng Zhijia Technology Co., Ltd., Beijing, China), a thermogravimetric analyzer (STA449F3, NETZSCH, Selb, Germany), a differential scanning calorimeter (2414Polyma, NETZSCH, Selb, Germany).

2.3. Raw Materials

2.3.1. Synthesis of PO3GCA

In this study, the PO3GCA copolymer was synthesized using a direct synthesis method [12,18–22]. Specifically, PO3G (300 g) and CA (21 g) were added to a 500 mL three-necked flask in a 1.3:1 molar ratio, with 0.5% (1.6 g) molybdenum trioxide used as the catalyst. After purging the reaction system with nitrogen gas three times, the system was evacuated to a pressure of 100 Pa using an oil pump. The system was gradually heated to 150 °C and held for 30 min, followed by heating at 170 °C for 8 h. Mechanical stirring was employed throughout the entire reaction process. After natural cooling to room temperature, the product was dissolved in dichloromethane (DCM) and stirred magnetically for 1 h. The solution was then left to stand for 10 h until most of the molybdenum trioxide precipitated at the bottom of the flask. The liquid above the precipitate was filtered and collected, and the filtrate was filtered 2–3 times to remove all molybdenum trioxide. The resulting filtrate was placed in the flask and distilled at 50 °C until no more vapor was produced. The second distillation was carried out at 110 °C until no more vapor was produced, yielding approximately 290 g of a light yellow to yellow, viscous liquid, with a yield of approximately 90.3%. This light yellow to yellow, viscous liquid is PO3GCA.

Figure 1 shows the reaction equation of PO3GCA synthesized from poly (1, 3-propylene glycol) and citric acid.

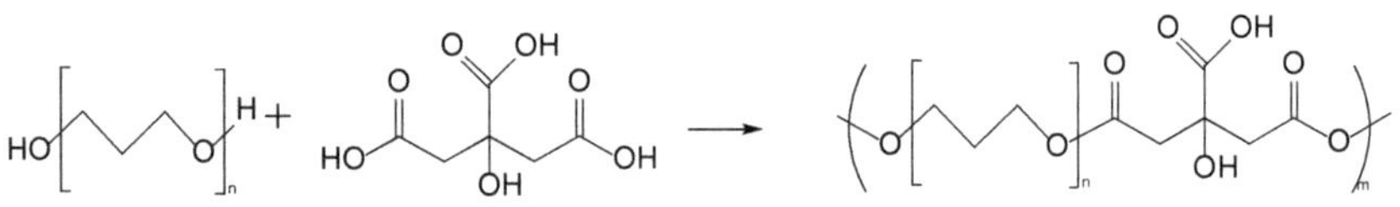

Figure 1. Reaction equation for the synthesis of PO3GCA from poly1,3-propylene glycol, and citric acid.

2.3.2. Preparation of PLLA/PO3GCA Films

The study involved the preparation of PLLA/PO3GCA composite films with a total mass concentration of 20%. The proportion of PO3GCA to solute ranged from 0% to 20%, with increments of 5% (The solvent used for the solution was 1,4-dioxane). The solution was stirred until it reached a temperature of 80 °C, followed by ultrasonic defoaming. The solution was then scraped onto a clean glass plate using a film scraper (750 μm) and allowed to dry after solvent volatilization. The resulting composite film samples had a uniform thickness of 0.07–0.08 mm and were naturally dried at room temperature for 48 h before being collected and stored for further use. The samples were labeled PLLA, PO3GCA5, PO3GCA10, PO3GCA15, and PO3GCA20, according to the concentration of PO3GCA.

2.4. Characterization of PO3GCA and the PLLA/PO3GCA Films

2.4.1. ^{1}H NMR Spectroscopy Characterization of PO3GCA

The samples for ^{1}H NMR spectroscopy analysis were prepared by placing 15 mg of PO3GCA in an NMR tube and dissolving it using 480 μL of Deuterium chloroform. Tetramethylsilane was used as the internal standard.

2.4.2. FTIR Spectroscopy Characterization of PO3GCA

KBr pellets were prepared by mixing and grinding dried PO3GCA powder and KBr in a mass ratio of 2:100, followed by pressing with a tablet press. The FTIR spectra were recorded over a range of 400–4000 cm^{-1}.

2.4.3. Tensile Test of the PLLA/PO3GCA Films

The PLLA/PO3GCA films were cut into smooth and nonwound samples with a size of 20 mm × 60 mm. At ambient temperature (25 °C), the initial length of the sample was 50 mm, the tensile rate was 10 mm/min, and the relative humidity was 60%. The sample was fixed on the fixture of the testing machine. The clamping length of the sample film was 50 mm, the force measurement accuracy was 0.01 cN, and the elongation accuracy was 0.01 mm. Stress–strain curves were obtained by testing each sample on a microcomputer-controlled electronic universal testing machine five times. The elongation at the break of each PLLA/PO3GCA film was determined using Equation (1).

$$\text{Elongation at break} = \frac{\text{Fracture displacement value(mm)}}{\text{The initial length of the membrane(mm)}} \times 100\% \quad (1)$$

2.4.4. Differential Scanning Calorimetry (DSC) Characterization of the PLLA/PO3GCA Films

The thermal properties of PLLA/PO3GCA thin films were analyzed using differential scanning calorimetry (DSC). Film samples weighing approximately 5–8 mg were placed in an aluminum crucible and heated from −40 °C to 200° C at a rate of 10 °C/min to obtain the DSC curve of the film samples [23].

The crystallinity of the composite films was calculated using Equation (2) [24,25], where X_c denotes crystallinity, ΔH_m denotes the enthalpy of melting, ΔH_{cc} denotes the enthalpy of cold crystallization, ΔH_{mPLLA} denotes the standard enthalpy of melting of PLLA (93.6 J·g), and ω_{PLLA} represents the mass fraction of PLLA in the composite film.

$$X_c = \frac{\Delta H_m - \Delta H_{cc}}{\Delta H_{mPLLA} \times \omega_{PLLA}} \times 100\% \quad (2)$$

2.4.5. Field-Emission Scanning Electron Microscopy (SEM) Characterization of the PLLA/PO3GCA Films

The morphology and fracture surface of the PLLA/PO3GCA films were examined using field-emission scanning electron microscopy (SEM). For surface morphology analysis, fully dried films were affixed to the sample table using conductive carbon glue and coated with a thin layer of gold for 30 s under vacuum conditions to enhance their conductivity. The samples were then observed at various magnifications at an accelerating voltage of 2 kV following vacuum extraction, and the corresponding images were captured for further analysis. For fracture surface analysis, the cross-sections of sample bars pulled at a rate of 10 mm/min on a universal tensile testing machine were used for SEM testing. Prior to testing, the surface of the samples was coated with a layer of conductive metal to enhance their conductivity, and then the samples were fractured to expose their internal structure. The SEM imaging process involved several steps, including surface treatment of the samples, vacuum extraction to remove air and moisture, and observation and capture of images at various magnifications in the SEM. This rigorous sample preparation and imaging protocol ensured the high-resolution imaging of the PLLA/PO3GCA films, allowing for a detailed examination of their microstructure, surface characteristics, and fracture behavior.

2.4.6. Thermogravimetric Analysis (TGA) of the PLLA/PO3GCA Films

The thermal stability of the PLLA/PO3GCA films was evaluated via TGA from room temperature (23 °C) to 600 °C at a heating rate of 10 °C/min under flowing nitrogen. The differential thermogravimetric (DTG) curves were obtained by differentiating the TGA curves.

3. Results and Discussion

3.1. Characterization of PO3GCA

Figure 2 shows the FTIR spectra of PO3G, PO3GCA, and CA. In the graph of PO3G, the peaks at 3500 and 1108 cm are the stretching vibrations of the terminal OH and ether bond C-O. The ester absorption peak at 1737 cm in the graph of PO3GCA shifted to the right compared to the carboxyl absorption peak at 1728 cm in the graph of CA, indicating that an esterification reaction occurred. The OH absorption peak near 3500 cm is significantly weaker than the OH absorption peak in the PO3G spectrum, which also indicates that OH participates in the esterification reaction.

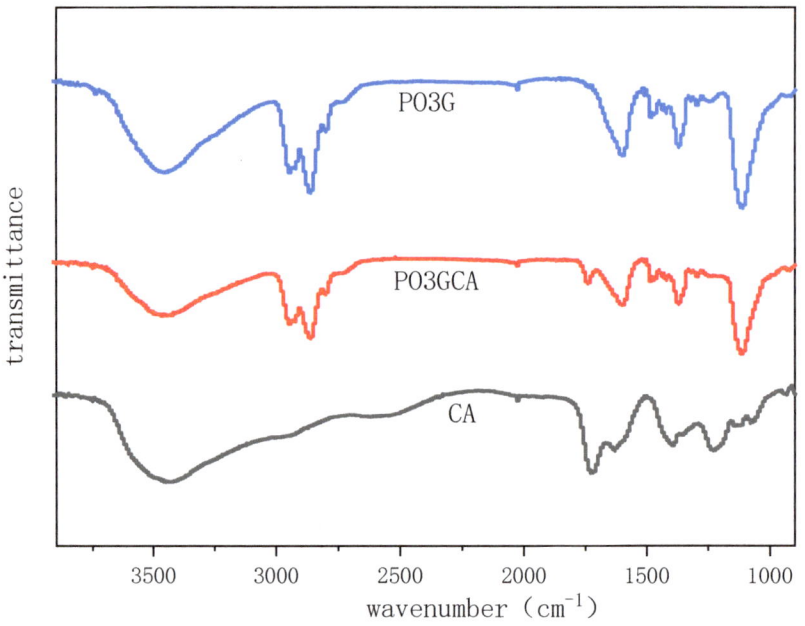

Figure 2. Fourier transform infrared (FTIR) spectra of poly(1,3-propylene glycol) (PO3G), poly(1,3-propylene glycol citrate) (PO3GCA), and citric acid (CA).

Figure 3 shows the ^{1}H NMR spectra of CA, PO3G, and PO3GCA. In the spectrum of CA, the two peaks at 2.7–2.7 ppm correspond to –CH_2–(C=O)OH. In the spectrum of PO3G, the peaks at 3.55, 3.5, and 1.8 ppm can be attributed to the –CH_2–OH–CH_2–CH_2–O, and –CH_2–CH_2– groups, respectively; the integral areas of these peaks were determined to be 1, 19.57, and 39.12, respectively. For PO3G, the average degree of polymerization was ~3 and Mn was 2260, which is close to the molecular weight of the raw material. The peak at 3.55 ppm in the spectrum of PO3G moves to 4.2 ppm in the spectrum of PO3GCA, indicating the formation of a–CH$_2$–O(C=O)– group as a result of the reaction of the hydroxyl group of PO3G with CA.

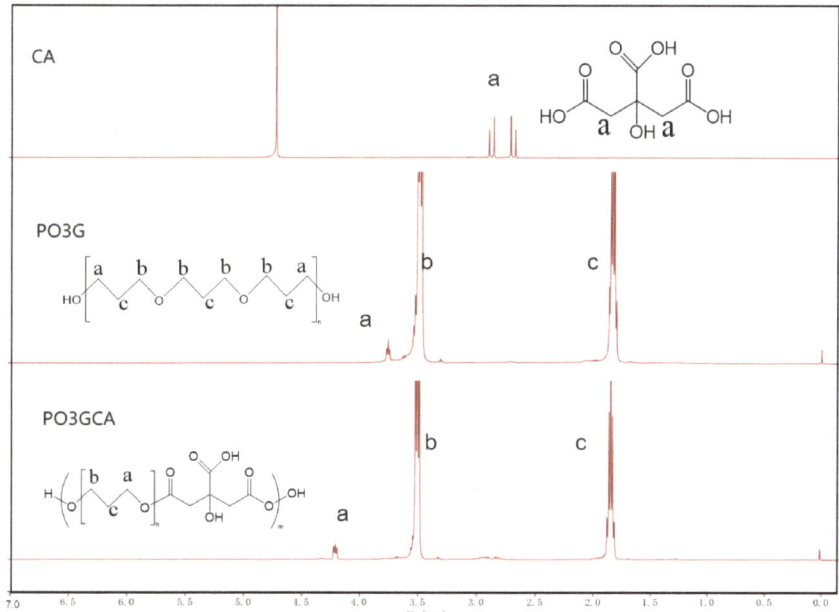

Figure 3. ^{1}H nuclear magnetic resonance (NMR) spectra of citric acid (CA), poly(1,3-propylene glycol) (PO3G), and poly(1,3-propylene glycol citrate) (PO3GCA).

3.2. Characterization of the PLLA/PO3GCA Films

3.2.1. Structure and Thermal Properties

TGA

Thermal properties of PLLA/PO3GCA composite films and PO3GCA were analyzed using thermogravimetric (TG) and derivative thermogravimetric (DTG) curves, as shown in Figures 4 and 5, respectively. The weight loss of PLLA and PLLA/PO3GCA composite membranes occurred in the temperature range of 75–140 °C, which was attributed to the evaporation of bonded water and residual solvents in the membranes. In contrast, the weight loss of PO3GCA was not evident in this temperature range. The PLLA thin film experienced a second weight loss at 295–390 °C, while the PLLA/PO3GCA composite thin film experienced a second weight loss at 300–430 °C. PO3GCA showed slight weight loss in the temperature range of 270–310 °C, which was partly due to the thermal degradation of low molecular weight PO3GCA. PO3GCA also showed a secondary weight loss at 310–450 °C. Both types of films exhibited 50% weight loss at 360 °C, while PO3GCA exhibited 50% weight loss at 380 °C. The maximum thermal decomposition rate of the film during the second weight loss was negatively correlated with the content of PO3GCA, as shown in Figure 5. With the increase of PO3GCA content, the maximum thermal decomposition rate decreased. These results indicate that the addition of PO3GCA can effectively improve the thermal stability of PLLA thin film. Overall, the TG and DTG analyses provided valuable insights into the thermal behavior of PLLA/PO3GCA composite films and highlighted the potential of PO3GCA as a thermal stabilizer for PLLA.

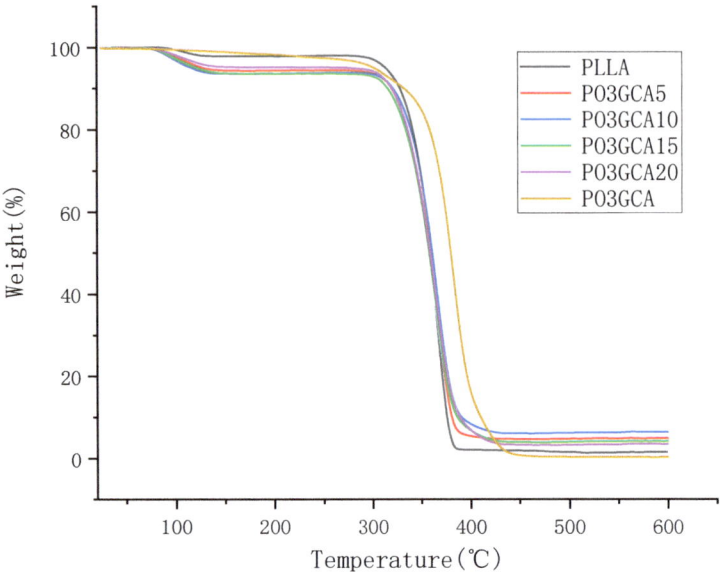

Figure 4. Thermogravimetric (TG) curves of polylactic-L-acid/poly(1,3-propylene glycol citrate) (PLLA/PO3GCA) films with different PO3GCA contents.

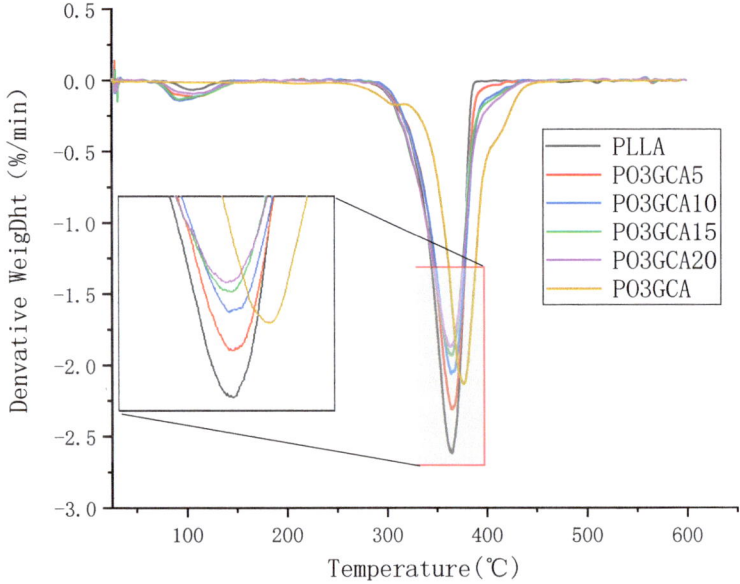

Figure 5. Differential thermogravimetric (DTG) curves of polylactic-L-acid/poly(1,3-propylene glycol citrate) (PLLA/PO3GCA) films with different PO3GCA contents.

DSC Analysis

The PLLA/PO3GCA composite films with varying PO3GCA content and DSC curves of PO3GCA are shown in Figure 6. The glass transition temperature (T_g) of PO3GCA is 8.2 °C. The T_g, cold crystallization temperature (T_{cc}), and melting temperature (T_m) of the polylactic acid (PLLA) film are 62.3 °C, 123.3 °C, and 165.4 °C, respectively. As can be seen from Figure 6 and Table 1, with increasing PO3GCA content, the Tg of the PLLA/PO3GCA composite films gradually decreases compared to that of the pure PLLA film. This indicates that the addition of PO3GCA improves the mobility of the PLLA segments, possibly due to the good compatibility between PLLA and PO3GCA. Moreover, when the PO3GCA content increases from 10% to 15%, a second glass transition temperature appears at 8.7 °C, which is close to the Tg of PO3GCA, indicating that slight phase separation occurs between PO3GCA and PLLA. Additionally, the second Tg at 43.7 °C is higher than that of PO3GCA10 at 37.7 °C. When the PO3GCA content increases to 20%, the phase separation becomes more severe, and the Tg increases to 45.3 °C.

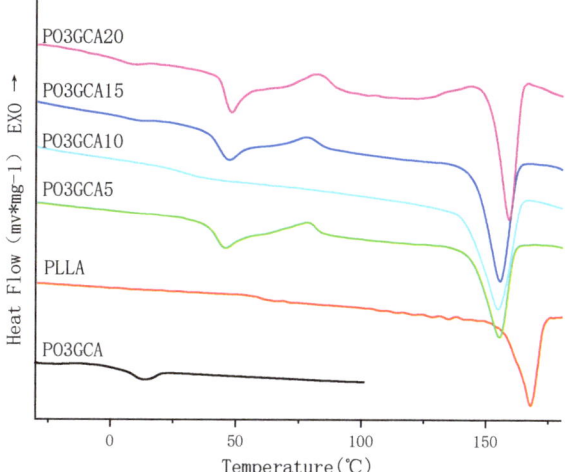

Figure 6. Differential scanning calorimetry (DSC) curves of polylactic-L-acid/poly(1,3-propylene glycol citrate) (PLLA/PO3GCA) films with different PO3GCA contents.

Table 1. Differential scanning calorimetry (DSC) data of PLLA thin film, PLLA/PO3GCA composite thin film, and PO3GCA.

Sample	Tg1 (°C)	Tg2 (°C)	Tcc (°C)	Tm (°C)	ΔHcc (J/g)	ΔHm (J/g)	Xc (%)
PLLA/PO3GCA5	-	41.6	78.0	155.2	5.76	27.98	24.99
PLLA/PO3GCA10	-	37.7	-	154.7	-	30.79	36.55
PLLA/PO3GCA15	8.7	43.7	78.6	155.4	3.74	28.53	31.15
PLLA/PO3GCA20	8.5	45.3	82.3	159.0	5.04	22.78	23.69
PLLA	-	62.3	123.3	165.4	35.6	39.7	4.4
PO3GCA	8.2	-	-	-	-	-	-

Furthermore, it is worth noting that when the PO3GCA content is less than 10%, the peak cold crystallization temperature of the composite film decreases with the increasing PO3GCA content. This indicates that the addition of PO3GCA effectively promotes the cold crystallization of PLLA, which is likely due to the nucleation effect of PO3GCA. Specifically, PO3GCA provides more nucleation sites for the formation of PLLA crystals, thereby enhancing the crystallization process. Additionally, as the PO3GCA content increases, the melting peak of the composite film slightly broadens and shifts to a lower temperature, indicating that the addition of PO3GCA affects the melting behavior of PLLA. Overall, the DSC analysis provides valuable insights into the thermal properties and compatibility of the PLLA/PO3GCA composite films.

As depicted in Table 1, it is evident that when the PO3GCA content is lower than 10%, the crystallinity of the composite film is positively correlated with the PO3GCA content, and the crystallinity of the film is significantly improved. This suggests that the addition of PO3GCA can effectively enhance the crystallization property of PLLA under suitable conditions. However, when the PO3GCA content exceeds 15%, the crystallinity of the composite film begins to decline. This is because the excessive addition of PO3GCA results in the separation of PO3GCA from PLLA, which is not conducive to the crystallization of PLLA. Therefore, it is crucial to maintain an appropriate PO3GCA content to achieve the desired crystallization improvement effect. In addition, it should be noted that the improvement in crystallinity is also closely related to the compatibility between PO3GCA and PLLA. When the compatibility is good, the nucleation effect of PO3GCA on PLLA crystallization is enhanced, resulting in higher crystallinity. Overall, the crystallinity analysis provides valuable insights into the crystallization behavior of the PLLA/PO3GCA composite films and highlights the importance of maintaining an appropriate PO3GCA content and good compatibility with PLLA.

Field-Emission SEM

Figures 6 and 7 present the surface micromorphology and micromorphology of stretched sections of PLLA/PO3GCA films with varying PO3GCA contents. The images in Figure 7 demonstrate that as the PO3GCA content increased, the surface of the films changed from a uniform phase to a dispersed phase with a sea-island structure. Additionally, the distribution of PO3GCA in PLLA shifted from the interior to the surface, particularly when the PO3GCA content increased from 10% to 15%. These observations suggest that the incorporation of PO3GCA into PLLA can significantly alter the morphology of the composite films, which could have implications for their mechanical and functional properties.

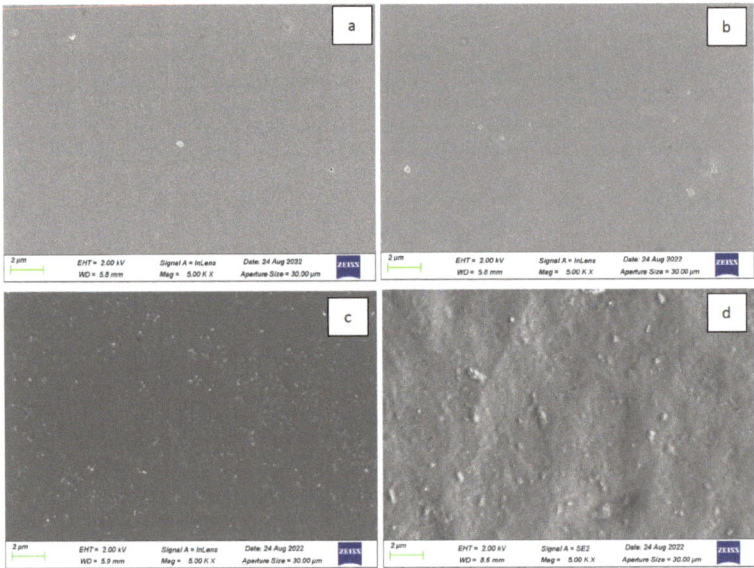

Figure 7. Scanning electron microscopy (SEM) images of polylactic-L-acid/poly(1,3-propylene glycol citrate) (PLLA/PO3GCA) films: (**a**) PO3GCA content, 5%; (**b**) PO3GCA content, 10%; (**c**) PO3GCA content, 15%; (**d**) PO3GCA content, 20%.

The tensile fracture sections shown in Figure 8 revealed that stratification occurred in the PLLA/PO3GCA15 and PLLA/PO3GCA20 films, whereas this phenomenon was not observed in the PLLA/PO3GCA5 and PLLA/PO3GCA10 films. This finding indicates that PO3GCA may be included in the cracks of the PLLA chains, filling the gaps between PLLA layers. However, excess PO3GCA cannot be accommodated in these gaps, resulting in the stratification of the film and the formation of a sea-island structure of dispersed phases on the surface. Furthermore, scattered fine particles were observed on the surface of the PLLA/PO3GCA5 and PLLA/PO3GCA10 films, which are most likely PO3GCA macromolecules that cannot be accommodated into the gaps between the PLLA layers. Another possibility is that the PO3GCA plasticizer crystallizes in advance during the solvent volatilization process owing to its high molecular weight, promoting the aggregation of the surrounding plasticizer and resulting in scattered particles on the film surface.

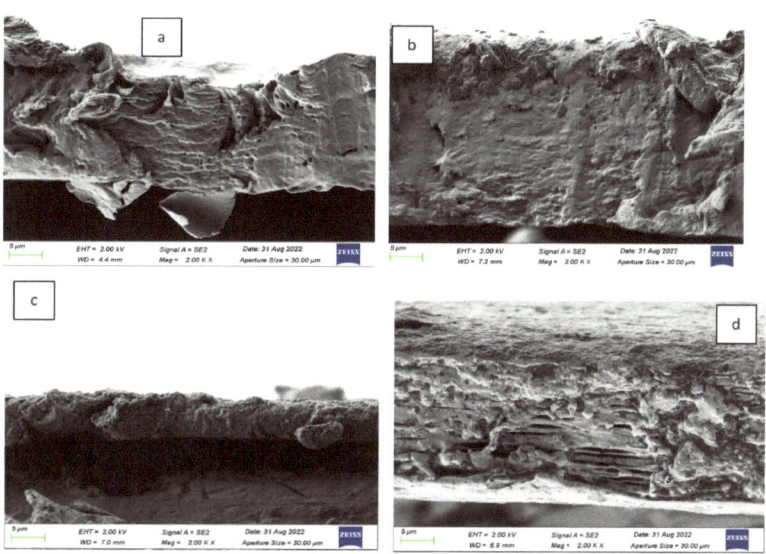

Figure 8. Scanning electron microscopy (SEM) images of the tensile fracture surface of polylactic acid/poly(1,3-propylene glycol citrate) (PLLA/PO3GCA) films: (**a**) PO3GCA content, 5%; (**b**) PO3GCA content, 10%; (**c**) PO3GCA content, 15%; (**d**) PO3GCA content, 20%.

The fracture surface of the PLLA/PO3GCA blend films was rough, with a distinct fibrous surface visible at a PO3GCA content of 15%, indicating a ductile fracture corresponding to a high elongation at break and high toughness of PLLA/PO3GCA15. As the PO3GCA content increased from 5% to 20%, burrs on the fracture surface became increasingly thin, indicating that PO3GCA may enhance the toughness of the PLLA film. However, the plasticizer overflowed between layers and accumulated on the surface as the PO3GCA content increased, thereby decreasing the tensile properties and toughness. These findings suggest that the optimal PO3GCA content for improving the mechanical properties of PLLA/PO3GCA films should be carefully considered.

3.2.2. Mechanical Properties of the PLLA/PO3GCA Films

Figure 9 illustrates the stress-strain curves obtained from the mechanical testing of PLLA/PO3GCA films with varying compositions. PLLA films are inherently rigid and brittle, with an elastic modulus of 1528 MPa and an elongation at a break of less than 10%. As shown in Table 2, the addition of PO3GCA plasticizer had a significant impact on the elongation at the break of the films. Among the different hybrid systems tested, the PLLA/PO3GCA15 film exhibited the highest elongation at break, which increased by 220% compared to pure PLLA.

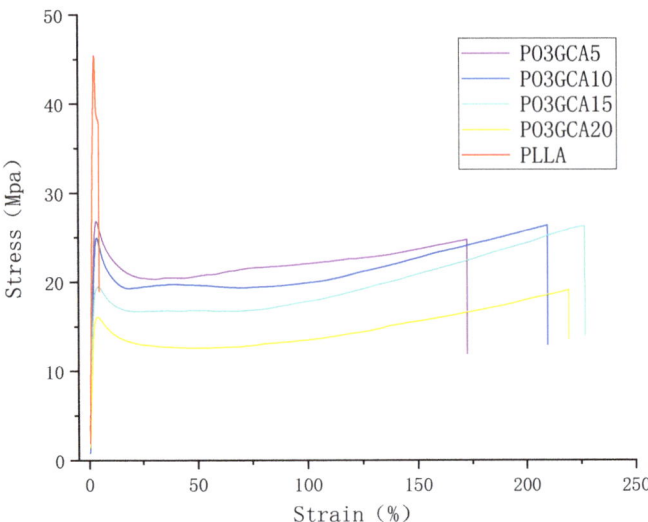

Figure 9. Stress–strain curves of polylactic-L-acid/poly(1,3-propylene glycol citrate) (PLLA/PO3GCA) films with different PO3GCA contents.

Table 2. Mechanical properties of PO3GCA/PLLA films.

PO3GCA Content (%)	Elongation AT Break (%)	Modulus of Elasticity (MPA)	Tensile Yield Stress (MPA)	Tensile Strength (MPA)	Tensile Toughness (MPA)
0	4.42	1528.00	38.12	27.28	1.47
5	172.40	1434.41	20.34	38.69	38.13
10	209.45	1339.90	19.30	35.64	44.89
15	230.03	1192.41	16.73	28.77	44.69
20	218.92	918.19	19.17	22.76	34.33

The stress-strain curves (Figure 9) indicate that the addition of PO3GCA resulted in plastic deformation of the PLLA/PO3GCA films due to changes in the interaction at the PLLA interface. The interfacial interaction between PO3GCA and the PLLA matrix facilitated the sliding of the interface of the hybrid matrix film, leading to a reduction in the tensile yield stress relative to the maximum force of the pure PLLA film. However, when the PO3GCA content reached 20%, the mechanical properties of the corresponding film decreased to varying degrees. This was attributed to excessive PO3GCA filling the gaps between the PLLA layers and aggregating on the surface of the film, causing the PLLA layer of the mixed films to slip easily and affecting the intralaminar structure of PLLA, ultimately leading to film fracture. As the elongation decreased, the elastic modulus, tensile yield stress, and maximum force also decreased.

In summary, the addition of PO3GCA significantly improved the mechanical properties of PLLA films, particularly in terms of elongation at break and toughness. However, excessive PO3GCA content can negatively impact the mechanical properties of PLLA/PO3GCA films. These findings have important implications for the use of PO3GCA as a plasticizer for PLLA in various applications, such as food packaging and medical devices.

The tensile toughness of the films was determined by integrating the stress-strain curves [26], and the results are presented in Table 2. The pure PLLA film exhibited a toughness of 1.47 MPa. However, upon the incorporation of PO3GCA, a substantial improvement in the toughness of the films was observed. Specifically, the toughness values for PLLA/PO3GCA5, PLLA/PO3GCA10, PLLA/PO3GCA15, and PLLA/PO3GCA20 were found to be 38.14, 44.89, 44.69, and 34.33 MPa, respectively. These values represent a remarkable increase of 2495%, 2953%, 2940%, and 2235% compared to that of pure PLLA. These findings indicate that PO3GCA is an excellent plasticizer for PLLA, and has the potential to significantly enhance the mechanical properties of PLLA-based materials.

4. Conclusions

In this study, poly(1,3-propylene glycol citrate) (PO3GCA) was synthesized from biologically derived citric acid (CA) and PO3G and was used as a toughening agent for Polylactic-L-acid (PLLA). The solution casting method was employed to prepare PLLA/PO3GCA films with varying PO3GCA contents while ensuring satisfactory compatibility between PO3GCA and PLLA.

The addition of PO3GCA to PLLA resulted in a slight increase in the thermal stability of the PLLA film and a significant improvement in its toughness. The elongation at the break of the composite film reached 230% when the PO3GCA content was 15%, compared to the pure PLLA film. These results suggest that PO3GCA may be a promising plasticizer for PLLA films, as it enhances the toughness of the film while maintaining its biodegradability and environmental-friendliness.

Overall, this study contributes to the development of sustainable and eco-friendly materials for the plastic industry. The use of biologically derived citric acid and poly(1,3-propylene glycol citrate) as toughening agents for PLLA can significantly enhance the properties of PLLA-based materials, while also reducing the environmental impact of plastic production. Further research is needed to investigate the long-term stability and biodegradability of PLLA/PO3GCA composite films, as well as to optimize the processing conditions and compatibility of the blended systems to achieve the desired properties of PLLA-based materials.

Author Contributions: Conceptualisation, D.J.; Data curation, J.C. and M.M.; Formal analysis, D.J. and X.S.; Funding acquisition, M.Y.; Investigation, D.J. and M.Y.; Methodology, D.J., H.A., W.Z., J.L., C.Z., W.Z. and Y.L.; Project administration, D.J. and M.Y.; Resources, Y.L. and M.Y.; Validation, J.C.; Visualisation, M.M.; Writing—original draft, J.C.; Writing—review and editing, D.J. All authors have read and agreed to the published version of the manuscript.

Funding: This research was funded by National Natural Science Foundation of China [grant numbers: U2002215 and 52163013], Yunnan Province "Thousand Talents Program" project training project [grant numbers: YNQR-CYRC-2018-012] and Yunnan Provincial Science and Technology Department (grant numbers: 202001AU070007).

Institutional Review Board Statement: Not applicable.

Informed Consent Statement: Not applicable.

Data Availability Statement: The data presented in this study are available on request from the corresponding author.

Acknowledgments: This work was supported by National Natural Science Foundation of China and Yunnan Provincial Science and Technology Department.

Conflicts of Interest: The authors declare no conflict of interest.

References

1. Fukushima, K.; Kimura, Y. An efficient solid-state polycondensation method for synthesizing stereocomplexed poly(lactic acid)s with high molecular weight. *J. Polym. Sci. Part A Polym. Chem.* **2008**, *46*, 3714–3722. [CrossRef]
2. Montané, X.; Montornes, J.M.; Nogalska, A.; Olkiewicz, M.; Tylkowski, B. Synthesis and synthetic mechanism of Polylactic acid. *Phys. Sci. Rev.* **2020**, *5*, 12. [CrossRef]
3. Nair, N.R.; Nampoothiri, K.M.; Banarjee, R. Simultaneous saccharification and fermentation (SSF) of jackfruit seed powder (JFSP) to l-lactic acid and to polylactide polymer. *Bioresour. Technol.* **2016**, *213*, 283–288. [CrossRef] [PubMed]
4. Tyler, B.; Gullotti, D.; Mangraviti, A. Polylactic acid (PLA) controlled delivery carriers for biomedical applications. *Adv. Drug Deliv. Rev.* **2016**, *107*, 163–175. [CrossRef] [PubMed]

5. Mohamad, N.; Mazlan, M.M.; Tawakkal, I.; Talib, R.A.; Jawaid, M. Development of active agents filled polylactic acid films for food packaging application. *Int. J. Biol. Macromol.* **2020**, *163*, 1451–1457. [CrossRef]
6. Guruprasad, R.; Prasad, G.K.; Prabu, G.; Raj, S.; Patil, P.G. Low-stress mechanical properties and fabric hand of cotton and polylactic acid fibre blended knitted fabrics. *Indian J. Fibre Text. Res.* **2018**, *43*, 381–384.
7. Malek, N.; Faizuwan, M.; Khusaimi, Z.; Bonnia, N.N.; Asli, N.A. Preparation and Characterization of Biodegradable Polylactic Acid (PLA) Film for Food Packaging Application: A Review. *J. Phys. Conf. Ser.* **2021**, *1892*, 012037. [CrossRef]
8. Shetty, S.D.; Shetty, N. Investigation of mechanical properties and applications of polylactic acids—A review. *Mater. Res. Express* **2019**, *6*, 112002. [CrossRef]
9. Puthumana, M.; Krishnan, P.; Nayak, S.K. Chemical modifications of PLA through copolymerization. *Int. J. Polym. Anal. Charact.* **2020**, *25*, 634–648. [CrossRef]
10. Akrami, M.; Ghasemi, I.; Azizi, H.; Karrabi, M.; Seyedabadi, M. A new approach in compatibilization of the poly(lactic acid)/thermoplastic starch (PLA/TPS) blends. *Carbohydr. Polym.* **2016**, *144*, 254–262. [CrossRef]
11. Su, S.; Kopitzky, R.; Tolga, S. Polylactide (PLA) and Its Blends with Poly(butylene succinate) (PBS): A Brief Review. *Polymers* **2019**, *11*, 1193. [CrossRef]
12. Krishnan, S.; Pandey, P.; Mohanty, S.; Nayak, S.K. Toughening of Polylactic Acid: An Overview of Research Progress. *Polym.-Plast. Technol. Eng.* **2016**, *55*, 1623–1652. [CrossRef]
13. Gui, Z.; Xu, Y.; Gao, Y.; Lu, C.; Cheng, S. Novel polyethylene glycol-based polyester-toughened polylactide. *Mater. Lett.* **2012**, *71*, 63–65. [CrossRef]
14. Ljungberg, N.; Andersson, T.; Wesslén, B. Film extrusion and film weldability of poly(lactic acid) plasticized with triacetine and tributyl citrate. *J. Appl. Polym. Sci.* **2003**, *88*, 3239–3247. [CrossRef]
15. Maiza, M.; Benaniba, M.T.; Massardier-Nageotte, V. Plasticizing effects of citrate esters on properties of poly(lactic acid). *J. Polym. Eng.* **2016**, *36*, 371–380. [CrossRef]
16. Butt, J.; Oxford, P.; Sadeghi-Esfahlani, S. Hybrid Manufacturing and Mechanical Characterization of Cu/PLA Composites. *Arab. J. Sci. Eng.* **2020**, *45*, 9339–9356. [CrossRef]
17. Yoo, H.M.; Jeong, S.Y.; Choi, S.W. Analysis of the Rheological Property and Crystallization Behavior of Polylactic Acid (Ingeo Biopolymer 4032D) at Different Process Temperatures. *e-Polymers* **2021**, *21*, 72–79. [CrossRef]
18. Vo, A.D.; Wei, J.C.; Mcauley, K.B. An Improved PO3G Model–Accounting for Cyclic Oligomers. *Macromol. Theory Simul.* **2020**, *29*, 2000023. [CrossRef]
19. Cong, Z.; Lua-Cheng, H.; Gui-You, W. A novel thermosensitive triblock copolymer from 100% renewably sourced poly(trimethylene ether) glycol. *J. Appl. Polym. Sci.* **2018**, *135*, 46112.
20. JJia-Heng, L.; Hui, L.; Xiao-Di, D.; An-Fu, Z. Effect of diester in esterification product of polyethylene glycol and acrylic acid on the performance of polycarboxylate superplasticizer. *Iran. Polym. J.* **2013**, *22*, 117–122.
21. Zoi, T.; Nejib, K.; Vasilios, T.; Nikolaos, D. Synthesis and Characterization of Bio-Based Polyesters: Poly(2-methyl-1,3-propylene-2,5-furanoate), Poly(isosorbide-2,5-furanoate), Poly(1,4-cyclohexanedimethylene-2,5-furanoate). *Materials* **2017**, *10*, 801.
22. Fang-Lian, Y.; Yun, B.; Yun-Tao, Z.; Hao, W. Synthesis and characterization of multiblock copolymers based on L-lactic acid, citric acid, and poly(ethylene glycol). *J. Polym. Sci. Part A Polym. Chem.* **2003**, *41*, 2073–2081.
23. Guo, Q.; Liu, X.; Liu, M.; Han, M.; Liu, Y.; Ji, S. Effect of Molecular Weight of Poly(Ethylene Glycol) on Plasticization of Poly(L-Lactic Acid). *Polymers* **2021**, *223*, 123720. [CrossRef]
24. Gray, A.P. Polymer crystallinity determinations by DSC. *Thermochim. Acta* **1970**, *1*, 563–579. [CrossRef]
25. Heather, S.; Praphulla, T.; Marianna, K. Improvements in the crystallinity and mechanical properties of PLA by nucleation and annealing. *Polym. Degrad. Stab.* **2019**, *166*, 248–257.
26. Rezgui, F.; Swistek, M.; Hiver, J.M.; G'Sell, C. Deformation and damage upon stretching of degradable polymers (PLA and PCL). *Polymer* **2005**, *46*, 7370–7385. [CrossRef]

Disclaimer/Publisher's Note: The statements, opinions and data contained in all publications are solely those of the individual author(s) and contributor(s) and not of MDPI and/or the editor(s). MDPI and/or the editor(s) disclaim responsibility for any injury to people or property resulting from any ideas, methods, instructions or products referred to in the content.

MDPI AG
Grosspeteranlage 5
4052 Basel
Switzerland
Tel.: +41 61 683 77 34

Polymers Editorial Office
E-mail: polymers@mdpi.com
www.mdpi.com/journal/polymers

Disclaimer/Publisher's Note: The statements, opinions and data contained in all publications are solely those of the individual author(s) and contributor(s) and not of MDPI and/or the editor(s). MDPI and/or the editor(s) disclaim responsibility for any injury to people or property resulting from any ideas, methods, instructions or products referred to in the content.

www.ingramcontent.com/pod-product-compliance
Lightning Source LLC
LaVergne TN
LVHW072347090526
838202LV00019B/2497